T0329717

Hybrid Dynamical Systems

Hybrid Dynamical Systems

Modeling, Stability, and Robustness

Rafal Goebel, Ricardo G. Sanfelice, and Andrew R. Teel

PRINCETON UNIVERSITY PRESS

PRINCETON AND OXFORD

ISBN: 978-0-691-15389-6

Library of Congress Control Number: 2011941674

British Library Cataloging-in-Publication Data is available

This book has been composed in LATEX

The publisher would like to acknowledge the author of this volume for providing
the digital files from which this book was printed

Printed on acid-free paper ∞

press.princeton.edu

Printed in the United States of America

10 9 8 7 6 5 4 3 2 1

To Ariana,
   Drew,
    Ellie,
     Julek,
      Melia,
       and Ola.

Without you, this book would have been possible much sooner.

Contents

Preface		**ix**
1	**Introduction**	**1**
	1.1 The modeling framework	1
	1.2 Examples in science and engineering	2
	1.3 Control system examples	7
	1.4 Connections to other modeling frameworks	15
	1.5 Notes	22
2	**The solution concept**	**25**
	2.1 Data of a hybrid system	25
	2.2 Hybrid time domains and hybrid arcs	26
	2.3 Solutions and their basic properties	29
	2.4 Generators for classes of switching signals	35
	2.5 Notes	41
3	**Uniform asymptotic stability, an initial treatment**	**43**
	3.1 Uniform global pre-asymptotic stability	43
	3.2 Lyapunov functions	50
	3.3 Relaxed Lyapunov conditions	60
	3.4 Stability from containment	64
	3.5 Equivalent characterizations	68
	3.6 Notes	71
4	**Perturbations and generalized solutions**	**73**
	4.1 Differential and difference equations	73
	4.2 Systems with state perturbations	76
	4.3 Generalized solutions	79
	4.4 Measurement noise in feedback control	84
	4.5 Krasovskii solutions are Hermes solutions	88
	4.6 Notes	94
5	**Preliminaries from set-valued analysis**	**97**
	5.1 Set convergence	97
	5.2 Set-valued mappings	101
	5.3 Graphical convergence of hybrid arcs	107

5.4   Differential inclusions 111
5.5   Notes 115

6   Well-posed hybrid systems and their properties   117
6.1   Nominally well-posed hybrid systems 117
6.2   Basic assumptions on the data 120
6.3   Consequences of nominal well-posedness 125
6.4   Well-posed hybrid systems 132
6.5   Consequences of well-posedness 134
6.6   Notes 137

7   Asymptotic stability, an in-depth treatment   139
7.1   Pre-asymptotic stability for nominally well-posed systems 141
7.2   Robustness concepts 148
7.3   Well-posed systems 151
7.4   Robustness corollaries 153
7.5   Smooth Lyapunov functions 156
7.6   Proof of robustness implies smooth Lyapunov functions 161
7.7   Notes 167

8   Invariance principles   169
8.1   Invariance and ω-limits 169
8.2   Invariance principles involving Lyapunov-like functions 170
8.3   Stability analysis using invariance principles 176
8.4   Meagre-limsup invariance principles 178
8.5   Invariance principles for switching systems 181
8.6   Notes 184

9   Conical approximation and asymptotic stability   185
9.1   Homogeneous hybrid systems 185
9.2   Homogeneity and perturbations 189
9.3   Conical approximation and stability 192
9.4   Notes 196

Appendix: List of Symbols   199

Bibliography   201

Index   211

Preface

A dynamical system is usually classified as either a continuous-time dynamical system or a discrete-time dynamical system. For example, classical mechanical systems and analog electronic circuits evolving in time according to principles of physics, such as Newton's and Kirchoff's laws, can be viewed naturally as continuous-time dynamical systems. Financial accounts, optimization algorithms, and digital systems can be viewed naturally as discrete-time dynamical systems.

Numerous dynamical systems escape such a clear-cut classification. In fact, there are dynamical systems that exhibit characteristics typical of both continuous-time systems and discrete-time systems. Examples are provided by circuits that combine analog and digital components and by mechanical devices controlled by digital computers. Such systems are called hybrid dynamical systems or just hybrid systems.

Modeling issues suggest an even broader understanding of hybrid systems. Many dynamical systems — including some that seem to fall into one of the two classical categories — are beyond the descriptive power of common modeling tools for continuous-time dynamical systems, such as differential equations, and common modeling tools for discrete-time dynamical systems, such as difference equations. For example, standard differential equations cannot describe changes of a logical variable that can take on only the values of 0 and 1. Hence, differential equations on their own are not able to model a continuous-time system controlled by an algorithm involving logic. Such a closed-loop system may be modeled, however, through a combination of differential equations with difference equations.

Another opportunity for combining the modeling tools for continuous-time and discrete-time dynamical systems comes in describing changes in a dynamical system that occur at dramatically different rates. For example, in a mechanical system with impacts the evolution of velocities during a collision can be modeled as instantaneous changes. Difference equations can model such changes, and differential equations may still describe the behavior in between collisions. While some concepts of generalized differential equations, involving time scales or measures that are not absolutely continuous, may treat such situations, a control student may find advantages in using the more familiar tools.

A hybrid dynamical system, or just a hybrid system, is then, for the purpose of this book, a dynamical system that exhibits characteristics of both continuous-time and discrete-time dynamical systems or a dynamical system that is modeled with a combination of common modeling tools for continuous-time and discrete-time dynamical systems. The goals of this book are

(i) To formulate a seemingly simple mathematical model of a hybrid system that is still extremely rich in descriptive capabilities;

(ii) To unify and generalize to the hybrid systems setting numerous results from stability theory for classical nonlinear dynamical systems;

(iii) To underline how "well-posedness" of a hybrid system — essentially, a reasonable dependence of solutions on initial conditions and the system's insensitivity to perturbations — makes some of the goals in (ii) attainable.

At the same time, this book aims at familiarizing the reader with some key mathematical concepts that do not fit in classical analysis but that are needed to meet the goals listed above. Attention is restricted to finite-dimensional hybrid systems, that is, systems where the state evolves in a finite-dimensional Euclidean space.

The mathematical model of hybrid systems used in this book goes beyond a combination of differential equations and difference equations. It combines differential equations or inclusions, difference equations or inclusions, and sets specifying where these equations or inclusions apply. The model is illustrated in Chapter 1 and rigorously developed in Chapter 2. The use of inclusions is justified to some extent by modeling needs, but is also deeply motivated by robustness considerations. The latter motivation is the topic of Chapter 4, where generalized solutions to hybrid systems and their relationship to perturbations are studied.

An initial discussion of asymptotic stability in a hybrid system and sufficient Lyapunov conditions for asymptotic stability are in Chapter 3. Further topics in asymptotic stability of hybrid systems, including the analysis of robustness of asymptotic stability and results showing the existence of smooth Lyapunov functions, are described in Chapter 7. Invariance principles and invariance-based sufficient conditions for asymptotic stability appear in Chapter 8. These further topics, in contrast to sufficient Lyapunov conditions in Chapter 3, rely on structural properties of the sets of solutions to hybrid systems, such as the dependence of solutions on initial conditions and other parameters. These structural properties are developed in Chapter 6. The mathematical concepts that are needed in Chapter 6, such as set convergence, graphical convergence, and continuity notions for set-valued mappings, are summarized in Chapter 5. Finally, more advanced topics in asymptotic stability of hybrid systems, for example, extending the concept of linearization, appear in Chapter 9.

The introduction of the mathematical modeling approach, the solution concept, as well as notions and sufficient conditions for stability in Chapters 1-3 do not insist on well-posedness of hybrid systems. However, we strongly advocate modeling that yields well-posed hybrid systems so that the tools developed in later chapters can be applied.

The sufficient background to follow the material is an undergraduate course in real analysis and in either differential equations or nonlinear systems. The order in which the material is presented is chosen with a control engineering student in mind and with additional necessary mathematical tools usually appearing just before they are needed.

A reader most interested in asymptotic stability theory for hybrid systems and its relevance in feedback control may choose to focus on Chapters 2, 3, 7, 8, and possibly 9 in the first reading. A reader more interested in elements of nonclassical mathematical analysis and their role in the study of solutions to hybrid systems may instead choose to focus on Chapters 2, 4, 5, 6, and 9. Chapter 4 is especially relevant to the reader interested in understanding the issues arising when the well-posedness assumptions of Chapter 6 do not hold.

Every chapter in this book concludes with a Notes section. The Notes include brief commentary on the development of some of the new concepts in the bookand list several references. The list of references is not meant to give a complete overview of the literature, but rather, it is meant to give the interested reader a good place to start further studies.

A list of symbols and general notation used in the book is compiled in Appendix 9.4.

We gratefully acknowledge the National Science Foundation, the Air Force Office of Scientific Research and the Army Research Office, for their support of research on analysis and control design tools for hybrid systems.

Finally, we wish to acknowledge our great debt to many colleagues and students who, in different ways, helped and educated us in writing this book. While we made every effort for this book to be free of typos, an errata list is available at http://www.u.arizona.edu/~sricardo/index.php?n=Main.Books. A list of suggested problems for inclusion in the classroom is also available at this website.

Rafal Goebel
Chicago, Illinois

Ricardo G. Sanfelice
Tucson, Arizona

Andrew R. Teel
Santa Barbara, California

Hybrid Dynamical Systems

Chapter One

Introduction

The model of a hybrid system used in this book is informally presented in this section. The focus is on the data structure and on modeling. Several examples are given, including models of hybrid control systems. The model of a hybrid system is then related to other modeling frameworks, such as hybrid automata, impulsive differential equations, and switching systems. A formal presentation of the model, together with a rigorous definition of the solution, is postponed until Chapter 2.

1.1 THE MODELING FRAMEWORK

The model of a hybrid system used in this book can be represented in the following form:

$$\begin{cases} x \in C & \dot{x} \in F(x) \\ x \in D & x^+ \in G(x). \end{cases} \tag{1.1}$$

A reader less familiar with set-valued mappings and differential or difference inclusions may choose to keep in mind a less general representation involving equations:

$$\begin{cases} x \in C & \dot{x} = f(x) \\ x \in D & x^+ = g(x). \end{cases} \tag{1.2}$$

This representation suggests that the state of the hybrid system, represented by x, can change according to a differential inclusion $\dot{x} \in F(x)$ or differential equation $\dot{x} = f(x)$ while in the set C, and it can change according to a difference inclusion $x^+ \in G(x)$ or difference equation $x^+ = g(x)$ while in the set D. The notation $\dot{x}$ represents the velocity of the state x, while x^+ represents the value of the state after an instantaneous change.

A rigorous statement of what constitutes a model of a hybrid system and what is a solution to the model is postponed until Chapter 2. This chapter focuses on modeling of various hybrid systems in the form (1.1) or (1.2).

To shorten the terminology, the behavior of a dynamical system that can be described by a differential equation or inclusion is referred to as *flow*. The behavior of a dynamical system that can be described by a difference equation or inclusion is referred to as *jumps*. This leads to the following names for the four objects involved in (1.1) or (1.2):

- C is the *flow set*.

- F (or f) is the *flow map*.

- D is the *jump set*.

- G (or g) is the *jump map*.

This book discusses hybrid systems in finite-dimensional spaces, that is, the flow set C and the jump set D are subsets of an n-dimensional Euclidean space $\mathbb{R}^n$. For consistency in the model, it will be required that the function f, respectively g, be defined on at least the set C, respectively D. In the case of set-valued flow and jump maps, it will be required that F, respectively G, have nonempty values on C, respectively D.

As the model in (1.2) or (1.1) suggests, the flow set, the flow map, the jump set, and the jump map can be specialized to capture the dynamics of purely continuous-time or discrete-time systems on $\mathbb{R}^n$. The former corresponds to a flow set equal to $\mathbb{R}^n$ and an empty jump set, while the latter can be captured with an empty flow set and a jump set defined as $\mathbb{R}^n$.

1.2 EXAMPLES IN SCIENCE AND ENGINEERING

Many mechanical systems experience impacts. Examples range from elaborate systems such as walking robots, through colliding billiard balls or the Newton's cradle, to a seemingly simple bouncing ball. Such systems flow in between impacts. A rough approximation of the impacts suggests considering them as instantaneous, and hence, as leading to jumps in the state of the system. Consequently, systems with impacts can be viewed as hybrid systems.

The first example is the mentioned bouncing ball. This example, and some of the later ones in this chapter, reappear throughout the book as illustrations of various properties and results.

Example 1.1. (Bouncing ball) Consider a point-mass bouncing vertically on a horizontal surface. In between impacts the point-mass flows, experiencing acceleration due to gravity. At impacts, when the point-mass hits the surface, the change in velocity is approximated as being instantaneously reversed and possibly diminished in magnitude due to dissipation of energy.

The state of the point-mass can be described with

$$x = \begin{pmatrix} x_1 \\ x_2 \end{pmatrix} \in \mathbb{R}^2,$$

where x_1 represents the height above the surface and x_2 represents the vertical velocity. It is natural to say that flow is possible when the point-mass is above the surface, or when it is at the surface and its velocity points up. Hence, the flow set is

$$C = \left\{ x \in \mathbb{R}^2 \, : \, x_1 > 0 \text{ or } x_1 = 0, x_2 \geq 0 \right\}.$$

The choice of a flow map is delicate at one point in C, that is, at $x = 0$. First, it is natural to say that

$$f(x) = \begin{pmatrix} x_2 \\ -\gamma \end{pmatrix} \qquad \text{when } x_1 > 0 \text{ or } x_1 = 0, x_2 > 0,$$

where $-\gamma$ is the acceleration due to gravity. Second, it is natural to say that $f(0) = 0$; it has to be accepted, though, that the resulting flow map f is not continuous at 0. Impacts happen when the point-mass is on the surface with negative velocity. Hence, the jump set is

$$D = \left\{ x \in \mathbb{R}^2 : x_1 = 0, x_2 < 0 \right\}.$$

The jump map is given, for some $\lambda \in (0, 1)$, by

$$g(x) = \begin{pmatrix} 0 \\ -\lambda x_2 \end{pmatrix}.$$

An alternative choice for g is the vector $-\lambda x$ since this function agrees with $g(x)$ on the set D. Figure 1.1 illustrates the data of the bouncing ball system.

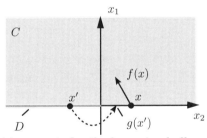

Figure 1.1: Flow and jump sets for the bouncing ball system in Example 1.1.

In the bouncing ball model above, every jump is followed by a period of flow. In other words, consecutive jumps do not happen. Consecutive jumps can happen in other systems with impacts, like in a model of Newton's cradle. Newton's cradle consists of at least three identical steel balls, each of which is suspended on a pendulum. At the stationary state, the balls are aligned along a horizontal line. Lifting a ball from one end of the alignment and releasing it leads to a collision of the lifted ball with the remaining balls. After the collision, the ball that was lifted and released becomes stationary and the ball on the other end of the alignment swings up. One way to model this interaction is to consider a sequence of collisions between pairs of adjacent balls.

A number of biological systems, such as groups of fireflies or crickets, are able to produce synchronized behavior, flashing or chirping, respectively, through a dynamical mechanism that can be viewed as hybrid.

Example 1.2. (Flashing fireflies) The timing of flashes of a firefly is determined by the firefly's internal clock. In between flashes, the internal clock gradually increases. When it reaches a threshold, a flash occurs and the clock is instantly reset to 0. In a group of fireflies, the flash of one firefly affects the internal clock of all other fireflies. That is, when a firefly witnesses a flash from another firefly, its internal clock instantly increases to a value closer to the threshold.

To model the internal clocks of n fireflies, normalize units so that each firefly's internal clock, denoted x_i, takes values in the interval $[0,1]$, i.e., every threshold is 1. The flow set is then

$$C = [0,1)^n := \{x \in \mathbb{R}^n \ : \ x_i \in [0,1), \ i = 1, 2, \ldots, n\}.$$

In between the flashes, every clock state flows toward the threshold according to the differential equation $\dot{x}_i = f_i(x_i)$, where $f_i : [0,1] \to \mathbb{R}_{>0}$, $i = 1, 2, \ldots, n$, is continuous. This defines the flow map f.

Jumps occur when one of the internal clocks reaches the threshold. Thus, the jump set is

$$D = \left\{x \in [0,1]^n \ : \ \max_i x_i = 1\right\}.$$

One method to model the (instantaneous) changes in internal clocks during a flash is through the jump map defined by

$$g(x) = \begin{pmatrix} g_1(x_1) \\ g_2(x_2) \\ \vdots \\ g_n(x_n) \end{pmatrix}, \qquad g_i(x_i) = \begin{cases} (1+\varepsilon)x_i, & \text{when } (1+\varepsilon)x_i < 1, \\ 0, & \text{otherwise,} \end{cases}$$

where $\varepsilon > 0$. This indicates that the internal clock x_i of a firefly witnessing a flash increases to $(1+\varepsilon)x_i$, unless this would result in reaching or exceeding the threshold, in which case the internal clock is reset to 0 together with the internal clock of the flashing firefly. Figure 1.2 illustrates the evolution of the clock variable x for $n = 2$ and $n = 10$ when $f_i \equiv 1$ for each i.

Example 1.3. (Power control with a thyristor) Consider the electric circuit in Figure 1.3(a) for controlling the power delivered to a load. The load consists of a resistor R and an inductance L that is connected to a power source through a thyristor with a gate control port. A simple model describing the operation of the thyristor is as follows. When in conduction mode, which can be triggered through the gate port, the thyristor allows flow of current from anode to cathode, which are the terminals denoted as $a+$ and $c-$ in Figure 1.3(a), respectively. It will turn off once the current from anode to cathode becomes zero. The load current is denoted by i_L, its voltage by v_L, and the capacitor's voltage by v_o. The sinusoidal input voltage with angular frequency ω is denoted by v_s and is generated by the output $v_s = z_1$ of the system

$$\dot{z}_1 = \omega z_2, \qquad \dot{z}_2 = -\omega z_1. \tag{1.3}$$

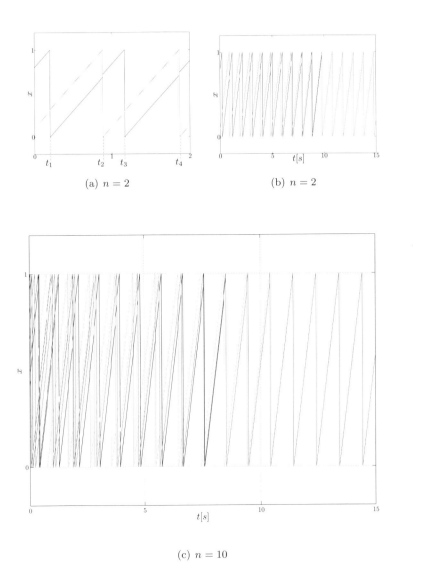

(a) $n = 2$ (b) $n = 2$

(c) $n = 10$

Figure 1.2: Evolution of coupled impulsive oscillators in fireflies with unitary threshold and $f_i \equiv 1$ for each i.

A discrete state $q \in \{0, 1\}$ is used to indicate whether the thyristor is *on* $(q = 1)$ or *off* $(q = 0)$, while a continuous state $\tau \in \mathbb{R}$ is used to model the firing events in the gate port, given as a function of the firing angle parameter $\alpha \in (0, \pi)$.

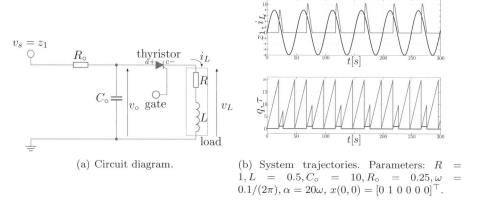

(a) Circuit diagram.

(b) System trajectories. Parameters: $R = 1, L = 0.5, C_o = 10, R_o = 0.25, \omega = 0.1/(2\pi), \alpha = 20\omega, x(0,0) = [0\ 1\ 0\ 0\ 0\ 0]^\top$.

Figure 1.3: Power control circuit with thyristor.

By defining the state of the system to be

$$x := (z_1, z_2, i_L, v_o, q, \tau) \in \mathbb{R}^6,$$

the continuous dynamics are defined by

$$F(x) = \begin{pmatrix} \omega z_2 \\ -\omega z_1 \\ q\left(\frac{v_o - R i_L}{L}\right) \\ -\frac{1}{C_o R_o} v_o + \frac{1}{C_o R_o} z_1 - \frac{1}{C_o} i_L \\ 0 \\ 1 \end{pmatrix}.$$

These equations can be derived applying electrical circuit theory for each mode of operation. Note that $\dot{q} = 0$ indicates that the discrete state remains constant during flows, and that $\dot{\tau} = 1$ enforces that τ counts the flow time in between switches. Assuming that when the thyristor is in off mode the load current is zero, two conditions trigger switches of the thyristor mode:

- When the thyristor is *off* ($q = 0$, $i_L = 0$), the firing angle has been reached ($\tau \geq \alpha/\omega$), and the capacitor voltage is positive ($v_0 > 0$), then switch to *on* ($q = 1$).

- When the thyristor is *on*, the load current is zero and decreasing ($i_L = 0$, $\dot{i}_L < 0$), then switch to *off*.

These conditions can be captured with the flow and jump sets

$$C \; := \; \left\{ x \; : \; q = 0, \tau < \frac{\alpha}{\omega}, i_L = 0 \right\} \cup \{ x \; : \; q = 1, i_L > 0 \},$$

$$D \; := \; \left\{ x \; : \; q = 0, \tau \geq \frac{\alpha}{\omega}, i_L = 0, v_0 > 0 \right\} \cup \{ x \; : \; q = 1, i_L = 0, v_0 < 0 \},$$

and the jump map

$$G(x) := (z_2 \;\; z_1 \;\; i_L \;\; v_\circ \;\; 1 - q \;\; 0)^\top.$$

At every jump, q is toggled and the timer is restarted to trigger the next jump to *on* mode at the programmed firing angle. The top plot in Figure 1.3(b) shows the input voltage with $\omega = 0.1/(2\pi)$ rad/sec and the resulting load's current with a firing angle of 20ω rad, while the bottom plot shows the associated logic and timer states.

1.3 CONTROL SYSTEM EXAMPLES

The control of a continuous-time system with state feedback faces both practical and theoretical obstacles: precise information about the state may not be available at all times, even if frequent measurements of the state are available; the behavior of the closed-loop system may be very sensitive to errors in the state measurements; or satisfactory performance of the closed-loop system may not be achievable by using just one state-feedback controller. These issues provide motivation for the use of hybrid control, several simple instances of which are described below.

Example 1.4. (Sample-and-hold control) Given a continuous-time control system and a state-feedback controller, associating with each state of the system the control to be applied there, a *sample-and-hold* implementation of the feedback is essentially as follows:

- *sample:* measure the state of the system, and use the feedback controller to obtain the control value based on the measurements;

- *hold:* apply the computed constant control value for certain amount of time;

and repeat the procedure infinitely many times. The processes of sampling and computing the control can be modeled as an instantaneous event. This leads to a continuous behavior of the closed-loop system in between the sampling times, according to the continuous-time dynamics of the control system and the constant value of the control, and an instantaneous change at every sampling time, when the control value is instantly updated.

A schematic example of a sample-and-hold control system is in Figure 1.4, where a digital device controls an analog plant. The basic operation of the system is as follows. The output of the plant is sampled by an *analog-to-digital*

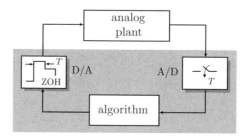

Figure 1.4: Digital control of a continuous-time nonlinear system with sample-and-hold devices.

converter, denoted A/D. The digitized output is processed by the algorithm, and the result is applied to the plant through a *digital-to-analog converter*, denoted D/A. For a periodic A/D sampler and a zero-order hold (ZOH) type of D/A, the output samples and control input updates occur at a fixed sampling period T.

To model such a system as a hybrid system, suppose that the control system is given by

$$\dot{z} = \widetilde{f}(z, u), \tag{1.4}$$

where $z \in \mathbb{R}^{n_p}$ is the state of the system, $u \in \mathbb{R}^{n_c}$ is the control variable, and $\widetilde{f} : \mathbb{R}^{n_p} \times \mathbb{R}^{n_c} \to \mathbb{R}^{n_p}$ is a function. Let the state-feedback controller be given by $u = \kappa(z)$. The standard closed-loop, without a sample-and-hold strategy, leads to a continuous-time closed-loop system

$$\dot{z} = \widetilde{f}(z, \kappa(z)).$$

A sample-and-hold implementation can be modeled as a hybrid system, with the state variable

$$x = \begin{pmatrix} z \\ u \\ \tau \end{pmatrix} \in \mathbb{R}^{n_p + n_c + 1}.$$

Note that, for simplicity, the control input u itself is taken to be a state variable for the closed-loop system resulting from sample-and-hold control. Suppose that the sampling period is T. Flow occurs when the timer variable τ belongs to the interval $[0, T)$. During flow, the variable u remains constant, τ keeps track of elapsed time, and the state of the plant z evolves according to the dynamics in (1.4). Thus, the flow set and the flow map can be taken to be

$$C = \mathbb{R}^{n_p} \times \mathbb{R}^{n_c} \times [0, T), \qquad f(x) = \begin{pmatrix} \widetilde{f}(z, u) \\ 0 \\ 1 \end{pmatrix}. \tag{1.5}$$

Jumps occur when the timer variable reaches T. At jumps, the variable u is updated to $\kappa(x)$, the timer is reset to 0, and the state of plant does not change. Hence the jump set and the jump map can be taken to be

$$D = \mathbb{R}^{n_p} \times \mathbb{R}^{n_c} \times \{T\}, \qquad g(x) = \begin{pmatrix} z \\ \kappa(z) \\ 0 \end{pmatrix}. \tag{1.6}$$

Example 1.5. (A quantized control system) Some control systems that use quantized measurements include a mechanism for adjusting quantization parameters on-line. These adjustments are made to vary the accuracy of the measurements at different locations in the state space. For example, consider the control system

$$\dot{\zeta} = \zeta + u \tag{1.7}$$

with measurements

$$y = \mu q(\zeta/\mu),$$

where $q : \mathbb{R} \to \mathbb{R}$ is a function that represents measurement quantization and μ is a positive parameter that can be adjusted discretely as part of a control algorithm. The main requirement on the function q is that there exist positive real numbers Δ and M with $\Delta \ll M$ such that

$$|z| \leq M \text{ implies } |q(z) - z| \leq \Delta$$

$$|q(z)| \leq M - \Delta \text{ implies } |z| \leq M.$$

In this way, the value $q(z)$ gives some rough information about the value of z. An adaptive, quantized hybrid feedback law could consist of

- a feedback rule $u = -ky$, where $k > 1$, designed to steer the state ζ of (1.7) to zero;

- a discrete-time update rule for the parameter μ;

- a specification of sets where flows are allowed because μ does not need to be adjusted;

- a specification of sets where jumps are allowed because the parameter μ should be increased or decreased to put the argument of q into an acceptable range.

For example, letting the positive real numbers ℓ_{in}, ℓ_{out}, λ_{in}, and λ_{out} satisfy $\ell_{\text{in}} < \ell_{\text{out}}$ and $\lambda_{\text{in}} < 1 < \lambda_{\text{out}}$, consider taking the flow set to be

$$C = \{(\zeta, \mu) \in \mathbb{R} \times (0, \infty) : |q(\zeta/\mu)| \in [\ell_{\text{in}}, \ell_{\text{out}}]\},$$

the jump set to be

$$\begin{aligned} D_{\text{in}} &= \{(\zeta, \mu) \in \mathbb{R} \times (0, \infty) : |q(\zeta/\mu)| < \ell_{\text{in}}\} \\ D_{\text{out}} &= \{(\zeta, \mu) \in \mathbb{R} \times (0, \infty) : |q(\zeta/\mu)| > \ell_{\text{out}}\} \\ D &= D_{\text{in}} \cup D_{\text{out}}, \end{aligned}$$

and the jump map to be

$$g(\zeta, \mu) = \left\{ \begin{array}{ll} \lambda_{\text{in}}\mu & \forall(\zeta, \mu) \in D_{\text{in}} \\ \lambda_{\text{out}}\mu & \forall(\zeta, \mu) \in D_{\text{out}}. \end{array} \right.$$

The hybrid control algorithm increases or decreases the size of μ in an attempt to drive the state to the flow set. Depending on the initial value of (ζ, μ), multiple consecutive jumps may be required to reach the flow set. Ideally, ℓ_{in} and ℓ_{out} are chosen based on M and Δ so that, after some point in time, the system no longer reaches D_{out}, it repeatedly reaches D_{in}, and $|q(\zeta/\mu)| \leq M - \Delta$ so that $|\zeta/\mu| \leq M$. In this case, μ repeatedly shrinks by the factor λ_{in} and the convergence of μ to zero implies that ζ also converges to zero.

Example 1.6. (Reset linear control systems) In classical control theory, the output of a controller of a continuous-time plant evolves continuously in time. *Reset control systems* differ from those traditional controllers as their output experiences jumps caused by resets of the controller state. These resets may depend on the value of the controller inputs. In some scenarios, in comparison to (non-reset) classical controllers, reset controllers lead to improved system performance.

The first reset controller that appeared in the literature is the so-called *Clegg integrator*, a single-input/single-output linear controller that resets its output to zero when its input and output do not have the same sign. Figure 1.5 shows the response of the Clegg integrator to a sinusoidal input. During flows, the Clegg integrator's output is the integral of its input. Since the Clegg integrator does not permit the signs of its input and output to differ from one another, it forces a jump in its state when the input changes sign. After such a jump, the system flows again.

A schematic example of a plant controlled by a reset control system is in Figure 1.6. If the controller's input, which in Figure 1.6 is the error between the plant output and the reference input, and the controller's output satisfies a *reset condition* then the controller state is reset to a pre-specified value. The closed-loop system is a hybrid system with flows interrupted by state-dependent jumps, which are triggered when the reset condition is satisfied.

Consider a reset linear control system where the plant state is x_p and the controller state is x_c. The closed-loop system state is

$$x = \begin{pmatrix} x_p \\ x_c \end{pmatrix} \in \mathbb{R}^{n_p + n_c}.$$

Since the closed-loop system without resets is linear, the flow map is a linear function

$$f(x) = A_f x.$$

The resetting mechanism is also linear, so that the jump map has the form

$$g(x) = A_g x.$$

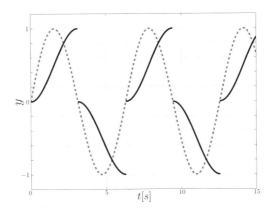

Figure 1.5: Output response y (solid) of Clegg integrator to a sinusoidal input (dashed).

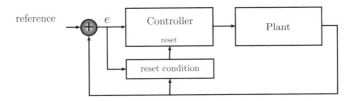

Figure 1.6: Closed-loop system with reset controller.

Resets typically occur when the state x satisfies some quadratic inequality, perhaps coming from insisting that two variables are always related by having the same sign. Thus, the jump set may have the form

$$D = \left\{ x \in \mathbb{R}^{n_p + n_c} : x^T M x \leq 0 \right\}$$

where $M = M^T$, that is, M is a symmetric matrix. One then may consider taking $C = \mathbb{R}^{n_p + n_c} \setminus D$. A particular construction of these sets is depicted in Figure 1.7.

Notice that the origin does not belong to the flow set but belongs to the jump set and that g maps the origin back to the origin. Thus, from the origin it is only possible to jump without ever flowing. To address this situation, one may consider forcing a small amount $\delta > 0$ of flow time between jumps. This can be done with a technique called "temporal regularization." In this case, one

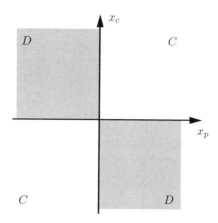

Figure 1.7: Examples of flow and jump sets for reset control with plant output $y = x_p$ and input $u = x_c$ ($n_p = n_c = 1$). The matrix $M \in \mathbb{R}^{2 \times 2}$ is given by $(0, 1; 1, 0)$, which enforces that flows occur when the components of $x = (x_p, x_c)$ have the same sign.

augments the state with a timer variable τ and takes the jump set to be

$$D := \left\{ x \in \mathbb{R}^{n_p + n_c} , \ \tau \in \mathbb{R} : x^T M x \leq 0 , \ \tau \geq \delta \right\}.$$

The flow set is taken to be

$$C := \left\{ x \in \mathbb{R}^{n_p + n_c} , \ \tau \in \mathbb{R} : x^T M x > 0 \text{ or } \tau \in [0, \delta] \right\}.$$

The jump map is augmented with the equation $\tau^+ = 0$ and the flow map is augmented with the equation $\dot{\tau} = 1$ for $\tau \in [0, 2\delta)$, $\dot{\tau} = 0$ for $\tau = 2\delta$, which, in particular, keeps τ bounded.

Example 1.7. (Combining local and global controllers) In several control applications, the design of a continuous-time feedback controller that performs a particular control task is not possible. For example, in the problem of globally stabilizing a multi-link pendulum to the upright position with actuation on the first link only, topological constraints rule out the existence of a continuous-time feedback controller that accomplishes this task globally and robustly. However, it is often possible to overcome such topological obstructions using hybrid feedback control to combine continuous-time feedback controllers that achieve certain subtasks.

To illustrate this idea, consider the task of combining a high-performance controller that works only near a reference point with a controller that is able to steer every trajectory toward the reference point, but does not have very good performance near that point. We refer to these controllers as *local* and *global* controllers, respectively.

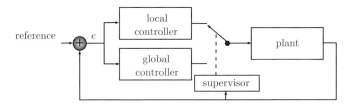

Figure 1.8: Closed-loop system combining local and global controllers.

Figure 1.8 depicts a block diagram of the control mechanism being described. Each controller measures the error signal given by the plant output and reference input. The controller selection is performed by a *supervisor* and is based on the plant's output and on the controller currently applied. Switching from one controller to the other results in a jump in the logic variable. In between the jumps, continuous evolution of the state of the system occurs.

More precisely, suppose that each of the two state feedback control laws, κ_1 and κ_2, asymptotically stabilizes the origin for the control system (1.4). Furthermore, suppose that κ_1 produces efficient transient responses, but works only near the origin, while κ_2 produces less efficient transients but works globally. The goal is to build a hybrid feedback law that globally asymptotically stabilizes the origin while using κ_1 near the origin and κ_2 far from the origin.

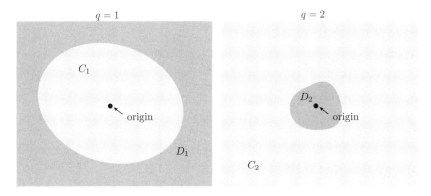

Figure 1.9: Sets for the hybrid controller combining control laws κ_1 and κ_2.

To eliminate the possibility of several instantaneous switches between controllers, a hysteresis mechanism is used. With the sets C_i, D_i, $i = 1, 2$ as in Figure 1.9, the switching idea is as follows: if κ_1 is being used and $z \in C_1$, do not switch, but if $z \in D_1$, switch to κ_2; while if κ_2 is being used and $z \in C_2$, do not switch, but if $z \in D_2$, switch to κ_1. Thus, when κ_1 is used, continuous evolution takes place when $z \in C_1$ and is described by $\dot{z} = \widetilde{f}(z, \kappa_1(z))$, while a jump takes place when $z \in D_1$ and results in q toggled to 2. When κ_2 is used, continuous evolution takes place when $z \in C_2$ and is described by $\dot{z} = \tilde{f}(z, \kappa_2(z))$, while a jump takes place when $z \in D_2$ and results in q toggled to 1. A general approach to modeling systems of this kind, where a logical variable (here equal to either 1 or 2) determines the hybrid dynamics, is given in Section 1.4.1. Here, it is illustrated that this can be modeled by including a logical variable q, taking values in $\{1, 2\}$, in the state

$$x = \begin{pmatrix} q \\ z \end{pmatrix} \in \mathbb{R}^{n_p+1},$$

and with the following sets and functions:

$$C = (\{1\} \times C_1) \cup (\{2\} \times C_2), \qquad f(x) = \begin{pmatrix} 0 \\ \tilde{f}(z, \kappa_q(z)) \end{pmatrix},$$

$$D = (\{1\} \times D_1) \cup (\{2\} \times D_2), \qquad g(x) = \begin{pmatrix} 3 - q \\ z \end{pmatrix}. \tag{1.8}$$

In fact, since $3 - q = 2$ when $q = 1$ and $3 - q = 1$ when $q = 2$, the state q is toggled when z enters the set D_q.

Finally, in order for the hybrid feedback law to work as intended, there should be a relationship between D_2 and C_1. In particular, if solutions to $\dot{z} = \tilde{f}(z, \kappa_1(z))$ start in D_2, they should remain in a closed set that is a strict subset of C_1; moreover, any trajectory of this system that starts in C_1 and remains in C_1 should converge to the origin. Since the local controller is locally asymptotically stabilizing, both of these properties can be induced by first picking C_1 to be a sufficiently small neighborhood of the origin and then picking D_2 to be another sufficiently small neighborhood of the origin strictly contained in C_1.

To illustrate this hybrid feedback construction, consider a positioning control system used for data read/write in hard disk drives. The objective of the control algorithm is to provide precise positioning of the magnetic heads to read and write information from the disk's tracks. A technique utilized in commercial devices for this purpose is called *mode-switching control*. It combines two controllers for stabilizing the position of the magnetic head to a desired position p^* on the disk with zero velocity: a controller (global) capable of steering it to a neighborhood of p^* and a controller (local) capable of stabilizing it to p^* with high precision. The plant can be modeled as a double integrator $\dot{p} = v$, $\dot{v} = u$ with state $z = (p, v)$, where $p \in \mathbb{R}$ is the position, $v \in \mathbb{R}$ the velocity of the magnetic head of the hard disk drive, and $z^* = (p^*, 0)$ is the point to stabilize. Suppose that the global controller is given by κ_2 and the local controller by κ_1. The hybrid control scheme leading to the closed-loop system in (1.8) can be employed to accomplish the control objective. The set C_1 can be taken to be a compact neighborhood of z^* that is contained in the basin of attraction for z^* when using κ_1, and D_2 can be taken to be a compact neighborhood of z^* such that solutions using κ_1 that start in D_2 do not reach the boundary of C_1. Then, $C_2 = \overline{\mathbb{R}^2 \setminus D_2}$ and $D_1 = \overline{\mathbb{R}^2 \setminus C_1}$.

1.4 CONNECTIONS TO OTHER MODELING FRAMEWORKS

The models (1.1) and (1.2) can describe several classes of hybrid systems that are frequently modeled in different frameworks. These different frameworks include hybrid automata, impulsive differential equations or inclusions, and switching systems. This section illustrates how models from these frameworks can be translated to (1.1) or (1.2). The benefit of passing to (1.1) or (1.2) is that the asymptotic stability theory developed in this book can then be applied to a broader class of systems. For example, as described in Section 8.5, invariance principles for hybrid systems can be applied to switching systems.

1.4.1 Systems with explicit "discrete states" or "logical modes"

The state in several hybrid systems can be decomposed into a "continuous state" and a "discrete state." The discrete state takes values in a discrete, often finite, set. It may represent a *mode* in which the system, or part of the system, is operating. For example, the discrete state can take values representing modes such as "on" or "off"; "first gear," "second gear," "third gear"; "controller 1" or "controller 2" as in Example 1.7; and so on. The discrete state, by its nature, can change only via a jump. The continuous state can change via flow and, sometimes, via a jump too. It may represent position, velocity, and other continuous-valued variables. For example, in a temperature control system, a discrete state can indicate whether a thermostat is "on" or "off" while a continuous state can indicate the temperature. In such a case, the continuous state may not change via a jump. If the discrete state represents whether a connection in an electrical circuit is "open" or "closed," as it does in Example 1.3, and the continuous variable represents the current in some part of the circuit, it may be natural to allow for instantaneous changes in the continuous variable that are simultaneous with changes in the discrete variable.

A system with continuous and discrete states usually can be represented by a set $Q = \{1, 2, \ldots, q_{\max}\}$, and for each $q \in Q$, a flow set $C_q \subset \mathbb{R}^n$, a flow map $F_q : \mathbb{R}^n \rightrightarrows \mathbb{R}^n$, a jump set $D_q \subset \mathbb{R}^n$, and a jump map $G_q : \mathbb{R}^n \rightrightarrows Q \times \mathbb{R}^n$. The suggestive form to represent such a system, parallel to (1.1), is

$$\begin{cases} z \in C_q & \dot{z} \in F_q(z) \\ z \in D_q & (q, z)^+ \in G_q(z). \end{cases} \tag{1.9}$$

When the discrete variable q has the value $q^* \in Q$ and the continuous variable z is in the flow set C_{q^*}, flow is possible according to the inclusion $\dot{z} \in F_{q^*}(z)$. During flow, the discrete variable remains constant. The condition $\dot{q} = 0$ is not explicitly mentioned in (1.9). When the discrete variable has the value $q^* \in Q$

and the continuous variable z is in the jump set D_{q^*}, a jump is possible, with both q and z changing values according to G_{q^*}. For systems where the continuous variable does not change via jumps, the inclusion $(q, z)^+ \in G_q(z)$ can be replaced by the simpler $q^+ \in G_q(z)$, in which case the equation $z^+ = z$ is usually not mentioned explicitly.

The system (1.9) can be formulated in the form (1.1). To this end one takes

$$x = \begin{pmatrix} q \\ z \end{pmatrix} \in \mathbb{R}^{n+1}$$

and

$$
\begin{aligned}
C &= \bigcup_{q \in Q} (\{q\} \times C_q) & F(x) &= (0, F_q(z)), \\
D &= \bigcup_{q \in Q} (\{q\} \times D_q) & G(x) &= G_q(z).
\end{aligned}
\tag{1.10}
$$

This construction covers the one used in Example 1.7.

Example 1.8. (Combining local and global controllers - revisited) The flow and jump maps of Example 1.7 can be written as in (1.10) by defining

$$F_q(z) := \tilde{f}(z, \kappa_q(z)), \quad G_q(z) := \begin{pmatrix} 3 - q \\ z \end{pmatrix}.$$

The flow and jump sets were already defined in Example 1.7 as in (1.10).

The following example illustrates the use of a discrete state to explicitly model an on/off mechanism.

Example 1.9. (Thermostat) On/off control of a heater for temperature control of a room can be modeled with an explicit discrete state. The evolution of the room's temperature z can be approximated by the differential equation

$$\dot{z} = -z + z_0 + z_\Delta q \,, \tag{1.11}$$

where z_0 represents the natural temperature of the room, z_Δ the capacity of the heater to raise the temperature in the room by always being on, and q the state of the heater, which can be either 1 ("on") or 0 ("off"). Typically, it is desired to keep the temperature between two specified values $z_{\min}$ and $z_{\max}$, given in Fahrenheit units, satisfying the following relationship

$$z_0 < z_{\min} < z_{\max} < z_0 + z_\Delta \,.$$

For purposes of illustration, consider the case when $z_{\min} = 70$ and $z_{\max} = 80$. A control algorithm that attempts to keep the temperature between such thresholds is the following:

```
if q=1 and z >= 80 then
  q = 0
elseif q = 0 and z <= 70 then
  q = 1
end
```

This algorithm implements the following logic: if the heater is "on" and the temperature is larger than 80, then turn off the heater, while if the heater is "off" and the temperature is smaller than 70, then turn on the heater. With this algorithm, the temperature of the system evolves as shown in Figure 1.10, where parameters $z_0 = 60$ and $z_\Delta = 30$ were used.

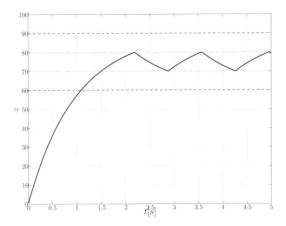

Figure 1.10: Temperature control. Evolution of temperature with control algorithm.

The control logic above results in a hybrid system of the form (1.10) with

$$
\begin{aligned}
F_q(z) &:= -z + z_0 + z_\Delta q, & G_q(z) &:= \begin{pmatrix} 1 - q \\ z \end{pmatrix}, \\
C_0 &:= \{z : z > 70\}, & C_1 &:= \{z : z < 80\}, \\
D_0 &:= \{z : z \leq 70\}, & D_1 &:= \{z : z \geq 80\}.
\end{aligned}
$$

1.4.2   Hybrid automata

Systems with explicit "discrete states" or "logical modes" where, in each logical mode, different jump maps are specified on different subsets of a jump set, or where the jumps are modeled by an automaton, can also be molded into the framework of (1.1). Such systems are usually given by

- a set of *modes* Q, which is identified here with $\{1, 2, \ldots, q_{max}\}$;

- a *domain* mapping Domain, giving for each $q \in Q$ a set Domain(q) in which the continuous state z may evolve;

- a *flow map* $f : Q \times \mathbb{R}^n \to \mathbb{R}^n$, which describes the continuous evolution of the continuous state variable z; in fact, it is enough that $f(q, \cdot)$ be defined on Domain(q), for each $q \in Q$;

- a *set of edges* Edges $\subset Q \times Q$, identifying pairs (q, q') such that a transition from q to q' is possible;

- *guard conditions* which identify, for each edge $(q, q') \in$ Edges, the set Guard(q, q') to which the continuous state z has to belong for transitions from q to q' to be enabled;

- *reset map* Reset : Edges$\times \mathbb{R}^n \to \mathbb{R}^n$, which describes, for each edge $(q, q') \in$ Edges and continuous state $z \in \mathbb{R}^n$, the jump of the continuous state during a transition from q to q'; in fact, it is enough for Reset$(q, q', \cdot)$ to be defined on Guard(q, q'). When the continuous variable z remains constant at jumps from q to q', the reset map Reset$(q, q', \cdot)$ can be taken to be the identity.

To capture the dynamics resulting from such a set of data in the format (1.9), for each $q \in Q$, consider

$$
\begin{aligned}
C_q &= \text{Domain}(q), \\
F_q(z) &= f(q, z) & \forall z \in C_q, \\
D_q &= \bigcup_{(q,q')\in \text{Edges}} \text{Guard}(q, q'), & (1.12) \\
G_q(z) &= \bigcup_{\{q' : z \in \text{Guard}(q,q')\}} \begin{pmatrix} q' \\ \text{Reset}(q, q', z) \end{pmatrix} & \forall z \in D_q.
\end{aligned}
$$

The values of F_q and G_q outside of C_q and D_q, respectively, can be taken to be empty. Such a definition of G_q naturally introduces set-valuedness. Indeed, $G_q(z)$ is a set whenever z is an element of two different guard sets Guard(q, q') and Guard(q, q''). In fact, $G_q(z)$ is a set in such a case even when all reset maps are identities, in other words, when z does not change during jumps.

Example 1.10. (Modeling a hybrid automaton) Consider the hybrid automaton shown in Figure 1.11, with the set of modes $Q = \{1, 2\}$; the domain map given by

$$\text{Domain}(1) = \mathbb{R}_{\geq 0} \times \mathbb{R}, \quad \text{Domain}(2) = \mathbb{R} \times \mathbb{R}_{\geq 0};$$

the flow map, for all $z \in \mathbb{R}^2$, given by

$$f(1, z) = \begin{pmatrix} 1 \\ 0 \end{pmatrix}, \quad f(2, z) = \begin{pmatrix} z_2 \\ -z_1 \end{pmatrix};$$

the set of edges given by Edges $= \{(1,1),(1,2),(2,1)\}$; the guard map given by

$$\text{Guard}(1,1) = \{0\} \times \mathbb{R}_{\geq 0}, \quad \text{Guard}(1,2) = \{0\} \times \mathbb{R}_{\leq 0}, \quad \text{Guard}(2,1) = [1,3] \times \{0\};$$

and the reset map, for all $z \in \mathbb{R}^2$, given by

$$\text{Reset}(1,1,z) = (-1,0), \quad \text{Reset}(1,2,z) = z, \quad \text{Reset}(2,1,z) = -z.$$

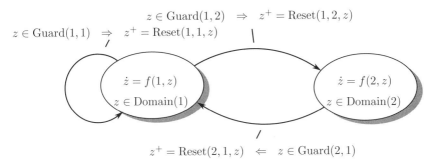

Figure 1.11: Two modes of the hybrid automaton.

The sets Guard$(1,1)$ and Guard$(1,2)$ overlap, indicating that in mode 1, a reset of the state z to $(-1,0)$ or a switch of the mode to 2 is possible from $z = 0$. Formulating this hybrid automaton as a hybrid system with explicitly shown modes (1.9) leads to $C_1 = \text{Domain}(1) = \mathbb{R}_{\geq 0} \times \mathbb{R}$, $C_2 = \text{Domain}(2) = \mathbb{R} \times \mathbb{R}_{\geq 0}$, $F_1(z) = f(1,z) = \begin{pmatrix} 1 \\ 0 \end{pmatrix}$, $F_2(z) = f(2,z) = \begin{pmatrix} z_2 \\ -z_1 \end{pmatrix}$, $D_1 = \text{Guard}(1,1) \cup \text{Guard}(1,2) = \{0\} \times \mathbb{R}$, a set-valued jump map G_1 given by

$$G_1(z) = \begin{cases} (1,-1,0), & \text{if } z_1 = 0, z_2 > 0, \\ (1,-1,0) \cup (2,z), & \text{if } z = 0, \\ (2,z), & \text{if } z_1 = 0, z_2 < 0, \end{cases}$$

and $G_2(z) = (1,-z)$. Formulating the system with explicitly shown modes just described as (1.1) leads to a hybrid system in $\mathbb{R}^3$, where x_1 corresponds to q, x_2

corresponds to z_1, x_3 corresponds to z_2, and the data is given by

$$
\begin{aligned}
C &= \left(\{1\} \times \mathbb{R}_{\geq 0} \times \mathbb{R}\right) \cup \left(\{2\} \times \mathbb{R} \times \mathbb{R}_{\geq 0}\right), \\[1mm]
F(x) &= \begin{cases} (0, -1, 0), & \text{if } x_1 = 1, \\ (0, x_3, -x_2), & \text{if } x_1 = 2, \end{cases} \\[1mm]
D &= \left(\{1\} \times \{0\} \times \mathbb{R}\right) \cup \left(\{2\} \times [1, 3] \times \{0\}\right), \\[1mm]
G(x) &= \begin{cases} \begin{cases} (1, -1, 0), & \text{if } x_2 = 0, x_3 > 0, \\ (1, -1, 0) \cup (2, x_2, x_3), & \text{if } x_2 = 0, x_3 = 0, \\ (2, x_2, x_3), & \text{if } x_2 = 0, x_3 < 0, \end{cases} & \text{if } x_1 = 1, \\[4mm] (1, -x_2, -x_3), & \text{if } x_1 = 2. \end{cases}
\end{aligned}
$$

Figure 1.12 gives a pictorial representation of the data of the hybrid automaton as a hybrid system.

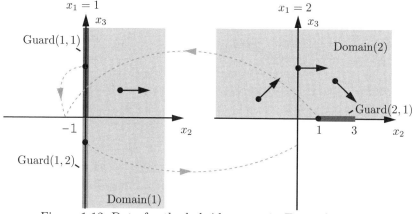

Figure 1.12: Data for the hybrid system in Example 1.10.

1.4.3 Impulsive differential equations

Consider the differential equation

$$
\dot{z} = f(z),
$$

for some $f : \mathbb{R}^n \to \mathbb{R}^n$, with impulses leading to instantaneous change at pre-determined times $t_1, t_2, t_3, \ldots$, according to

$$
\Delta z(t_i) = g(z, t_i),
$$

for some $g : \mathbb{R}^n \times \mathcal{T} \to \mathbb{R}^n$ and $\mathcal{T} = \{t_1, t_2, \dots\}$. For simplicity, suppose that $\{t_i\}_{i=1}^{\infty}$ is an increasing and divergent to ∞ sequence of positive numbers. Such impulsive differential equations can be modeled in the format (1.1). The straight-forward approach is to consider $x = (z, \tau) \in \mathbb{R}^{n+1}$ and the hybrid system

$$C = \mathbb{R}^n \times (\mathbb{R}_{\geq 0} \setminus \mathcal{T}) \qquad F(x) = \begin{pmatrix} f(z) \\ 1 \end{pmatrix}$$

$$D = \mathbb{R}^n \times \mathcal{T} \qquad G(x) = \begin{pmatrix} z + g(z, \tau) \\ \tau \end{pmatrix}$$

and consider initial conditions with $\tau = 0$, so that the τ-variable represents time. The discussion in Chapter 4 will show that such a formulation is not robust to perturbations. A preferred approach may be to consider $x = (z, \tau_1, \tau_2) \in \mathbb{R}^{n+2}$ and the hybrid system

$$C = \mathbb{R}^n \times \mathbb{R}_{\geq 0} \times \mathbb{R}_{\geq 0} \qquad F(x) = \begin{pmatrix} f(z) \\ 1 \\ -1 \end{pmatrix}$$

$$D = \mathbb{R}^n \times \mathcal{T} \times \{0\} \qquad G(x) = \begin{pmatrix} z + g(z, \tau_1) \\ \tau_1 \\ t_{i+1} - t_i \end{pmatrix} \quad \text{when} \quad \tau_1 = t_i.$$

Here the variable τ_1 represents time and variable τ_2 is a timer that ensures, robustly, that flow does not occur when $\tau_1 \in \mathcal{T}$.

1.4.4 Switching systems

Broadly speaking, switching systems are continuous-time systems given by a family of differential equations, where the particular differential equation that governs the evolution of the state at any given time instant is determined by a switching rule/signal. That is, consider a set Q and, for each $q \in Q$, a function $f_q : \mathbb{R}^n \to \mathbb{R}^n$. When the switching signal σ, which takes on values in Q, remains constant, the variable z evolves continuously according to the differential equation

$$\dot{z} = f_\sigma(z).$$

When a switch in the switching signal occurs, it can be described by $\sigma^+ \in Q$. This can be modeled by the following hybrid system:

$$\begin{cases} z \in \mathbb{R}^n, q \in Q & \dot{z} = f_q(z) \\ z \in \mathbb{R}^n, q \in Q & q^+ \in Q \end{cases} \tag{1.13}$$

with the state

$$x = \begin{pmatrix} z \\ q \end{pmatrix} \in \mathbb{R}^{n+1}.$$

In (1.13), q remains constant during flow and z remains constant during jumps. More rigorously, (1.13) fits the framework of hybrid systems with

$$\begin{pmatrix} \dot{z} \\ \dot{q} \end{pmatrix} = F(x) := \begin{pmatrix} f_q(z) \\ 0 \end{pmatrix}, \quad \begin{pmatrix} z^+ \\ q^+ \end{pmatrix} \in G(x) := \begin{pmatrix} z \\ Q \end{pmatrix},$$

while $C = D = \mathbb{R}^n \times Q$.

The model (1.13) is not very helpful in practice. For example, when a switching signal is given, a switching system is just a time-varying differential equation. Similarly, questions about behavior of a switching system under all possible switching signals are better handled in a framework of differential inclusions and not through the analysis of (1.13). This is further justified by Corollary 4.24 and the discussion surrounding it.

On the other hand, a hybrid systems approach to the analysis of switching systems is useful when only switching signals from certain classes are allowed. For example, when the frequency of switching is limited, a clock state can be introduced to limit how many switches occur in a given time interval. Modeling such cases is presented in Section 2.4, where the relationships between solutions to switching systems and to hybrid systems that model them are also discussed.

1.5 NOTES

The model (1.1) or (1.2), identifying the data of a hybrid system as consisting of a flow set, flow map, jump set, and jump map, was proposed in Goebel et al. [37] and more formally stated in Goebel and Teel [40]. Models closely related to (1.1), also involving set-valued dynamics, appeared previously in Aubin and Haddad [8] and Aubin et al. [9], and concurrently in Collins [29]. Early consideration of set-valued dynamics in hybrid systems is found in Puri and Varaiya [98] and Aubin [5].

Notable early references with models of hybrid systems that distinguish between "continuous states" and "discrete states" or use the language of hybrid automata include Witsenhausen [128], Tavernini [116], Alur et al. [1], Henzinger [51], doctoral dissertations by Branicky [18] and Lygeros [78], and the book by van der Schaft and Schumacher [123]. A thorough discussion of numerous early models of hybrid systems is included in [18]. References for impulsive differential equations, as summarized here, include several books: Bainov and Simeonov [13], Lakshmikantham et al. [66], Yang [129], and Haddad et al. [45]. The standard reference for switching systems is Liberzon [73].

A different approach to modeling the behavior of mechanical systems with friction, unilateral constraints, and impacts is visible in Moreau [93], Monteiro Marques [86], and Brogliato [20], with an extensive review of mathematical literature in Stewart [114]. The approach often leads to dynamical complementarity systems, which mix differential equations and complementarity systems common in optimization. Relation of such dynamical systems to hybrid systems is discussed by van der Schaft and Schumacher [122] and, in the linear case, by Heemels et al. [50]. See also Heemels and Brogliato [49]. Numerous references in the area are listed by Brogliato [21]. Closely related is the framework of measure-driven differential equations and inclusions; see Dal Maso and Rampazzo [30] and Silva and Vinter [111].

A detailed discussion of the Newton's cradle can be found in [123]. The flashing fireflies model in Example 1.2 draws inspiration from the hybrid model used by Strogatz and Mirollo [92]. The adjustment mechanism in the quantized control system and the typical conditions on the quantizer used in Example 1.5 are taken from Liberzon [72]. Further analysis of reset systems can be found in Beker et al. [15] and Nešić et al. [95]. The idea behind the hybrid control strategy in Example 1.7 applies to arbitrary nonlinear control systems and state-feedback laws by Prieur [97], and also motivated the hybrid control strategy in Sanfelice and Teel [106] combining state-feedback and open-loop laws. The illustration of the hybrid control strategy in this example on mode-switching control algorithms for hard disk drives follows the algorithms reported in Goh et al. [43], Venkataramanana et al. [124], and Taghirad and Jamei [115].

Chapter Two

The solution concept

A rigorous development of the concept of a solution to a hybrid system is the topic of this chapter. The data of a hybrid system is defined, and a generalized concept of time is introduced. Solutions to a hybrid system are defined and basic properties of solutions, like their existence and uniqueness, are addressed. The concept of a solution is further illustrated by hybrid models of switching systems under broad families of switching signals.

2.1 DATA OF A HYBRID SYSTEM

From now on, a hybrid system is identified with the model describing it, in the form (1.1) or (1.2). Data of a hybrid system is formally defined below, after an introduction of some basic terminology regarding set-valued mappings. Solutions to a hybrid system are defined in Section 2.3.

A *set-valued mapping* from $\mathbb{R}^m$, or from a subset S of $\mathbb{R}^m$, associates, with every point $x \in \mathbb{R}^m$, or every point $x \in S$, a subset of $\mathbb{R}^n$. The double arrow notation $M : \mathbb{R}^m \rightrightarrows \mathbb{R}^n$ or $M : S \rightrightarrows \mathbb{R}^n$ distinguishes a set-valued mapping M from a function. The notation $M : \mathbb{R}^m \rightrightarrows S$, for $S \subset \mathbb{R}^n$, indicates that $M : \mathbb{R}^m \rightrightarrows \mathbb{R}^n$ is a set-valued mapping with $M(x) \subset S$ for all $x \in \mathbb{R}^m$.

Definition 2.1. (Domain of a set-valued mapping) *Given a set-valued mapping* $M : \mathbb{R}^m \rightrightarrows \mathbb{R}^n$, *the* domain *of* M *is the set*

$$\operatorname{dom} M = \{x \in \mathbb{R}^m \: : \: M(x) \neq \emptyset\}.$$

As suggested in Chapter 1, a hybrid system in this book is represented by four elements of data.

Definition 2.2. (Data of a hybrid system) Data of a hybrid system *in* $\mathbb{R}^n$ *consists of four elements:*

- *a set* $C \subset \mathbb{R}^n$, *called the* flow set;

- *a set-valued mapping* $F : \mathbb{R}^n \rightrightarrows \mathbb{R}^n$ *with* $C \subset \operatorname{dom} F$, *called the* flow map;

- *a set* $D \subset \mathbb{R}^n$, *called the* jump set;

- *a set-valued mapping* $G : \mathbb{R}^n \rightrightarrows \mathbb{R}^n$ *with* $D \subset \operatorname{dom} G$, *called the* jump map.

A hybrid system with the data as above will be represented by the notation $\mathcal{H} = (C, F, D, G)$ or, briefly, by $\mathcal{H}$.

Data of a hybrid system in $\mathbb{R}^n$ given by a flow set C, jump set D, and functions $f : C \to \mathbb{R}^n$ and $g : D \to \mathbb{R}^n$ as the flow map and jump map, respectively, does fall under Definition 2.2. Indeed, it suffices to identify the function $f : C \to \mathbb{R}^n$ with a set-valued mapping on $\mathbb{R}^n$ defined on C by f and having empty values outside of C, and to repeat this identification for the jump map.

More generally, given a set $S \subset \mathbb{R}^m$, a set-valued mapping $M : S \rightrightarrows \mathbb{R}^n$ can be trivially extended to a mapping (with some abuse of notation) $M : \mathbb{R}^m \rightrightarrows \mathbb{R}^n$, by setting $M(x) = \emptyset$ for $x \notin S$. The concept of the domain, defined for mappings from $\mathbb{R}^m$ to $\mathbb{R}^n$, when applied to a mapping from $S \subset \mathbb{R}^m$, should be understood as the domain of the trivial extension. Hence, for $M : S \rightrightarrows \mathbb{R}^n$, $\operatorname{dom} M = \{x \in \mathbb{R}^m : M(x) \neq \emptyset\} = \{x \in S : M(x) \neq \emptyset\}$.

Set-valued analysis suggests a reduction — at least a visual one — from four to two in the number of the elements of data of a hybrid system. That is, the flow set and the jump set information can be captured by a new flow map and a new jump map, thanks to the following trick: augmenting F and G to be empty-valued outside C and D, respectively. More specifically, given C, F, D, G as in Definition 2.2, let the set-valued mappings $F_C : \mathbb{R}^n \rightrightarrows \mathbb{R}^n$, $G_D : \mathbb{R}^n \rightrightarrows \mathbb{R}^n$ given by

$$F_C(x) = \begin{cases} F(x) & \text{if } x \in C, \\ \emptyset & \text{if } x \notin C, \end{cases} \qquad G_D(x) = \begin{cases} G(x) & \text{if } x \in D, \\ \emptyset & \text{if } x \notin D. \end{cases}$$

Then, $\operatorname{dom} F_C = C$, $\operatorname{dom} G_D = D$, and the inclusions $\dot{x} \in F_C(x)$ and $x^+ \in G_D(x)$ implicitly require that $x \in C$ and $x \in D$, respectively, provided that $\dot{x}$ and x^+ exist. Hence, only two elements of data could be sufficient. Such a simplification is not done in this book for a couple of reasons. One reason is that the full set of data, including the flow and the jump sets, yields a more explicit description. Another reason is that in many aspects of the analysis, the geometry of the flow and of the jump sets needs to be analyzed, and it is more convenient to have them given explicitly.

2.2 HYBRID TIME DOMAINS AND HYBRID ARCS

In continuous-time systems, solutions are parameterized by $t \in \mathbb{R}_{\geq 0}$, in other words, by time, and in discrete-time systems, solutions are parameterized by $j \in \mathbb{N}$, that is, by the number of jumps or discrete steps. For hybrid systems, it is natural to suggest that solutions be parameterized by both t, the amount of time passed, and j, the number of jumps that have occurred. Of course, it is impossible to parameterize a particular evolution of a hybrid system with all $(t, j) \in \mathbb{R}_{\geq 0} \times \mathbb{N}$. For example, for an evolution in which three jumps occur before the total time of flow reaches two seconds, it makes no sense to ask what is happening after four seconds of flow and before any jumps. More precisely, only certain subsets of $\mathbb{R}_{\geq 0} \times \mathbb{N}$ can correspond to evolutions of hybrid systems. Such sets are called *hybrid time domains*.

Definition 2.3. (Hybrid time domains) *A subset $E \subset \mathbb{R}_{\geq 0} \times \mathbb{N}$ is a compact hybrid time domain if*

$$E = \bigcup_{j=0}^{J-1} ([t_j, t_{j+1}], j)$$

for some finite sequence of times $0 = t_0 \leq t_1 \leq t_2 \leq \ldots \leq t_J$. It is a hybrid time domain if for all $(T, J) \in E$, $E \cap ([0, T] \times \{0, 1, \ldots, J\})$ is a compact hybrid domain.

Equivalently, E is a compact hybrid time domain if E is a union of a finite sequence of intervals $[t_j, t_{j+1}] \times \{j\}$, while E is a hybrid time domain if it is a union of a finite or infinite sequence of intervals $[t_j, t_{j+1}] \times \{j\}$, with the last interval (if existent) possibly of the form $[t_j, T)$ with T finite or $T = \infty$.

Figure 2.1 shows an example of a hybrid time domain E given by the sequence of times $0 = t_0 < t_1 < t_2 = t_3 < t_4$. Note that for $(T, J) \in E$ in Figure 2.1, $E \cap ([0, T] \times \{0, 1, \ldots, J\})$ is a compact hybrid domain. The figure suggests that for each hybrid time domain E, there is a natural (lexicographical) way of ordering its points: given $(t, j), (t', j') \in E$, $(t, j) \preceq (t', j')$ if $t < t'$ or $t = t'$ and $j \leq j'$. Equivalently, as long as the points are taken from the same time domain E, $(t, j) \preceq (t', j')$ if $t + j \leq t' + j'$. Points in two different hybrid time domains need not be comparable. For example, points $(1, 0)$ and $(0, 1)$ — which cannot belong to the same hybrid time domain — are not comparable: it is not the case that either $(1, 0) \preceq (0, 1)$ or $(1, 0) \preceq (0, 1)$.

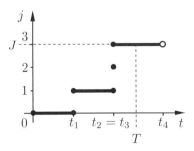

Figure 2.1: A hybrid time domain E. For the point $(T, J) \in E$, the set $E \cap ([0, T] \times \{0, 1, \ldots, J\})$ is a compact hybrid time domain.

Given a hybrid time domain E,

$$\sup_t E = \sup \{t \in \mathbb{R}_{\geq 0} \: : \: \exists \, j \in \mathbb{N} \text{ such that } (t, j) \in E\},$$
$$\sup_j E = \sup \{j \in \mathbb{N} \: : \: \exists \, t \in \mathbb{R}_{\geq 0} \text{ such that } (t, j) \in E\}.$$

That is, the operations $\sup_t$ and $\sup_j$ on a hybrid time domain E return the supremum of the t and j coordinates, respectively, of points in E. Furthermore, $\sup E = (\sup_t E, \sup_j E)$, and finally, $\text{length}(E) = \sup_t E + \sup_j E$.

Definition 2.4. (Hybrid arc)   *A function* $\phi : E \to \mathbb{R}^n$ *is a* hybrid arc *if E is a hybrid time domain and if for each $j \in \mathbb{N}$, the function $t \mapsto \phi(t, j)$ is locally absolutely continuous on the interval $I^j = \{t : (t, j) \in E\}$.*

In the definition above, the absolute continuity requirement is only relevant for those intervals I^j that have nonempty interiors. In general, intervals I^j may be empty or consist of only one point. On each I^j with a nonempty interior, local absolute continuity of $t \mapsto \phi(t, j)$ means that $t \mapsto \phi(t, j)$ is absolutely continuous on each compact subinterval of I^j. On each such I^j, $t \mapsto \phi(t, j)$ is differentiable almost everywhere, and $\dot{\phi}(t, j)$ denotes the time derivative of $\phi(t, j)$, whenever it exists. In short,

$$\dot{\phi}(t, j) = \frac{d}{dt}\phi(t, j).$$

Given a hybrid arc ϕ, the notation $\text{dom}\,\phi$ represents its domain, which is a hybrid time domain. Such notation is consistent with the following "set-valued interpretation" of a hybrid arc. A hybrid arc ϕ can be defined as a set-valued mapping $\phi : \mathbb{R}^2 \rightrightarrows \mathbb{R}^n$ that is single-valued on its domain $\text{dom}\,\phi$ (i.e., on the set of (t, j) on which $\phi(t, j) \neq \emptyset$; recall Definition 2.1), which is a hybrid time domain and for which $t \mapsto \phi(t, j)$ is locally absolutely continuous for each fixed $j \in \mathbb{N}$. Such an interpretation makes it more natural to think that the hybrid time domain $\text{dom}\,\phi$ is determined by the hybrid arc ϕ. This is particularly relevant when talking about hybrid arcs that are solutions to a hybrid system. Then, it is certainly not appropriate to consider an arbitrary hybrid time domain E first, and then try to find a solution with E as a domain. Rather, it is necessary to find a solution ϕ first, and say that its domain $\text{dom}\,\phi$ is determined by ϕ. Furthermore, the "set-valued interpretation" above helps carry over some concepts of convergence and closeness of mappings from the set-valued analysis realm to hybrid arcs and to solutions of hybrid systems in Chapters 4 and 5.

Figure 2.2 shows a graph of a hybrid arc ϕ with hybrid time domain $\text{dom}\,\phi$ that happens to coincide with the hybrid time domain in Figure 2.1.

Certain classes of hybrid arcs can be defined based on the structure of their domains.

Definition 2.5. (Types of hybrid arcs)   *A hybrid arc ϕ is called*

- nontrivial *if $\text{dom}\,\phi$ contains at least two points;*

- complete *if $\text{dom}\,\phi$ is unbounded, i.e., if $\text{length}(E) = \infty$;*

- Zeno *if it is complete and $\sup_t \text{dom}\,\phi < \infty$;*

- eventually discrete *if $T = \sup_t \text{dom}\,\phi < \infty$ and $\text{dom}\,\phi \cap (\{T\} \times \mathbb{N})$ contains at least two points;*

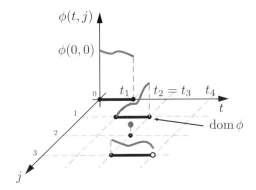

Figure 2.2: Hybrid arc ϕ.

- discrete *if nontrivial and* $\operatorname{dom}\phi \subset \{0\} \times \mathbb{N}$;

- eventually continuous *if* $J = \sup_j \operatorname{dom}\phi < \infty$ *and* $\operatorname{dom}\phi \cap (\mathbb{R}_{\geq 0} \times \{J\})$ *contains at least two points;*

- continuous *if nontrivial and* $\operatorname{dom}\phi \subset \mathbb{R}_{\geq 0} \times \{0\}$;

- compact *if* $\operatorname{dom}\phi$ *is compact.*

The hybrid time domains associated with some of the classes defined above are shown in Figure 2.3.

Note that completeness of a hybrid arc ϕ does not imply that $\sup_t \operatorname{dom}\phi = \infty$. Symmetrically, it does not imply that $\sup_j \operatorname{dom}\phi = \infty$ either. Every discrete hybrid arc is eventually discrete. A complete and eventually discrete hybrid arc is Zeno, but Zeno describes a far more general class than complete and eventually discrete hybrid arcs. Sometimes, Zeno arcs that are not eventually discrete are referred to as "genuinely Zeno" while complete and discrete arcs are referred to as "instantaneously Zeno."

2.3   SOLUTIONS AND THEIR BASIC PROPERTIES

Given a hybrid system (C, F, D, G), its solutions are hybrid arcs ϕ that satisfy certain conditions determined by the hybrid time domain $\operatorname{dom}\phi$ and the data of the hybrid system. Figure 2.4 illustrates the definition below.

Definition 2.6. *(Solution to a hybrid system) A hybrid arc ϕ is a solution to the hybrid system (C, F, D, G) if $\phi(0,0) \in \overline{C} \cup D$, and*

(S1) for all $j \in \mathbb{N}$ such that $I^j := \{t : (t,j) \in \operatorname{dom}\phi\}$ has nonempty interior

$$
\begin{aligned}
\phi(t,j) &\in C && \text{for all} && t \in \operatorname{int} I^j, \\
\dot{\phi}(t,j) &\in F(\phi(t,j)) && \text{for almost all} && t \in I^j;
\end{aligned}
\qquad (2.1)
$$

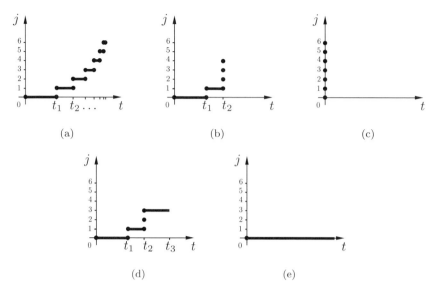

Figure 2.3: Hybrid time domains associated with various arc types: (a) Zeno, (b) eventually discrete, (c) discrete, (d) eventually continuous, and (e) continuous hybrid arcs.

(S2) for all $(t, j) \in \operatorname{dom} \phi$ such that $(t, j + 1) \in \operatorname{dom} \phi$,

$$
\begin{aligned}
\phi(t, j) &\in D, \\
\phi(t, j + 1) &\in G(\phi(t, j)).
\end{aligned}
\tag{2.2}
$$

The definition is quite broad. It does not require that $\phi(t, j) \in C$ at the endpoints of I^j, nor does it insist that $\phi(t, j) \notin D$ when $t \in \operatorname{int} I^j$. Figure 2.4 and Examples 2.8 and 2.9 illustrate the variety of solutions admitted by Definition 2.6.

As in Definition 2.5, solutions to hybrid systems are classified based on their hybrid time domains as nontrivial, complete, Zeno, eventually discrete, discrete, eventually continuous, and continuous. Additionally, solutions that cannot be extended are said to be maximal.

Definition 2.7. (Maximal solutions) *A solution ϕ to $\mathcal{H}$ is maximal if there does not exist another solution ψ to $\mathcal{H}$ such that $\operatorname{dom} \phi$ is a proper subset of $\operatorname{dom} \psi$ and $\phi(t, j) = \psi(t, j)$ for all $(t, j) \in \operatorname{dom} \phi$.*

Clearly, complete solutions are maximal, but the converse statement is not true.

The next two examples illustrate the notion of the solution to a hybrid system. They also verify that maximal solutions with quite different domains can appear in a hybrid system. In fact, Example 2.8 shows that even a hybrid

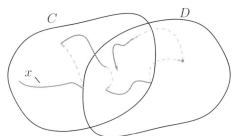

Figure 2.4: Evolution of a solution to a hybrid system. Flows and jumps of the solution x are allowed only on the flow set C and from the jump set D, respectively. The solid curves indicate flow. The dashed arcs indicate jumps. The solid curves must belong to the flow set C. The dashed arcs must originate from the jump set D.

system that is simple in appearance can have maximal solutions from each of the following categories: continuous, eventually continuous but not continuous, eventually discrete but not discrete, and Zeno but not discrete. Example 2.9 shows different kinds of maximal but not complete solutions.

Example 2.8. (Solutions) Consider a hybrid system in $\mathbb{R}^2$ given by

$$C = \mathbb{R}^2 \setminus D \qquad\qquad f(x) = \begin{pmatrix} 1 \\ 1 \end{pmatrix}$$

$$D = \left\{ x \in \mathbb{R}^2 : 0 \le x_2 \le -\frac{1}{5}x_1 + \frac{2}{5} \right\} \qquad g(x) = \begin{pmatrix} \frac{3}{4}x_1 \\ \frac{1}{4}x_1 \end{pmatrix}.$$

The maximal solution from $(2, -1)$, denoted ϕ_a, has the domain $\operatorname{dom}\phi_a = \mathbb{R}_{\ge 0} \times \{0\}$ and is given by

$$\phi_a(t, 0) = \begin{pmatrix} t + 2 \\ t - 1 \end{pmatrix}.$$

The maximal solution from $(1, -1)$, denoted ϕ_b, has the domain $\operatorname{dom}\phi_b = [0, 1] \times \{0\} \cup [1, \infty) \times \{1\}$ and is given by

$$\phi_b(t, 0) = \begin{pmatrix} t + 1 \\ t - 1 \end{pmatrix}, \quad \phi_b(t, 1) = \begin{pmatrix} (t - 1) + \frac{3}{2} \\ (t - 1) + \frac{1}{2} \end{pmatrix}.$$

In simple words, ϕ_b flows from the initial point for one unit of time, reaches $(2, 0)$ and jumps to $\left(\frac{3}{2}, \frac{1}{2}\right)$ from there, and flows afterwards. One maximal solution from $(0, -1)$, denoted ϕ_c, has the domain $\operatorname{dom}\phi_c = \operatorname{dom}\phi_b$ and is given by

$$\phi_c(t, 0) = \begin{pmatrix} t \\ t - 1 \end{pmatrix}, \quad \phi_c(t, 1) = \begin{pmatrix} (t - 1) + \frac{3}{4} \\ (t - 1) + \frac{1}{4} \end{pmatrix}.$$

In simple words, ϕ_c behaves similarly to ϕ_b. Another maximal solution from $(0, -1)$, denoted ϕ_d, has the domain $\operatorname{dom} \phi_d = ([0, 1] \times \{0\}) \cup (\{1\} \times \mathbb{N})$ and is given by

$$\phi_d(t, 0) = \begin{pmatrix} t \\ t - 1 \end{pmatrix}, \quad \phi_d(1, j) = \begin{pmatrix} \left(\frac{3}{4}\right)^j \\ \left(\frac{1}{4}\right)^j \end{pmatrix}.$$

In simple words, ϕ_d flows from the initial point for one unit of time, reaches $(1, 0)$ and jumps to $\left(\frac{3}{4}, \frac{1}{4}\right)$ from there, and then keeps on jumping infinitely many times. The maximal solution from $(0, 0)$, denoted ϕ_e, has the domain $\operatorname{dom} \phi_e = \{0\} \times \mathbb{N}$ and is given by

$$\phi_e(0, j) = \begin{pmatrix} 0 \\ 0 \end{pmatrix}.$$

In simple words, $(0, 0)$ is an equilibrium point, despite the fact that f is nonzero there, and the solution ϕ_e jumps infinitely many times from $(0, 0)$ to $(0, 0)$. The maximal solution from $(-1, 0)$, denoted ϕ_f, has the domain $\operatorname{dom} \phi_f = \bigcup_{j=0}^{\infty}([t_j, t_{j+1}] \times \{j\})$, where $t_0 = 0$, $t_1 = 0$, $t_2 = \frac{1}{4}$, $t_3 = \frac{1}{4} + \frac{1}{8}$, $t_4 = \frac{1}{4} + \frac{1}{8} + \frac{1}{16}$, etc., so that $t_{j+1} - t_j = \frac{1}{4} \left(\frac{1}{2}\right)^{j-1}$ for $j = 1, 2, \ldots$, and is given by

$$\phi_f(0, 0) = \begin{pmatrix} -1 \\ 0 \end{pmatrix}, \quad \phi_f(t, j) = \begin{pmatrix} (t - t_j) - \frac{3}{4}\left(\frac{1}{2}\right)^{j-1} \\ (t - t_j) - \frac{1}{4}\left(\frac{1}{2}\right)^{j-1} \end{pmatrix} \quad \text{for } j = 1, 2, \ldots.$$

Note that solutions ϕ_a, ϕ_b, ϕ_c, ϕ_d, and ϕ_e are maximal and complete. The solution ϕ_a is continuous, ϕ_b and ϕ_c are eventually continuous, ϕ_d is eventually discrete, ϕ_e is discrete, and ϕ_f is Zeno. Solutions ϕ_a, ϕ_b, and ϕ_d are depicted in Figure 2.5.

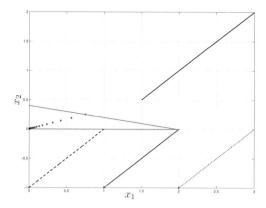

Figure 2.5: Solutions ϕ_a, ϕ_b, and ϕ_d for the system in Example 2.8. The black lines denote the boundary of the jump set.

Example 2.9. (Solutions) Consider a hybrid system in $\mathbb{R}$ given by

$$C = \{x \in \mathbb{R} : |x| \geq 1\} \qquad f(x) = x^2$$
$$D = [0, 1] \qquad g(x) = x - 1.$$

One maximal solution from 1, denoted ϕ_a, has domain $\operatorname{dom} \phi_a = [0, 1) \times \{0\}$ and is given by $\phi_a(t, 0) = (1 - t)^{-1}$. Another maximal solution from 1, denoted ϕ_b, has domain $\operatorname{dom} \phi_b = \{0\} \times \{0, 1, 2\}$ and is given by $\phi_b(0, 0) = 1$, $\phi_b(0, 1) = 0$, $\phi_b(0, 2) = -1$; see Figure 2.6(a). The maximal solution from -2, denoted ϕ_c, has domain $\operatorname{dom} \phi_c = [0, \frac{1}{2}] \times \{0\}$ and is given by $\phi_c = -(\frac{1}{2} + t)^{-1}$; see Figure 2.6(b). Note that solutions ϕ_a, ϕ_b, and ϕ_c are maximal but not complete.

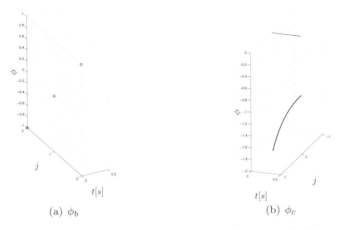

(a) ϕ_b (b) ϕ_c

Figure 2.6: Solutions ϕ_b and ϕ_c for the system in Example 2.9.

Throughout the book, $\mathcal{S}_\mathcal{H}(S)$ denotes the set of all maximal solutions ϕ to $\mathcal{H}$ with $\phi(0, 0) \in S$. For example, writing $\phi \in \mathcal{S}_\mathcal{H}(\xi)$ means that ϕ is a maximal solution to $\mathcal{H}$ with $\phi(0, 0) = \xi$. If no set S is mentioned, $\phi \in \mathcal{S}_\mathcal{H}$ means that ϕ is a maximal solution to $\mathcal{H}$. The following proposition gives natural conditions for the existence of nontrivial solutions to hybrid systems. Furthermore, it characterizes maximal solutions.

Proposition 2.10. (Basic existence) *Consider the hybrid system $\mathcal{H} = (C, F, D, G)$. Let $\xi \in \overline{C} \cup D$. If $\xi \in D$ or*

(VC) *there exists $\varepsilon > 0$ and an absolutely continuous function $z : [0, \varepsilon] \to \mathbb{R}^n$ such that $z(0) = \xi$, $\dot{z}(t) \in F(z(t))$ for almost all $t \in [0, \varepsilon]$ and $z(t) \in C$ for all $t \in (0, \varepsilon]$,*

then there exists a nontrivial solution ϕ to $\mathcal{H}$ with $\phi(0, 0) = \xi$. If (VC) holds for every $\xi \in \overline{C} \setminus D$, then there exists a nontrivial solution to $\mathcal{H}$ from every point of $\overline{C} \cup D$, and every $\phi \in \mathcal{S}_\mathcal{H}$ satisfies exactly one of the following:

(a) ϕ is complete;

(b) dom ϕ is bounded and, with $J = \sup_j \text{dom } \phi$, the interval I^J has nonempty interior and is open to the right, and there does not exist an absolutely continuous function $z : [a, b] \to \mathbb{R}^n$ satisfying $\dot{z}(t) \in F(z(t))$ for almost all $t \in [a, b]$, $z(t) \in C$ for all $t \in (a, b)$, and such that $I^J \subset [a, b)$ and $z(t) = \phi(t, J)$ for all $t \in I^J$;

(c) dom ϕ is bounded and $\phi(T, J) \notin \overline{C} \cup D$, where $(T, J) = \sup \text{dom } \phi$.

Furthermore, if $G(D) \subset \overline{C} \cup D$, then (c) above does not occur.

PROOF. The first conclusion follows from the definition of a solution to $\mathcal{H}$. To see the second conclusion, suppose that ϕ is a maximal solution that is not complete, i.e., dom ϕ is bounded. Let $(T, J) = \sup \text{dom } \phi$. If $(T, J) \in \text{dom } \phi$ and $\phi(T, J) \in \overline{C} \cup D$, then either $\phi(T, J) \in D$ in which case ϕ can be extended via a jump, or $\phi(T, J) \in \overline{C} \setminus D$ in which case ϕ can be extended via flow, thanks to (VC). This contradicts maximality of ϕ. Thus either (c) holds or $(T, J) \notin \text{dom } \phi$. If the latter holds, then the interior of I^J is nonempty, and (b) must hold to ensure maximality of ϕ. Indeed, if (b) failed, ϕ could be extended to a solution to $\mathcal{H}$ on $\overline{\text{dom } \phi}$. ☐

Case (b) above essentially says that the interval I^J has nonempty interior and $t \mapsto \phi(t, J)$ is a maximal solution, in an appropriate sense, to the constrained inclusion $\dot{z} \in F(z)$, $z \in C$. In particular, $t \mapsto \phi(t, J)$ could be a solution to the inclusion which blows up in finite time.

In Example 2.8, condition (VC) holds at every point in $\overline{C}$ except $x = (x_1, 0)$ with $x_1 < 2$, which are points that belong to D. Consequently, nontrivial solutions exist for every initial point in $\overline{C} \cup D = \mathbb{R}^2$. Since $G(D) \subset \overline{C} \cup D$, case (c) in Proposition 2.10 does not occur. Further careful analysis reveals that (b) does not occur, and thus all maximal solutions for the system in Example 2.8 are complete.

In Example 2.9, condition (VC) fails at $x = -1$, and there does exist a maximal solution, ϕ_c, that does not fall into either of the three cases in Proposition 2.10. Furthermore, the maximal solution ϕ_a blows up in finite time and falls into case (b), while the maximal solution ϕ_b jumps out of $\overline{C} \cup D$ and falls into case (c).

Conditions for uniqueness of solutions, stated below, are straightforward. The three conditions in Proposition 2.11 say, respectively, that from no point there exist two flowing solutions, from no point there exists a flowing solution and a jumping solution, and from no point there exist two jumping solutions.

Proposition 2.11. (Basic uniqueness) *Consider the hybrid system* $\mathcal{H} = (C, F, D, G)$. *For every* $\xi \in \overline{C} \cup D$ *there exists a unique maximal solution* ϕ *with* $\phi(0, 0) = \xi$ *provided that the following conditions hold:*

(a) for every $\xi \in \overline{C} \backslash D$, $T > 0$, if two absolutely continuous $z_1, z_2 : [0,T] \to \mathbb{R}^n$ are such that $\dot{z}_i(t) \in F(z_i(t))$ for almost all $t \in [0,T]$, $z_i(t) \in C$ for all $t \in (0,T]$, and $z_i(0) = \xi$, $i = 1, 2$, then $z_1(t) = z_2(t)$ for all $t \in [0,T]$;

(b) for every $\xi \in \overline{C} \cap D$, there does not exist $\varepsilon > 0$ and an absolutely continuous $z : [0,\varepsilon] \to \mathbb{R}^n$ such that $z(0) = \xi$, $\dot{z}(t) \in F(z(t))$ for almost all $t \in [0,\varepsilon]$ and $z(t) \in C$ for all $t \in (0,\varepsilon]$;

(c) for every $\xi \in D$, $G(\xi)$ consists of one point.

Example 2.12. (Bouncing ball — solutions) Consider the bouncing ball model of Example 1.1. The conditions for existence of solutions, as in Proposition 2.10, are satisfied. Indeed, initial points ξ with $\xi_1 > 0$ are in the interior of C, f is continuous on the interior of C, and (VC) follows from classical results for differential equations. For initial points ξ with $\xi_1 = 0$ and $\xi_2 > 0$, $f(\xi)$ points into the interior of C and this suggests that (VC) does hold. A rigorous justification is possible via Lemma 5.26. Finally, from the initial point $\xi = 0$, there is the obvious solution $\phi(t,0) = 0$ for all $t \in \mathbb{R}_{\geq 0}$.

Maximal solutions are complete. This can be shown by inspection, but also through Proposition 2.10. Since $g(D) \subset C$, situation (c) in the proposition is excluded. Furthermore, every solution to $\dot{z} \in F(z)$, $z \in C$ defined on an interval open to the right can be extended to an interval including the right endpoint, which excludes situation (b) in the proposition. Indeed, every solution is bounded, which can be shown using the total energy function; see Example 3.19 for details. Then, it is enough to note that the maximal solution from $(0,0)$ is complete while continuous solutions from other points evolve in a set on which the flow map is continuous, and hence can be extended. Hence, only situation (a) in Proposition 2.10 is possible.

Every maximal solution from a nonzero initial condition is Zeno. Given an initial condition ξ with $\xi_1 > 0$ or $\xi_1 = 0$, $\xi_2 \neq 0$, the first jump of the solution ϕ from ξ, corresponding to the first bounce of the ball, is

$$t_1 = \frac{\xi_2 + \sqrt{\xi_2^2 + 2\gamma\xi_1}}{\gamma} . \tag{2.3}$$

Recursive use of (2.3) shows that

$$\sup_t \text{dom}\,\phi = \frac{\xi_2 + \sqrt{\xi_2^2 + 2\gamma\xi_1}}{\gamma} + \frac{2\lambda\sqrt{\xi_2^2 + 2\gamma\xi_1}}{\gamma(1-\lambda)} . \tag{2.4}$$

As long as $\lambda < 1$, $\sup_t \text{dom}\,\phi < \infty$ and since it was already shown that ϕ is complete, ϕ is Zeno. Figure 2.7 depicts two such solutions and corresponding hybrid time domains for different initial conditions.

2.4 GENERATORS FOR CLASSES OF SWITCHING SIGNALS

Broadly speaking, switching systems are continuous-time systems given by a family of differential equations, where the particular differential equation that

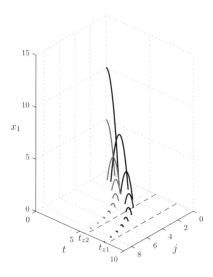

(a) Solutions and their $\sup_t \operatorname{dom} \phi$ denoted by t_z.

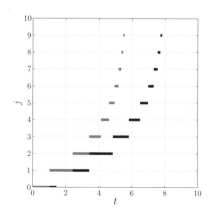

(b) Hybrid time domains of each of the solutions.

Figure 2.7: The height variable for two solutions to the bouncing ball system.

governs the evolution of the state at a given time instant is determined by a switching rule/signal. An early discussion and modeling of such systems in the hybrid system framework was given in Section 1.4.

Often, there is interest not in the behavior of a switching system under a particular switching signal, but rather, under all switching signals belonging to some class. Switching systems under various classes of signals can be modeled in the hybrid framework. A great benefit of this modeling effort is that information about switching signals can be embedded in the data of a hybrid system; such an embedding is impossible in the switching systems framework.

More precisely, let $Q = \{1, 2, \ldots, q_{max}\}$, and for each $q \in Q$, let $f_q : \mathbb{R}^n \to \mathbb{R}^n$ be continuous. A switching system can be represented by

$$\dot{z} = f_\sigma(z). \tag{2.5}$$

Typically, a solution to (2.5) is considered to consist of a piecewise constant function σ taking values in Q and a continuous and piecewise differentiable function z, with σ being right continuous and having left limits (such functions are called CADLAG, from the French "continue à droite, limite à gauche"), and satisfying

$$\dot{z}(t) = f_{\sigma(t)}(z(t))$$

at all times t except the "switching instants."

As long as σ has finitely many discontinuities in every compact subinterval of its domain of definition, a solution to (2.5) can be identified with a solution to the hybrid system (1.13) with the state $x = (z, q) \in \mathbb{R}^{n+1}$. In (1.13), q remains constant during flow and z remains constant during jumps. If σ in a solution of (2.5) is such that its discontinuities have accumulation points other than possibly the right endpoint of the domain of definition of σ, then hybrid time domains are not rich enough to describe such a solution while keeping track of the values of q. In the reverse direction, there are many solutions to (1.13) that do not directly correspond to a solution to the switching system; this includes all solutions that undergo many jumps at the same time instants, and in particular, the discrete solutions.

Many applications call for consideration of only particular types of switching signals. Suppose $z : [0, \infty) \to \mathbb{R}^n$ and $\sigma : [0, \infty) \to Q$ form a solution to the switching system (2.5). Let $t_0 = 0$, and $t_1, t_2, \ldots$ be the consecutive (positive) times at which σ is discontinuous (i.e., the switching times). Then

- σ is a *dwell-time signal* and the solution is a dwell-time solution with dwell time $\tau_D > 0$ if $t_{i+1} - t_i \geq \tau_D$ for $i = 1, 2, \ldots$. That is, jumps are separated by at least τ_D amount of time; see Figure 2.8(a).

- σ is a *persistent dwell-time signal* with persistent dwell time $\tau_D > 0$ and period of persistence $T > 0$ if there exists a subsequence $0 = t_{i_0}, t_{i_1}, t_{i_2}, \ldots$ of the sequence $\{t_i\}$ such that $t_{i_k+1} - t_{i_k} \geq \tau_D$ for $k = 1, 2, \ldots$ and $t_{i_{k+1}} - t_{i_k+1} \leq T$ for $k = 0, 1, \ldots$. (That is, at most T amount of time passes between two consecutive intervals of length at least τ_D on which there are no jumps.)

- σ is a *weak dwell-time signal* with dwell time $\tau_D > 0$ if there exists a subsequence $0 = t_{i_0}, t_{i_1}, t_{i_2}, \ldots$ of the sequence $\{t_i\}$ such that $t_{i_1+1} - t_{i_1} \geq \tau_D$ for $k = 1, 2, \ldots$. (That is, there are infinitely many intervals of length τ_D with no switching.)

- σ is an *average dwell-time signal* with dwell time $\tau_D > 0$ and offset $N_o \in \mathbb{N}$ if, for all $0 \le s < t$, $N(t,s) \le \frac{1}{\tau_D}(t-s) + N_o$ where $N(t,s)$ denotes the number of switching times in the interval $[s,t]$. Figure 2.8(b) depicts a hybrid time domain for an average dwell-time solution for parameters τ_D and N_o satisfying such condition with, for example, $\tau_D = \frac{t_2}{4}$ and $N_o = 4$. When $N_o = 1$, this class of signals agrees with the class of dwell-time signals with dwell time τ_D.

The switching system (2.5) with switching restricted to some of the above classes can be represented in the hybrid framework with a close correspondence between the solutions of (2.5) and the solutions of the associated hybrid system.

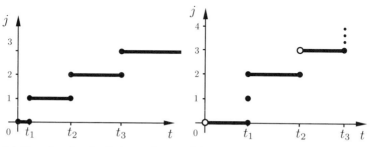

(a) Domain of a dwell-time solution with constant τ_D larger or equal than $\min\{t_2 - t_1, t_3 - t_2, \ldots\}$.

(b) Domain of an average dwell-time solution. The domain repeats periodically, as denoted by the white dots.

Figure 2.8: Hybrid time domain of solutions to a switched system.

Example 2.13. (Dwell-time signals and hybrid systems) Consider a hybrid system

$$x = \begin{pmatrix} z \\ q \\ \tau \end{pmatrix} \in \mathbb{R}^{n+2} \qquad \begin{cases} \begin{aligned} \dot{z} &= f_q(z) \\ \dot{\tau} &\in [0,1] \end{aligned} & \tau \in [0, \tau_D] \\ \\ \begin{aligned} q^+ &\in Q \\ \tau^+ &= 0 \end{aligned} & \tau = \tau_D. \end{cases} \tag{2.6}$$

As usual, not mentioning $\dot{q}$ or z^+ explicitly suggests that q remains constant during flow and z remains constant during jumps. More precisely, the data for the hybrid system (2.6) is given by

$$F(x) = \begin{pmatrix} f_q(\phi) \\ 0 \\ [0,1] \end{pmatrix}, \quad C = \mathbb{R}^n \times Q \times [0, \tau_D], \quad G(x) = \begin{pmatrix} z \\ Q \\ 0 \end{pmatrix}, \quad D = \mathbb{R}^n \times Q \times \{\tau_D\}.$$

Then, to each dwell-time solution z and q, with dwell-time $\tau_D > 0$, there corresponds a solution to the hybrid system (2.6), and viceversa. For simplicity, this correspondence is explained only for solutions with infinitely many switches/jumps.

Let $t_0 = 0$ and $t_1, t_2, \ldots$ be the consecutive switching times for the dwell-time solution z and q. Then a hybrid arc x with $\operatorname{dom} x = \bigcup_{j=0}^{\infty} [t_j, t_{j+1}] \times \{j\}$ and defined, for each $j \in \mathbb{N}$, by

$$
x(t, j) = \begin{pmatrix} z(t) \\ q(t) \\ \min\{t - t_j, \tau_D\} \end{pmatrix} \quad \forall t \in [t_j, t_{j+1}), \qquad x(t_{j+1}, j) = \lim_{t \nearrow t_{j+1}} x(t, j)
$$

is a solution to (2.6). Now consider a solution $x = \begin{pmatrix} z \\ q \\ \tau \end{pmatrix}$ to (2.6) on $\operatorname{dom} x = \bigcup_{j=0}^{\infty} [t_j, t_{j+1}] \times \{j\}$. Then $z'(t) = z(t, j)$, $q'(t) = q(t, j)$ for $t \in [t_j, t_{j+1})$, $j \in \mathbb{N}$ defines a dwell-time solution to (2.5), with dwell time τ_D.

Example 2.14. (Persistent dwell time and hybrid systems) Persistent dwell-time solutions to switching signals feature intervals of arbitrary switching between the functions f_q. It is convenient, and accurate, to model this with a differential inclusion. Consider a set-valued mapping $\Phi : \mathbb{R}^n \rightrightarrows \mathbb{R}^n$ defined at each $x \in \mathbb{R}^n$ by

$$
\Phi(x) := \overline{\operatorname{con}} \bigcup_{q \in Q} f_q(x),
$$

in other words, $\Phi(x)$ is the closed convex hull of $\bigcup_{q \in Q} f_q(x)$. In particular, for each $x \in \mathbb{R}^n$ and each $q \in Q$, $f_q(x) \in \Phi(x)$, and consequently, if z and q is a solution to the switching system (2.5), then z is a solution to the differential inclusion $\dot{z} \in \Phi(z)$. On the other hand, it can be shown (see Corollary 4.24) that every solution to the differential inclusion can be approximated by a solution to $\dot{z} = f_q(x)$ with appropriately chosen switching signal q.

Now, a switching system under persistent dwell-time switching, with dwell time τ_D and period of persistence T can be modeled by a hybrid system (2.7), given below, with the variable $x = \begin{pmatrix} z \\ q \\ \tau_1 \\ \tau_2 \end{pmatrix} \in \mathbb{R}^{n+3}$. The variable q can take values in Q as well as the value 0, representing a period of arbitrary switching. The timer τ_1 keeps track of the length of intervals where z flows according to $\dot{z} = f_q(z)$ for some $q \in Q$; the timer τ_2 keeps track of the length of intervals where z flows according to the differential inclusion, representing periods of arbitrary switching.

$$\begin{cases} \begin{aligned} \dot{z} &= f_q(z) \\ \dot{\tau}_1 &\in [0,1] \end{aligned} & \qquad q \in Q, \ \tau_1 \in [0, \tau_D] \\[2ex] \begin{aligned} \dot{z} &\in \Phi(z) \\ \dot{\tau}_2 &= 1 \end{aligned} & \qquad q = 0, \ \tau_2 \in [0, T] \\[2ex] \begin{aligned} q^+ &\in Q \setminus \{q\} \cup \{0\} \\ \tau_1^+ &= 0 \end{aligned} & \qquad q \in Q, \ \tau_1 = \tau_D \\[2ex] \begin{aligned} q^+ &\in Q \\ \tau_2^+ &= 0 \end{aligned} & \qquad q = 0, \ \tau_2 = T. \end{cases} \qquad (2.7)$$

Modeling this system in the standard format gives

$$F(x) = \begin{pmatrix} f_q(z) \\ 0 \\ [0,1] \\ 0 \end{pmatrix} \text{ if } q \in Q, \qquad F(x) = \begin{pmatrix} \Phi(z) \\ 0 \\ 0 \\ 1 \end{pmatrix} \text{ if } q = 0,$$

$$C = (\mathbb{R}^n \times Q \times [0, \tau_D] \times \{0\}) \cup (\mathbb{R}^n \times \{0\} \times \{0\} \times [0, T]),$$

$$G(x) = \begin{pmatrix} z \\ Q \setminus \{q\} \cup \{0\} \\ 0 \\ \tau_2 \end{pmatrix} \text{ if } q \in Q, \tau_1 = \tau_D, \qquad G(x) = \begin{pmatrix} z \\ Q \\ \tau_1 \\ 0 \end{pmatrix} \text{ if } q = 0, \tau_2 = T,$$

and

$$D = (\mathbb{R}^n \times Q \times \{\tau_D\} \times \{0\}) \cup (\mathbb{R}^n \times \{0\} \times \{0\} \times \{T\}).$$

To each persistent dwell-time solution to (2.5) with dwell time τ_D and period of persistence T, there corresponds one solution to (2.7) initialized with $\tau_1 = \tau_2 = 0$ and for which τ_1 increases at the rate 1 until it reaches τ_D. (In general, different behaviors of τ_1 can lead to the same length of the interval with no switching of $q \in Q$.) Furthermore, the z coordinate of a solution to (2.7) with the properties just described, parameterized by time only, can be approximated, with arbitrary precision on compact time intervals, with a z coordinate of a persistent dwell-time solution to (2.5).

Example 2.15. (Average dwell-time signals and hybrid systems) To each average dwell-time solution with dwell time $\tau_D > 0$ and offset N_o there corresponds a hybrid time domain such that, for each of its elements (s, i) and (t, j) with $(s, i) \preceq (t, j)$, the following bound holds:

$$j - i \leq \frac{1}{\tau_D}(t - s) + N_o. \qquad (2.8)$$

Such hybrid time domains can be generated by the simple hybrid system

$$\tau \in \mathbb{R} \qquad \begin{cases} \dot{\tau} \in [0, 1/\tau_D] & \tau \in [0, N_\circ] \\ \tau^+ = \tau - 1 & \tau \in [1, N_\circ] . \end{cases} \qquad (2.9)$$

The time domain for each solution of this hybrid system satisfies the constraint (2.8), and viceversa: for every hybrid time domain satisfying (2.8) there exists a solution of (2.9), starting at $\tau = N_\circ$, having the said hybrid time domain. In turn, switching systems with average dwell-time signals with parameters τ_D and $N_\circ$ can be modeled with the hybrid system

$$\begin{pmatrix} z \\ q \\ \tau \end{pmatrix} \in \mathbb{R}^{n+2} \qquad \begin{cases} \begin{aligned} \dot{z} &= f_q(z) \\ \dot{\tau} &\in [0, 1/\tau_D] \end{aligned} & q \in Q, \ \tau \in [0, N_\circ] \\[2ex] \begin{aligned} q^+ &\in Q \setminus \{q\} \\ \tau^+ &= \tau - 1 \end{aligned} & q \in Q, \ \tau \in [1, N_\circ]. \end{cases} \qquad (2.10)$$

There is a one-to-one correspondence between solutions to (2.10) starting with $\tau(0,0) = N_\circ$ for which τ flows at the rate of $1/\tau_D$ when less than $N_\circ$ and solutions to the switching system (2.5) with switching restricted to average dwell-time signals with parameters τ_D and $N_\circ$.

2.5   NOTES

Early references, such as Witsenhausen [128] and Tavernini [116], parameterize the solutions to hybrid systems by $t \in \mathbb{R}$. Solutions are then piecewise absolutely continuous or piecewise smooth functions, possessing both left and right limits at discontinuities, satisfying continuous-time dynamics in between the discontinuities and with discrete-time dynamics specifying the relationship between the left and right limits at discontinuities. Such an approach to solutions is still common in the impulsive differential equations and switching system literature. Parameterization by $t \in \mathbb{R}$ is also a feature of measure-drive differential equations or inclusions and of dynamical complementarity systems, where solutions are functions of bounded variation. More abstract approaches to solutions of hybrid automata often describe a solution as an alternating sequence of continuous and discrete evolutions. Sometimes, for example, Lynch et al. [81], the discrete evolution is allowed to consist, in the current terminology, of multiple jumps.

Recognition that parameterization of a solution to a hybrid dynamical system by $t \in \mathbb{R}$ is not sufficient for modeling and analytical purposes is evidenced as early as the doctoral theses by Deshpande [33] and Lygeros [78]. The object used there to parameterize a solution is a finite or infinite sequence of intervals $[t_j, t_{j+1}]$, where $t_j \leq t_{j+1}$, $j = 1, 2, \ldots$. Note the nonstrict inequality. This object is featured in later works, under the name "hybrid time trajectory," in Lygeros et al. [80] and [79] and others. Another, essentially equivalent, approach to "hybrid time trajectories" by specifying a nondecreasing sequence of impulse times, is visible in Aubin and Haddad [8]. A hybrid time trajectory is equivalent to a hybrid time domain, as in Definition 2.3, with the distinction that the former is a sequence of intervals in $\mathbb{R}$ while the latter is a subset of $\mathbb{R}^2$, which provides several advantages for analytical purposes. Hybrid time domains appeared first in Goebel et al. [37] and concurrently in Collins [29] under the name "hybrid time set."

The definition of a solution to a hybrid system, Definition 2.6, is similar, subject to translating the terminology of hybrid time trajectories and hybrid time sets and the data given in the format of a hybrid automaton to the current format, to the concepts of a "run" and an "execution" of a hybrid system or a hybrid automaton, as used in [78], [79], and many other works. When C is a closed set, Definition 2.6 is equivalent to the definition of solution proposed in [37].

An approach to unification of continuous-time and discrete-time dynamical system theories, different from what is suggested here, is possible by considering dynamical systems on time scales; see the book by Bohner and Peterson [17]. A time scale is an arbitrary closed subset of the real line, but it is fixed a priori. Modeling capabilities of systems on time scales are hence quite different from those of hybrid inclusions.

Formulation of different classes of switching signals for switching systems is in Hespanha [54]. Modeling of switching systems as hybrid systems was used in asymptotic stability analysis in Sanfelice et al. [104], Cai et al. [25], and Goebel et al. [38].

Chapter Three

Uniform asymptotic stability, an initial treatment

This chapter focuses on uniform asymptotic stability of a closed set. Studying this property provides another opportunity to illustrate the concept of a solution to a hybrid system. Asymptotic stability is a fundamental property of dynamical systems, one that is usually desired in natural and engineered systems. It provides qualitative information about solutions, especially a characterization of the solutions' long-term trends. Asymptotic stability of a closed set, rather than of an equilibrium point, is significant since the solutions of a hybrid system often do not settle down to an equilibrium point. In a sample-and-hold control system, for example, the controlled state is expected to settle down to an equilibrium, while the timer variable does not converge to a point but rather to an interval. Asymptotic stability of an equilibrium point is a special case of asymptotic stability of a closed set. Namely, an equilibrium point is a closed set containing a single point.

This chapter defines uniform asymptotic stability, provides equivalent characterizations, and gives various sufficient conditions for uniform asymptotic stability, focusing especially on Lyapunov functions, which are a staple of stability analysis for classical dynamical systems. Since this chapter is an initial treatment of asymptotic stability, only uniform *global* asymptotic stability is considered. Chapter 7 contains a more detailed treatment of asymptotic stability for a compact (closed and also bounded) set, including a discussion of local asymptotic stability and basins of attraction. The results of this chapter hold without assuming any of the well-posedness properties described in Chapter 6. In contrast, the stability results in Chapter 7, which are in many ways more versatile than the results of this chapter, rely heavily on the well-posedness properties of Chapter 6.

3.1 UNIFORM GLOBAL PRE-ASYMPTOTIC STABILITY

Uniform global pre-asymptotic stability (UGpAS) of a closed set entails the property that the distance of each solution to the set is bounded by a function of two quantities: the initial condition's distance to the set and the amount of elapsed time at which the solution is evaluated; moreover, this bound tends to zero as the initial condition's distance to the set tends to zero or the amount of elapsed hybrid time tends to infinity. "Pre" indicates that maximal solutions are not required to be complete. This aspect will be clarified subsequently.

43

A typical situation is when the closed set is a single point. In this situation, the point is an equilibrium since stability of a set implies that if a solution starts in the set then it remains in the set forever. Examples of sets that consist of many points and which may be of interest in asymptotic stability analysis are now given.

Example 3.1. (Sample-and-hold and compact attractors)   Example 1.4 outlined a sample-and-hold implementation of a feedback controller $u = \kappa(z)$ for a continuous-time control system $\dot{z} = f_p(z, u)$. The standard goal of the feedback controller is uniform global asymptotic stability of the origin for the continuous-time closed-loop system $\dot{z} = f_p(z, \kappa(z))$. The sample-and-hold implementation uses a timer variable τ and results in a hybrid closed-loop system with state $x = (z, u, \tau)$; the stability goal translates to uniform global asymptotic stability of the set

$$\mathcal{A} = \{0\} \times \mathbb{R}^{n_c} \times [0, T]$$

for the system with data given in (1.5)-(1.6). Indeed, uniform global asymptotic stability of $\mathcal{A}$ for the hybrid closed-loop system places no conditions on the components u and τ of the state x of the closed-loop. Note that when the control design enforces boundedness of u, which is a property guaranteeing that the implementation of the controller is feasible, then the set $\mathcal{A}$ can be chosen to be bounded. For example, if u is picked from a compact set of controls $U \in \mathbb{R}^{n_c}$, then it is possible to consider $\mathcal{A} = \{0\} \times U \times [0, T]$.

Example 3.2. (Switching systems and compact attractors)   Section 2.4 discussed how switching systems under various classes of switching signals can be modeled by hybrid systems. Questions of stability and asymptotic stability in switching systems, where the interest is usually only in the behavior of the "continuous" variable z (recall that a solution to a switching system Σ given by $\dot{z} = f_q(z)$ consists of a switching signal q and a resulting solution z), can be translated to the hybrid setting as well. For example, questions of appropriately understood (uniform global) asymptotic stability of 0 for Σ under all dwell-time switching signals with dwell time τ_D translate to questions of asymptotic stability of the set

$$\mathcal{A} = \{0\} \times Q \times [0, \tau_D]$$

for the hybrid system (2.6). Similarly, questions of asymptotic stability of 0 for Σ under all persistent dwell-time signals with parameters τ_D and T translate to questions of asymptotic stability of the set

$$\mathcal{A} = \{0\} \times Q \times [0, \tau_D] \times [0, T]$$

for the hybrid system (2.7).

Example 3.3. (Time-varying systems) In some situations, the conditions for flowing or jumping as well as the flow map and jump map depend on a variable like time, which typically does not remain bounded. For example, consider a system with state $x = (z, \tau) \in \mathbb{R}^{n+1}$, flow set $C \subset \mathbb{R}^{n+1}$, and $D \subset \mathbb{R}^{n+1}$. Suppose the flow map and jump map are given as

$$F(x) = \begin{pmatrix} f(z, \tau) \\ 1 \end{pmatrix}, \quad G(x) = \begin{pmatrix} g(z, \tau) \\ \tau + 1 \end{pmatrix}.$$

Since the variable τ satisfies $\dot{\tau} = 1$ during flows and $\tau^+ = \tau + 1$ at jumps, $\tau(t, j) = \tau(0, 0) + t + j$ so that $\tau(t, j) \to \infty$ when $t + j \to \infty$. Therefore, if a closed set $\mathcal{A} \subset \mathbb{R}^{n+1}$ is uniformly globally asymptotically stable for this (C, F, D, G), it typically is not bounded.

The rigorous definition of uniform asymptotic stability uses class-$\mathcal{K}_\infty$ functions and the distance of a vector $x \in \mathbb{R}^n$ to a closed set $\mathcal{A} \subset \mathbb{R}^n$.

Definition 3.4. (Class-$\mathcal{K}_\infty$ functions) *A function $\alpha : \mathbb{R}_{\geq 0} \to \mathbb{R}_{\geq 0}$ is a class-$\mathcal{K}_\infty$ function, also written $\alpha \in \mathcal{K}_\infty$, if α is zero at zero, continuous, strictly increasing, and unbounded.*

Definition 3.5. (Distance to a closed set) *Given a vector $x \in \mathbb{R}^n$ and a closed set $\mathcal{A} \subset \mathbb{R}^n$, the distance of x to $\mathcal{A}$ is denoted $|x|_\mathcal{A}$ and is defined by $|x|_\mathcal{A} := \inf_{y \in \mathcal{A}} |x - y|$.*

The definition of uniform global pre-asymptotic stability is as follows.

Definition 3.6. (Uniform global pre-asymptotic stability (UGpAS)) *Consider a hybrid system $\mathcal{H}$ on $\mathbb{R}^n$. Let $\mathcal{A} \subset \mathbb{R}^n$ be closed. The set $\mathcal{A}$ is said to be*

- *uniformly globally stable for $\mathcal{H}$ if there exists a class-$\mathcal{K}_\infty$ function α such that any solution ϕ to $\mathcal{H}$ satisfies $|\phi(t, j)|_\mathcal{A} \leq \alpha(|\phi(0, 0)|_\mathcal{A})$ for all $(t, j) \in \operatorname{dom} \phi$;*

- *uniformly globally pre-attractive for $\mathcal{H}$ if for each $\varepsilon > 0$ and $r > 0$ there exists $T > 0$ such that, for any solution ϕ to $\mathcal{H}$ with $|\phi(0, 0)|_\mathcal{A} \leq r$, $(t, j) \in \operatorname{dom} \phi$ and $t + j \geq T$ imply $|\phi(t, j)|_\mathcal{A} \leq \varepsilon$;*

- *uniformly globally pre-asymptotically stable for $\mathcal{H}$ if it is both uniformly globally stable and uniformly globally pre-attractive.*

The term "pre-attractive," as opposed to "attractive," indicates the possibility of a maximal solution that is not complete, even though it may be bounded. Allowing this phenomenon separates conditions for completeness, which is closely related to existence, from conditions for stability and attractivity. This separation is reasonable since many classical conditions for asymptotic stability, like Lyapunov functions discussed later, do not guarantee existence or completeness of solutions. This fact is especially true in hybrid systems, but can already be seen in purely continuous-time systems too, if the right-hand side is not sufficiently regular.

Example 3.7. (Pre-attractivity in continuous-time systems) On $\mathbb{R}^2$, consider a differential equation $\dot{z} = f(z)$ with a discontinuous right-hand side, given by

$$f(z) = \begin{cases} -z & \text{if} \quad z_2 \neq 0, \\ -\begin{pmatrix} z_1 \\ z_1 \end{pmatrix} & \text{if} \quad z_2 = 0, \end{cases}$$

and the version of Definition 2.6 for continuous-time setting: an absolutely continuous function $z : I \to \mathbb{R}^2$ is a solution to $\dot{z} = f(z)$ if $I \subset \mathbb{R}_{\geq 0}$ has nonempty interior and

$$\dot{z}(t) = f(z(t)) \qquad \text{for almost all} \ \ t \in I. \tag{3.1}$$

It is easy to verify that from any initial point ξ with $\xi_2 \neq 0$, the unique solution to the differential equation is given by $z(t) = \xi e^{-t}$. For initial points ξ with $\xi_2 = 0$, except the origin itself, no solution exists. Figure 3.1 shows the right-hand side of (3.1) in the plane. The origin of the differential equation under discussion is uniformly globally pre-asymptotically stable, in the sense of Definition 3.6. However, it is not asymptotically stable in the common sense of this term, as there are initial points arbitrarily close to 0 from which no nontrivial solutions exist. Note that this does not preclude the existence of a Lyapunov function, as presented in Section 3.2. Indeed, for $V(z) = \frac{1}{2}|z|^2$, the equality $\langle \nabla V(z), f(z) \rangle = -|z|^2$ holds for all $z \in \mathbb{R}^2$.

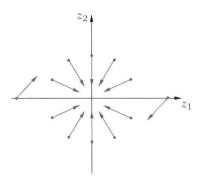

Figure 3.1: Vector field of (3.1).

The use of "pre-attractivity" is partly motivated by the fact that a system with a uniformly pre-asymptotically stable set but without complete solutions can often be "completed" so that uniform pre-asymptotic stability is preserved and maximal solutions are complete. Not insisting on complete solutions opens the door to calling some systems stable that might appear on the surface to be unstable. Both of these situations are illustrated in the following example.

Example 3.8. (Stability that looks unstable and completing a stable system) On $\mathbb{R}^2$, consider the hybrid system with flow set $C = \mathbb{R} \times [0, M]$ where

$M > 0$, flow map

$$f(x) = \begin{pmatrix} x_1 \\ 1 \end{pmatrix},$$

and jump set $D = \emptyset$. The compact set $\mathcal{A} = \{0\} \times [0, M]$ is UGpAS even though the distance to this set grows exponentially with time. The reason is that the amount of time is bounded. To verify UGpAS, note that the solution from each point in C is unique and its time domain is given as $[0, M - x_2(0,0)] \times \{0\}$. For all $t \in \mathrm{dom}\, x$, $x_2(t, 0) = x_2(0, 0) + t$ and $x_1(t, 0) = x_1(0, 0)e^t$. In particular, $|x(t, 0)|_{\mathcal{A}} = |x_1(t, 0)| \leq |x_1(0, 0)|e^M = |x(0, 0)|_{\mathcal{A}}e^M$. This calculation establishes uniform global stability of the set $\mathcal{A}$ with $\alpha \in \mathcal{K}_\infty$ given as $\alpha(s) = e^M s$ for all $s \geq 0$. Uniform global pre-attractivity follows by taking $T = M + 1$ for each $r > 0$ and $\varepsilon > 0$. Indeed, since there are no times $(t, 0) \in \mathrm{dom}\, x$ such that $t \geq M + 1$, there are no times that need to satisfy $|x(t, 0)|_{\mathcal{A}} \leq \varepsilon$. Figure 3.2 depicts a solution starting from $\mathcal{A} + \delta\mathbb{B}$ and staying in $\mathcal{A} + \varepsilon\mathbb{B}$.

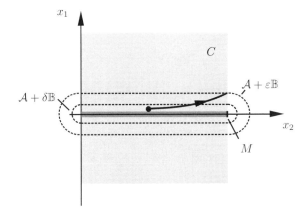

Figure 3.2: Sets and a solution to Example 3.8.

Now consider "completing" the system by changing the jump set to $D = \mathbb{R} \times \{M\}$ and using the jump map

$$g(x) = \begin{pmatrix} \lambda x_1 \\ 0 \end{pmatrix}$$

where $0 \leq \lambda < e^{-M}$. Solutions are still unique and now each maximal solution is complete. The time domain of a solution is such that each interval in the domain has length bounded by M. That is, $(t, j) \in \mathrm{dom}\, x$ implies that $t \leq M(j + 1)$, from which it follows that $j \geq (M + 1)^{-1}(t + j - M)$. Uniform global stability can again be verified with $\alpha(s) = e^M s$. Also, it can be verified that $|x(t, j)|_{\mathcal{A}} \leq e^M(\lambda e^M)^j |x(0, 0)|_{\mathcal{A}}$. Thus, given $r > 0$ and $\varepsilon > 0$, it suffices to pick $T \geq M + (M + 1)k$ where the integer k satisfies $e^M(\lambda e^M)^k r \leq \varepsilon$. Such an integer k exists since $0 \leq \lambda e^M < 1$.

Some additional observations about uniform global pre-asymptotic stability appear next. The first additional observation is that, in the case where $\mathcal{A}$ is not bounded, finite escape times are not incompatible with uniform global pre-asymptotic stability.

Example 3.9. (Uniform pre-asymptotic stability with finite escape times) On $\mathbb{R}^2$, consider the system with flow set $C = \mathbb{R}^2$, flow map

$$f(x) = \begin{pmatrix} x_1^3 \\ -x_2 \end{pmatrix},$$

and jump set $D = \emptyset$. The closed set $\mathcal{A} = \mathbb{R} \times \{0\}$ is UGpAS. Indeed, $|x|_{\mathcal{A}} = |x_2|$ and $\phi_2(t,j) = \phi_2(0,0)e^{-t}$ for all $(t,j) \in \operatorname{dom} \phi$. On the other hand, each maximal solution starting from a point with $\phi_1(0,0) \neq 0$ has a domain of the form $[0, \bar{t}) \times \{0\}$ where $\bar{t} \in [0, \infty)$. In particular, it can be established that

$$\phi_1^2(t,j) = \frac{\phi_1^2(0,0)}{1 - 2t\phi_1^2(0,0)} ,$$

and thus

$$\bar{t} = (2\phi_1^2(0,0))^{-1} .$$

It is possible for a closed set $\mathcal{A}$ to be pre-attractive, meaning that all complete solutions converge to $\mathcal{A}$, but not *uniformly* pre-attractive. This phenomenon requires one of three conditions: (1) the set $\mathcal{A}$ is not uniformly stable, (2) the set $\mathcal{A}$ is not bounded, or (3) the system does not satisfy regularity conditions — requiring, for example, that the flow and the jump sets be closed — that are discussed in Chapter 6. Compare with Theorem 7.12.

Example 3.10. (Nonuniform pre-attractivity due to $\mathcal{A}$ not stable) On $\mathbb{R}$, consider the system with flow set $C = \mathbb{R}$, flow map $f(x) = -x(x-1)^2$, and jump set $D = \emptyset$. Consider the case where $\mathcal{A}$ is the two-point set $\{0,1\}$. Every solution is complete and converges to $\mathcal{A}$. Solutions from initial points arbitrarily close to 1, and to the left of it, converge to 0; hence, $\mathcal{A}$ is not stable. The time required for such solutions to reach and remain within a small neighborhood of 0 grows to infinity when the initial point approaches 1 from the left. Note though that the interval $[0, 1]$, which contains $\mathcal{A}$, is UGpAS.

Example 3.11. (Nonuniform pre-attractive due to $\mathcal{A}$ not compact) On $\mathbb{R}^2$, consider the system with flow set $C = \mathbb{R} \times \mathbb{R}_{\geq 1}$, flow map

$$f(x) = \begin{pmatrix} -\dfrac{x_1}{x_2} \\ 1 \end{pmatrix} \qquad \forall x \in C ,$$

and jump set $D = \emptyset$. Let $\mathcal{A} = \{0\} \times \mathbb{R}$, so that $|x|_{\mathcal{A}} = |x_1|$. Every solution is complete, satisfies

$$\phi_1(t,j) = \phi_1(0,0) \exp\left(-\int_0^t \frac{1}{\phi_2(0,0) + s} ds \right) = \phi_1(0,0) \frac{\phi_2(0,0)}{\phi_2(0,0) + t} ,$$

and consequently, converges to $\mathcal{A}$. However, with $|\phi(0,0)|_{\mathcal{A}} = r$, the time required for the solution to satisfy $|\phi(t,0)|_{\mathcal{A}} = \varepsilon < r$ is given as

$$t - \phi_2(0,0) \left(\frac{r}{\varepsilon} - 1 \right) .$$

This time is unbounded in $\phi_2(0,0)$. Thus, $\mathcal{A}$ is not uniformly asymptotically stable.

Discontinuities in the dynamics can result in attractivity that is not uniform in differential or difference equations. For example, consider a differential equation $\dot{x} = f(x)$ on $\mathbb{R}^2$ with $f(x) = -\left(\frac{x_2}{x_1} \right) x$ if $x_1, x_2 > 0$, $f(x) = -x$ otherwise. Solutions flow exponentially along half-lines towards 0, with rates that can be arbitrarily small, when half-lines with small $\frac{x_2}{x_1}$ are considered. Thus $\mathcal{A} = \{0\}$ is attractive, in fact asymptotically stable, but the attractivity is not uniform. Now, consider a difference equation $x^+ = g(x)$ on $\mathbb{R}$, where $g(x) = x^2$ for $x \in (0,1)$, $g(x) = 0$ otherwise. For initial conditions arbitrarily close to but less than one, the number of jumps required to reach the interval $[0, 1/2]$ becomes arbitrarily large. Therefore, the origin is not uniformly pre-attractive, but it is asymptotically stable. A more exotic example can be given, where the origin in $\mathbb{R}$ is asymptotically stable, where every solution reaches it in a finite number of jumps, and where every neighborhood of every point in $\mathbb{R}$ contains initial points from which the solutions take an arbitrarily large number of jumps to reach the origin.

Example 3.12. ("Uniformly nonuniform attractivity") Let $g : \mathbb{R} \to \mathbb{R}$ be an odd function, given for $x > 0$ by

$$g(x) = \begin{cases} \frac{p-1}{q} & \text{if } x = \frac{p}{q} \text{ where } p, q \in \mathbb{N}, \; p, q > 1, \; q \text{ is prime, } \gcd(p,q) = 1 \\ 0 & \text{otherwise,} \end{cases}$$

where $\gcd(p,q)$ stands for the greatest common divisor of p and q. Every solution to $x^+ = g(x)$ reaches 0 in a finite number of jumps. It takes 1 jump if the initial point is not of the form $\pm \frac{p}{q}$, with p, q as above. If the initial point is of the form $\pm \frac{p}{q}$, then the number of jumps is r, if $\frac{p}{q} < 1$, or $r + 1$ if $\frac{p}{q} > 1$, where r is the remainder from the division of p by q. Indeed, after r jumps, the solution reaches an integer, and if that integer is not 0, it takes one more jump to reach 0. Since there exist arbitrarily large prime numbers, every open set in $\mathbb{R}$ contains a point of the form $\pm \frac{p}{q}$, moreover, with an arbitrarily large p.

As it was just illustrated, discontinuities in the flow map or the jump map can lead to attractivity of a set being not uniform, even for quite simple systems and, in fact, when the set is asymptotically stable (but not uniformly). In hybrid systems, lack of uniformity can also be caused by the flow set or the jump set being not closed. This is illustrated below.

Example 3.13. (Nonuniform pre-attractivity due to not closed jump set) On $\mathbb{R}$, consider the system with flow set $C = [1,2]$, flow map $f(x) = 1$, jump set $D = (-\infty, 1) \cup [2, \infty)$, and jump map $g(x) = (\max \{0, x\})^2$ for $x \in (-\infty, 1)$ and $g(x) = 0$ for $x \in [2, \infty)$, as depicted in Figure 3.3. Let $\mathcal{A}$ be the origin. All solutions are complete and converge to the origin. However, solutions starting in D but arbitrarily close to one take an arbitrarily large amount of time (jumps) to reach the interval $[0, 1/2]$. Therefore, $\mathcal{A}$ is attractive but not uniformly attractive.

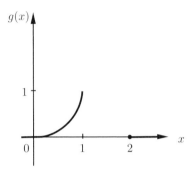

Figure 3.3: The map g for Example 3.13.

Example 3.14. (Nonuniform pre-attractivity due to not closed flow set) On $\mathbb{R}$, consider the system with flow set $C = (-\infty, 1)$, flow map $f(x) = x(x - 1)$, jump set $D = [1, \infty)$, and jump map $g(x) = 0$. Let $\mathcal{A}$ be the origin. All solutions are complete and converge to the origin. However, solutions starting in C but arbitrarily close to one take an arbitrarily large amount of time to reach the interval $[0, 1/2]$. Therefore, $\mathcal{A}$ is attractive but not uniformly attractive.

Example 3.15. (Nonuniform pre-attractivity due to discontinuous flow map) On $\mathbb{R}$, consider the system with flow set $C = \mathbb{R}$, flow map

$$f(x) = \begin{cases} x(x - 1) & x \in (-\infty, 1) \\ -1 & x \in [1, \infty) \end{cases}$$

as depicted in Figure 3.4, and jump set $D = \emptyset$. Let $\mathcal{A}$ be the origin. There is no solution from the initial condition $x = 1$. For initial conditions arbitrarily close to but less than one, the time required for the solution to reach the interval $[0, 1/2]$ becomes arbitrarily large. Therefore, the origin is not uniformly pre-attractive. Nevertheless, each complete solution converges to the origin.

3.2   LYAPUNOV FUNCTIONS

This section addresses the Lyapunov function as a sufficient condition for uniform global pre-asymptotic stability. This sufficient condition is also a necessary condition for UGpAS when the set $\mathcal{A}$ is compact and the data of the hybrid system satisfies basic conditions that are enumerated in Chapter 6.

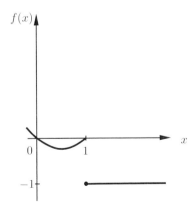

Figure 3.4: Map f for Example 3.15.

Lyapunov functions for differential equations

A fundamental tool in stability analysis for classical dynamical systems is the Lyapunov function. In the setting of differential equations $\dot{z} = f(z)$ with the state evolving in $\mathbb{R}^n$ and $f : \mathbb{R}^n \to \mathbb{R}^n$ continuous and stability of the origin, Lyapunov's method for stability analysis can be summarized as follows. If there exists a continuously differentiable function $V : \mathbb{R}^n \to \mathbb{R}_{\geq 0}$ with $V(0) = 0$ and such that, for all $z \neq 0$, $V(z) > 0$ and $\langle \nabla V(z), f(z) \rangle \leq 0$, then, for the differential equation $\dot{z} = f(z)$, the origin is

- Lyapunov stable, in the sense that for each $\varepsilon > 0$ there exists $\delta > 0$ such that $|z(0)| \leq \delta$ implies $|z(t)| \leq \varepsilon$ for each solution z and each $t \geq 0$.

The key to this conclusion is that along solutions to the differential equation, the value $V(z(t))$ is nonincreasing as a function of time t. If $\langle \nabla V(z), f(z) \rangle < 0$ for all $z \neq 0$ then $V(z(t))$ decreases with time t, as long as $z(t) \neq 0$, and hence the origin is also

- locally attractive, in the sense that all solutions z with $z(0)$ sufficiently close to the origin are such that $z(t) \to 0$ as $t \to \infty$.

Local attractivity implicitly relies on continuity of f, which together with stability guarantees that solutions from sufficiently close to 0 are complete, and hence talking about $z(t)$ as $t \to \infty$ is possible. Continuity can be weakened, though, to any property ensuring existence of solutions for each initial point sufficiently close to 0. Local stability together with local attractivity amounts to a property referred to as local asymptotic stability. Finally, if V also has compact sublevel sets, in other words, for each $r > 0$, $\{z : V(z) \leq r\}$ is compact, then the origin is globally asymptotically stable, in the sense that it is Lyapunov stable and each solution to the differential equation converges to the origin. A function V possessing the properties as above is referred to as a Lyapunov function.

The conclusions just stated can be strengthened to uniform global stability and uniform global attractivity. Compactness of the sublevel sets of V and the continuity of V imply upper and lower bounds on $V(z)$ expressible as $\mathcal{K}_\infty$-functions of $|z|$. These uniform bounds, together with the fact that $V(z(t))$ is not increasing with t, imply uniform global stability. Subsequently, the key to uniform global attractivity is the fact that, due to the continuity of f and ∇V, the values of $\langle \nabla V(z), f(z) \rangle$ are actually bounded from above, by a negative number, on each compact subset of $\mathbb{R}^n$ that does not include the origin. In the absence of continuity of f, the uniform negative bound on $\langle \nabla V(z), f(z) \rangle$ on each compact set that does not include the origin must be assumed.

Lyapunov conditions for hybrid systems

Lyapunov functions are also useful in analyzing stability for hybrid systems. Due to the constraints that are given in a hybrid system that limit where jumping and flowing is possible, it is not necessarily a requirement that a Lyapunov function be defined on all of $\mathbb{R}^n$ nor is it a requirement that it is continuously differentiable on all of $\mathbb{R}^n$. For now, a Lyapunov function is assumed to be continuously differentiable on a neighborhood of the flow set. The existence of smooth Lyapunov functions is addressed in Chapter 7. The definition below makes rigorous the conditions required for a function V to be considered a Lyapunov function candidate for establishing uniform global pre-asymptotic stability of a closed set for a hybrid system.

Definition 3.16. (Lyapunov function candidate) *A function $V : \operatorname{dom} V \to \mathbb{R}$ is said to be a Lyapunov function candidate for the hybrid system $\mathcal{H} = (C, F, D, G)$ if the following conditions hold:*

1. *$\overline{C} \cup D \cup G(D) \subset \operatorname{dom} V$;*

2. *V is continuously differentiable on an open set containing $\overline{C}$;*

where $\overline{C}$ denotes the closure of C.

Definition 3.17. (Positive definite functions) *A function $\rho : \mathbb{R}_{\geq 0} \to \mathbb{R}_{\geq 0}$ is positive definite, also written $\rho \in \mathcal{PD}$, if $\rho(s) > 0$ for all $s > 0$ and $\rho(0) = 0$.*

The following theorem provides conditions on a Lyapunov function candidate that guarantee uniform global pre-asymptotic stability.

Theorem 3.18. (Sufficient Lyapunov conditions) *Let $\mathcal{H} = (C, F, D, G)$ be a hybrid system and let $\mathcal{A} \subset \mathbb{R}^n$ be closed. If V is a Lyapunov function candidate for $\mathcal{H}$ and there exist $\alpha_1, \alpha_2 \in \mathcal{K}_\infty$, and a continuous $\rho \in \mathcal{PD}$ such that*

$$\alpha_1(|x|_{\mathcal{A}}) \leq V(x) \leq \alpha_2(|x|_{\mathcal{A}}) \qquad \forall x \in C \cup D \cup G(D) \tag{3.2a}$$

$$\langle \nabla V(x), f \rangle \leq -\rho(|x|_{\mathcal{A}}) \qquad \forall x \in C, \ f \in F(x) \tag{3.2b}$$

$$V(g) - V(x) \leq -\rho(|x|_{\mathcal{A}}) \qquad \forall x \in D, \ g \in G(x) \tag{3.2c}$$

then $\mathcal{A}$ is uniformly globally pre-asymptotically stable for $\mathcal{H}$.

PROOF. Let $\phi \in \mathcal{S}_{\mathcal{H}}$. Pick any $(t, j) \in \operatorname{dom} \phi$ and let $0 = t_0 \leq t_1 \leq \cdots \leq t_{j+1} = t$ satisfy

$$\operatorname{dom} \phi \cap ([0, t] \times \{0, \ldots, j\}) = \bigcup_{i=0}^{j} [t_i, t_{i+1}] \times \{i\} \ .$$

For each $i \in \{0, \ldots, j\}$ and almost all $s \in [t_i, t_{i+1}]$, $\phi(s, i) \in C$. Then, (3.2b) implies that, for each $i \in \{0, \ldots, j\}$ and for almost all $s \in [t_i, t_{i+1}]$,

$$\frac{d}{ds} V(\phi(s, i)) \leq -\rho(|\phi(s, i)|_{\mathcal{A}}) \ .$$

Integrating both sides of this inequality yields

$$V(\phi(t_{i+1}, i)) - V(\phi(t_i, i)) \leq - \int_{t_i}^{t_{i+1}} \rho(|\phi(s, i)|_{\mathcal{A}}) ds \qquad \forall i \in \{0, \ldots, j\} \ .$$

Similarly, for each $i \in \{1, \ldots, j\}$, $\phi(t_i, i-1) \in D$, and thus

$$V(\phi(t_i, i)) - V(\phi(t_i, i-1)) \leq -\rho(|\phi(t_i, i-1)|_{\mathcal{A}}) \qquad \forall i \in \{1, \ldots, j\} \ .$$

The last two displayed inequalities imply that

$$V(\phi(t, j)) + \sum_{i=0}^{j} \int_{t_i}^{t_{i+1}} \rho(|\phi(s, i)|_{\mathcal{A}}) ds + \sum_{i=1}^{j} \rho(|\phi(t_i, i-1)|_{\mathcal{A}}) \leq V(\phi(0, 0)) \ . \quad (3.3)$$

By the definition of solution, $\phi(t, j) \in \overline{C} \cup D \cup G(D)$ for all $(t, j) \in \operatorname{dom} \phi$. Due to the continuous differentiability of V on a neighborhood of $\overline{C}$, the bound (3.2a) holds on $\overline{C} \cup D \cup G(D)$. Positive definiteness of ρ, combined with (3.2a), yields

$$\alpha_1(|\phi(t, j)|_{\mathcal{A}}) \leq V(\phi(t, j)) \leq V(\phi(0, 0)) \leq \alpha_2(|\phi(0, 0)|_{\mathcal{A}}),$$

and consequently, $|\phi(t, j)|_{\mathcal{A}} \leq \alpha_1^{-1}(\alpha_2(|\phi(0, 0)|_{\mathcal{A}}))$. Since $\alpha_1^{-1} \circ \alpha_2 \in \mathcal{K}_{\infty}$ and because the bound holds for every $\phi \in \mathcal{S}_{\mathcal{H}}$ and every $(t, j) \in \operatorname{dom} \phi$, uniform stability is established.

To establish uniform pre-attractivity, pick any $\varepsilon, r > 0$. With α_1, α_2 as above, let $\delta = \alpha_2^{-1}(\varepsilon)$, and $R = \alpha_2(r)$, so that if $|\phi(0, 0)|_{\mathcal{A}} \leq \delta$ then $|\phi(t, j)|_{\mathcal{A}} \leq \varepsilon$ for all $(t, j) \in \operatorname{dom} \phi$, and if $|\phi(0, 0)|_{\mathcal{A}} \leq r$ then $|\phi(t, j)|_{\mathcal{A}} \leq R$ for all $(t, j) \in \operatorname{dom} \phi$, for any $\phi \in \mathcal{S}_{\mathcal{H}}$. Let $m = \min \rho([\delta, R])$ and

$$T = 1 + (\alpha_2(r) - \alpha_1(\delta))/m. \quad (3.4)$$

Suppose that $\phi \in \mathcal{S}_{\mathcal{H}}$ with $|\phi(0, 0)|_{\mathcal{A}} \leq r$ is such that $|\phi(t, j)|_{\mathcal{A}} \geq \delta$ for all $(t, j) \in \operatorname{dom} \phi$ with $t + j \leq T$. Then, for such (t, j), $\rho(|\phi(t, j)|_{\mathcal{A}}) \geq m$, and inequality (3.3) implies that

$$V(\phi(t, j)) \leq V(\phi(0, 0)) - (t + j)m$$

and, consequently, that

$$|\phi(t,j)|_{\mathcal{A}} \leq \alpha_1^{-1}\left(\alpha_2(|\phi(0,0)|_{\mathcal{A}}) - (t+j)m\right). \tag{3.5}$$

Unless length $\phi \leq T - 1$, there exist $(t',j') \in \operatorname{dom}\phi$ with $T - 1 < t' + j' \leq T$. Then (3.5) yields

$$|\phi(t',j')_{\mathcal{A}}| \leq \alpha_1^{-1}\left(\alpha_2(r) - (t'+j')m\right) < \alpha_1^{-1}\left(\alpha_2(r) - (T-1)m\right) = \delta.$$

However, $|\phi(t',j')|_{\mathcal{A}} < \delta$ is a contradiction. Hence, for every $\phi \in \mathcal{S}_{\mathcal{H}}$ with $|\phi(0,0)|_{\mathcal{A}} \leq r$, there exists $(t',j') \in \operatorname{dom}\phi$ with $t'+j' \leq T$ such that $|\phi(t',j')|_{\mathcal{A}} < \delta$. By the choice of δ, $|\phi(t,j)|_{\mathcal{A}} \leq \varepsilon$ for all $(t,j) \in \operatorname{dom}\phi$ with $t+j \geq t'+j'$, and, in particular, for all $(t,j) \in \operatorname{dom}\phi$ with $t+j \geq T$. In the case of length $\phi \leq T-1$, $|\phi(t,j)|_{\mathcal{A}} \leq \varepsilon$ for all $(t,j) \in \operatorname{dom}\phi$ with $t + j \geq T$ holds vacuously, and hence the uniform pre-attractivity has been proven. ◻

For historical reasons, a function V satisfying the conditions of Theorem 3.18 is called a *Lyapunov function* with respect to $\mathcal{A}$ for $\mathcal{H}$, and the conditions (3.2) are called *Lyapunov conditions*.

When the set $\mathcal{A}$ is compact and the Lyapunov function V is continuous, the condition (3.2a) holds if and only if $V(x) = 0$ for all $x \in \mathcal{A}$, $V(x) > 0$ for all $x \notin \mathcal{A}$, and the values $V(x)$ grow unbounded as $x \in C \cup D \cup G(D)$ grows unbounded.

There is no loss of generality in using the same function ρ in both the second and third inequality of the Lyapunov conditions since, if the two functions were different, they could each be replaced by the point-wise minimum of the two functions, which would be another function belonging to class-$\mathcal{PD}$.

When F and G are continuous (for example, single-valued continuous functions) and the Lyapunov conditions hold, they necessarily hold also when C and D are replaced by their closures. Thus, when relying on Lyapunov conditions for asymptotic stability, it is also possible to assert uniform global pre-asymptotic stability even for the hybrid system obtained by expanding the flow and jump sets to include their boundaries.

The next examples illustrate the construction of Lyapunov functions.

Example 3.19. (Bouncing ball) Consider the bouncing ball model as in Example 1.1, where

$$f(x) = \begin{pmatrix} x_2 \\ -\gamma \end{pmatrix}, \qquad C = \{x \in \mathbb{R}^2 : x_1 > 0\}$$

$$g(x) = \begin{pmatrix} 0 \\ -\lambda x_2 \end{pmatrix}, \qquad D = \{x \in \mathbb{R}^2 : x_1 = 0, x_2 < 0\}$$

where $\lambda \in [0,1)$ and $\gamma > 0$. First, consider the Lyapunov function candidate

$$V_1(x) = \frac{1}{2}x_2^2 + \gamma x_1$$

which satisfies (3.2a) with, for all $s \in \mathbb{R}_{\geq 0}$,

$$\alpha_1(s) := \min\left\{\frac{1}{2}\left(\frac{s}{\sqrt{2}}\right)^2, \gamma\left(\frac{s}{\sqrt{2}}\right)\right\}, \qquad \alpha_2(s) := \frac{1}{2}s^2 + \gamma s \qquad \forall s \in \mathbb{R}_{\geq 0}.$$

The function V_1 satisfies

$$\langle \nabla V_1(x), f(x) \rangle = 0 \qquad \forall x \in C$$

and

$$\begin{aligned} V_1(g(x)) - V_1(x) &= -\frac{1}{2}\left(1 - \lambda^2\right) x_2^2 \\ &= -\frac{1}{2}\left(1 - \lambda^2\right)\left(x_2^2 + x_1^2\right) \qquad \forall x \in D. \end{aligned}$$

The last equality comes from the fact that $x \in D$ implies $x_1 = 0$. Thus, the inequality (3.2c) is satisfied with

$$\rho(s) := \frac{1}{2}\left(1 - \lambda^2\right) s^2.$$

However, inequality (3.2b) is not satisfied. This situation is addressed directly later in Proposition 3.24 of Section 3.3. In the meantime, a small modification to the given V_1 results in all of the Lyapunov conditions being satisfied. In particular, it can be verified that the Lyapunov function candidate

$$V_2(x) := (1 + \theta \arctan(x_2)) V_1(x), \qquad \theta = \frac{1 - \lambda^2}{\pi(1 + \lambda^2)}$$

satisfies the conditions of Theorem 3.18, thus establishing uniform global pre-asymptotic stability for the origin of the bouncing ball model. Alternatively, it can be verified that the function

$$V_3(x) = \left(V_1(x) + \varepsilon x_2 |V_1(x)|^{1/2}\right)^3$$

is a Lyapunov function candidate that satisfies the conditions of Theorem 3.18 for $\varepsilon > 0$ sufficiently small.

Example 3.20. (Stability that looks unstable) As in Example 3.8, on $\mathbb{R}^2$ let the flow set be $C = \mathbb{R} \times [0, M]$ with $M > 0$, let the flow map be

$$f(x) = \begin{pmatrix} x_1 \\ 1 \end{pmatrix},$$

and let the jump set be $D = \emptyset$. The UGpAS of the compact set $\mathcal{A} = \{0\} \times [0, M]$ can be established using the Lyapunov function candidate $V(x) = x_1^2 e^{-2\sigma x_2}$ where $\sigma > 1$. For all $x \in C$,

$$e^{-2\sigma M} |x|_{\mathcal{A}}^2 \leq V(x) \leq |x|_{\mathcal{A}}^2$$

and

$$\langle \nabla V(x), f(x) \rangle = 2V(x) - 2\sigma V(x) = 2(1 - \sigma)V(x) \ .$$

Thus, the conditions of Theorem 3.18 are satisfied with $\alpha_1(s) = e^{-2\sigma M} s^2$, $\alpha_2(s) = s^2$, and $\rho(s) = 2(\sigma - 1)\alpha_1(s)$.

For the case of the "completed" system in Example 3.8, where the jump set D is changed to $\mathbb{R} \times \{M\}$ and the jump map is

$$g(x) = \left(\begin{array}{c} \lambda x_1 \\ 0 \end{array} \right)$$

where $0 \leq \lambda < e^{-M}$, the same Lyapunov function candidate is a Lyapunov function as long as $\sigma > 1$ is chosen to satisfy $\lambda e^{\sigma M} < 1$. Indeed, under this condition,

$$V(g(x)) = \lambda^2 |x_1|^2 = \lambda^2 |x_1|^2 e^{-2\sigma M} e^{2\sigma M} = V(x)(\lambda e^{\sigma M})^2 \ .$$

Thus, the conditions of Theorem 3.18 are satisfied by replacing ρ determined above by the pointwise minimum of ρ and the function $(1 - (\lambda e^{\sigma M})^2)\alpha_1(s)$.

Example 3.21. (Linear sampled-data systems) Consider the linear sampled-data system

$$\left\{ \begin{array}{rcl} \dot{z} & = & Az + Bu \\ \dot{u} & = & 0 \\ \dot{\tau} & = & 1 \\ z^+ & = & z \\ u^+ & = & Kz \\ \tau^+ & = & 0 \end{array} \right\} \begin{array}{l} \tau \in [0, T] \\ \\ \\ \tau = T \ . \end{array}$$

This corresponds to a hybrid system with state x where $x_1 = \left(\begin{array}{c} z \\ u \end{array} \right)$, $x_2 = \tau$,

$$f(x) = \left(\begin{array}{c} A_f x_1 \\ 1 \end{array} \right) , \quad C = \{x : x_2 \in [0, T]\}$$
$$g(x) = \left(\begin{array}{c} A_g x_1 \\ 0 \end{array} \right) , \quad D = \{x : x_2 = T\}$$

where

$$A_f := \left(\begin{array}{cc} A & B \\ 0 & 0 \end{array} \right) , \quad A_g := \left(\begin{array}{cc} I & 0 \\ K & 0 \end{array} \right) \ .$$

Define $H := \exp(A_f T) A_g$ and note that the matrix H indicates the evolution of the variable x_1 at sampling times just before jumps. In particular if $x(t, j) \in D$ then $x_1(t + T, j + 1) = H x_1(t, j)$. Suppose there exists a positive definite symmetric matrix P such that $H^T P H - P$ is negative definite; equivalently, the eigenvalues of H all have magnitude less than one. Under this condition, the compact set $\mathcal{A} := \{x : x_1 = 0 , \ x_2 \in [0, T]\}$ will be shown to be uniformly globally asymptotically stable using Theorem 3.18. Define $W(x_1) := x_1^T P x_1$ and note that there exists $\varepsilon > 0$ such that

$$W(Hx_1) - W(x_1) = x_1^T (H^T P H - P)x_1 \leq -\varepsilon |x_1|^2 .$$

Now consider the Lyapunov function candidate

$$V_1(x) := W\left(\exp(A_f(T - x_2))x_1\right) .$$

It can be verified that, given $T > 0$, there exist $\underline{c} > 0$ and $\overline{c} > 0$ such that

$$\underline{c} \, |x|_{\mathcal{A}}^2 \leq V_1(x) \leq \overline{c} \, |x|_{\mathcal{A}}^2 \qquad \forall x \in C \cup D .$$

It can also be established that

$$\langle \nabla V_1(x), f(x) \rangle = 0 \qquad \forall x \in C$$

and, for all $x \in D$,

$$
\begin{aligned}
V_1(g(x)) - V_1(x) &= W(\exp(A_f T)A_g x_1) - W(x_1) \\
&= W(Hx_1) - W(x_1) \\
&\leq -\varepsilon |x_1|^2 \\
&= -\varepsilon |x|_{\mathcal{A}}^2 .
\end{aligned}
$$

Like for the bouncing ball example, the inequality (3.2b) is not satisfied, a situation that is addressed in Section 3.3. However, a simple modification can be made to V_1 in order to have all of the Lyapunov conditions satisfied. In particular, it can be verified that, for $\sigma > 0$ sufficiently small, the Lyapunov function candidate

$$V_2(x) := \exp(-\sigma x_2)V_1(x)$$

satisfied the conditions of Theorem 3.18.

Example 3.22. (Switching systems) Consider a system that switches between a finite number of continuous vector fields, $f_q : \mathbb{R}^p \to \mathbb{R}^p$, $q \in \{1, \dots, q_{max}\} =: Q$, $q_{max} \in \mathbb{N}$, where the number of switches in a given ordinary time interval is limited by an "average dwell-time" condition. In particular, letting $N(s,t)$ denote the number of switches in the time period $[s, t]$, the constraint

$$N(s,t) \leq \eta(t - s) + N_\circ ,$$

where $\eta > 0$ and $N_\circ \in \mathbb{N}$, is imposed. As indicated in Section 2.4 of Chapter 2, the prescribed behavior is covered by the hybrid system

$$
\left\{
\begin{array}{rcl}
\left.\begin{array}{rcl}
\dot{z} &=& f_q(z) \\
\dot{q} &=& 0 \\
\dot{\tau} &\in& [0, \eta]
\end{array}\right\} & \tau \in [0, N_\circ] \\[2ex]
\left.\begin{array}{rcl}
z^+ &=& z \\
q^+ &\in& Q \\
\tau^+ &=& \tau - 1
\end{array}\right\} & \tau \in [1, N_\circ] .
\end{array}
\right.
$$

The overall state x combines z, q, and τ. The set $\mathcal{A}$ that should be uniformly globally (pre-)asymptotically stable is given by

$$\mathcal{A} := \{\, x : \quad z = 0\,,\ q \in Q\,,\ \tau \in [0, N_\circ]\,\}\,.$$

Note that $|x|_{\mathcal{A}} = |z|$ for all $x \in C \cup D$.

Suppose that each continuous-time system $\dot{z} = f_q(z)$, $q \in Q$, has the origin globally exponentially stable and let $W_q : \mathbb{R}^p \to \mathbb{R}^p$ for $q \in Q$ denote continuously differentiable functions satisfying the following properties:

1. There exist real numbers $\underline{c}, \overline{c}$ such that

$$\exp(\underline{c})|z|^2 \le W_q(z) \le \exp(\overline{c})|z|^2 \qquad \forall z \in \mathbb{R}^p\,,\ q \in Q\,.$$

2. There exists $\lambda > 0$ such that

$$\langle \nabla W_q(z), f_q(z) \rangle \le -\lambda W_q(z) \qquad \forall z \in \mathbb{R}^p\,,\ q \in Q\,.$$

3. $\overline{c} - \underline{c} < \lambda/\eta$.

Consider the Lyapunov function candidate

$$V(x) := \exp(\mu\tau) W_q(z)$$

where $\mu > 0$ is chosen so that

$$\overline{c} - \underline{c} < \mu < \lambda/\eta\,. \tag{3.6}$$

It follows that

$$\exp(\underline{c})|z|^2 \le V(x) \le \exp(\overline{c} + \mu N_\circ)|z|^2 \qquad \forall x \in C \cup D$$

and

$$\langle \nabla V(x), f(x) \rangle \le (\mu\eta - \lambda)\, V(x) \qquad \forall x \in C$$

and

$$
\begin{aligned}
\max_{g \in G(x)} V(g) &= \exp(-\mu) \exp(\mu\tau) \max_{q^+ \in Q} W_{q^+}(z) \\
&\le \exp(-\mu + \overline{c} - \underline{c}) V(x) \qquad \forall x \in D\,.
\end{aligned}
$$

It follows from (3.6) that V is a Lyapunov function and then it follows from Theorem 3.18 that the set $\mathcal{A}$ is uniformly globally pre-asymptotically stable.

Example 3.23. (Uniting local and global controllers) Recall, from Example 1.7, the continuous control system

$$\dot{z} = \widetilde{f}(z, u)$$

and two continuous functions $\kappa_i : \mathbb{R}^{n_p} \to \mathbb{R}^m$, $i = 1, 2$, representing local and global feedback controllers, and suppose that there exist two continuously differentiable functions $W_i : \mathbb{R}^{n_p} \to \mathbb{R}_{\ge 0}$ satisfying the following properties:

1. There exist $\alpha_1, \alpha_2 \in \mathcal{K}_\infty$ such that, for $i = 1, 2$,

$$\alpha_1(|z|) \le W_i(z) \le \alpha_2(|z|) \qquad \forall z \in \mathbb{R}^{n_p}.$$

2. There exist an open neighborhood $\mathcal{U}_1$ of the origin and a function $\rho_1 \in \mathcal{PD}$ such that

$$\langle \nabla W_1(z), \widetilde{f}(z, \kappa_1(z)) \rangle \le -\rho_1(|z|) \qquad \forall z \in \mathcal{U}_1 .$$

3. There exists a function $\rho_2 \in \mathcal{PD}$ such that

$$\langle \nabla W_2(z), \widetilde{f}(z, \kappa_2(z)) \rangle \le -\rho_2(|z|) \qquad \forall z \in \mathbb{R}^{n_p} .$$

Let c_1 be such that

$$L_{W_1}(c_1) := \{z \in \mathbb{R}^{n_p} : W_1(z) \le c_1\} \subset \mathcal{U}_1 .$$

Let $D_2 \subset L_{W_1}(c_1)$ be a compact set containing the origin in its interior. Take $C_2 = \mathbb{R}^{n_p} \backslash D_2$. Also, take $C_1 = \mathcal{U}_1$, $D_1 = \mathbb{R}^{n_p} \backslash \mathcal{U}_1$. Take

$$C := \{(q, z) : q \in \{1, 2\} \ , \ z \in C_q\} \ , \quad D := \{(q, z) : q \in \{1, 2\} \ , \ z \in D_q\} \ .$$

Also, use the flow map and jump map given in Example 1.7, which correspond to

$$\begin{aligned} \dot{q} &= 0 \\ \dot{z} &= \widetilde{f}(z, \kappa_q(z)) \end{aligned}$$

and

$$\begin{aligned} q^+ &= 3 - q \\ z^+ &= z . \end{aligned}$$

To establish uniform global (pre-)asymptotic stability for the set

$$\mathcal{A} := \{(q, z) : q = 1 \ , \ z = 0\} \ ,$$

let $\sigma_1 \in \mathcal{K}_\infty$ be smooth with a positive definite derivative and such that

$$\sigma_1(s) \le \alpha_1 \circ \alpha_2^{-1}(s) \qquad \forall s \in [0, c_1]$$

and

$$\alpha_2 \circ \alpha_1^{-1}(s) \le \sigma_1(s) \qquad \forall s \in [c_2, \infty)$$

where $c_2 > c_1$ is such that $L_{W_1}(c_2) \subset \mathcal{U}_1$, define $\sigma_2(s) := s$ for all $s \in \mathbb{R}_{\ge 0}$, and define $V(x) := \sigma_q(W_q(z))$. It can be verified that this function satisfies the conditions of Theorem 3.18 and thus the set $\mathcal{A}$ is uniformly globally pre-asymptotically stable for $\mathcal{H}$.

3.3   RELAXED LYAPUNOV CONDITIONS

This section gives several sufficient conditions for uniform global pre-asymptotic
stability, in which the strict decrease assumptions of Theorem 3.18 are weakened.
For the reader's convenience, the assumptions are now recalled:

$$\alpha_1(|x|_{\mathcal{A}}) \leq V(x) \leq \alpha_2(|x|_{\mathcal{A}}) \qquad \forall x \in C \cup D \cup G(D) \tag{3.7a}$$

$$\langle \nabla V(x), f \rangle \leq -\rho(|x|_{\mathcal{A}}) \qquad \forall x \in C,\ f \in F(x) \tag{3.7b}$$

$$V(g) - V(x) \leq -\rho(|x|_{\mathcal{A}}) \qquad \forall x \in D,\ g \in G(x). \tag{3.7c}$$

The assumptions of strict decrease of the Lyapunov function during both flows
and jumps can be weakened in several ways. First, note that uniform stability
is guaranteed as long as the Lyapunov function is not increasing; this is visible
in the proof of Theorem 3.18. Uniform pre-attractivity can then be deduced as
long as the Lyapunov function decreases, along solutions, over sufficiently long
hybrid time intervals. For example, this can be the case if the Lyapunov function
is nonincreasing during flows, strictly decreasing during jumps, and the jumps
occur frequently enough, as stated in the proposition below. Further sufficient
conditions, in which the Lyapunov function candidate is allowed to increase,
as long as the increase is balanced out by decrease, are presented later in the
section.

Proposition 3.24. (Sufficient Lyapunov conditions: persistent jumping)
Let $\mathcal{H} = (C, F, D, G)$ be a hybrid system and let $\mathcal{A} \subset \mathbb{R}^n$ be closed. Suppose
that V is a Lyapunov function candidate for $\mathcal{H}$ and there exist $\alpha_1, \alpha_2 \in \mathcal{K}_\infty$,
and a continuous $\rho \in \mathcal{PD}$ such that (3.7a), (3.7c) hold and

$$\langle \nabla V(x), f \rangle \leq 0 \quad \forall x \in C, f \in F(x). \tag{3.8}$$

If, for each $r > 0$, there exists $\gamma_r \in \mathcal{K}_\infty$, $N_r \geq 0$ such that for every solution ϕ
to $\mathcal{H}$, $|\phi(0,0)|_{\mathcal{A}} \in (0, r]$, $(t, j) \in \operatorname{dom} \phi$, $t + j \geq T$ imply $j \geq \gamma_r(T) - N_r$, then $\mathcal{A}$
is uniformly globally pre-asymptotically stable.

PROOF. The proof is very similar to the proof of Theorem 3.18. Uniform
stability is shown in exactly the same way. The proof of uniform pre-attractivity
requires minor changes. To define T, replace (3.4) with

$$T = 1 + \gamma_r^{-1}((\alpha_2(r) - \alpha_1(\delta))/m + N_r).$$

Because (3.8) is weaker than (3.7b), in place of the bound 3.5, the bound

$$|\phi(t,j)|_{\mathcal{A}} \leq \alpha_1^{-1}(\alpha_2(|\phi(0,0)|_{\mathcal{A}}) - jm)$$

is used. As in the proof of Theorem 3.18, unless length $\phi \leq T - 1$, there exist
$(t', j') \in \operatorname{dom} \phi$ with $T - 1 < t' + j' \leq T$. Here, $j' \geq \gamma_r(T-1) - N_r$. Then

$$|\phi(t',j')_{\mathcal{A}}| \leq \alpha_1^{-1}(\alpha_2(r) - j'm) < \alpha_1^{-1}(\alpha_2(r) - (\gamma_r(T-1) - N_r)m) = \delta.$$

The rest of the proof of Theorem 3.18 applies. □

Example 3.25. (Bouncing ball) Consider the Lyapunov function candidate

$$V_1(x) = \frac{1}{2}x_2^2 + \gamma x_1.$$

Example 3.19 showed that this function satisfies the inequalities (3.7a), (3.7c), and (3.8). Moreover, it follows from Example 2.12, in particular from (2.4), that for each $r > 0$ there exists $\tau_r > 0$ such that $|\phi(0,0)| \leq r$ and $(t,j) \in \operatorname{dom}\phi$ imply $t \leq \tau_r$. Thus $t + j \geq T$ implies $j \geq T - t \geq T - \tau_r$. Thus, all of the conditions of Proposition 3.24 are satisfied.

Example 3.26. (Sampled-data systems) Consider the system in Example 3.21 and the Lyapunov candidate V_1 given there. It was shown that V_1 satisfies the inequalities (3.7a), (3.7c), and (3.8). By construction, $(t,j) \in \operatorname{dom}\phi$ implies $t \leq (j+1)T_s$ where $T_s > 0$ is the sampling period. For $t + j \geq T$, this inequality yields $j \geq \frac{T}{T_s+1} - \frac{T_s}{T_s+1}$. Thus, Proposition 3.24 applies.

Similarly to the result in Proposition 3.24, pre-asymptotic stability can be established if the Lyapunov function is nonincreasing during jumps, strictly decreasing during flow, and the duration of flow is sufficiently large for every solution. This is stated formally in the proposition below. The proof is almost identical to the one given for Proposition 3.24, and so it is omitted.

Proposition 3.27. (Sufficient Lyapunov conditions: persistent flowing) *Let $\mathcal{H} = (C, F, D, G)$ be a hybrid system and let $\mathcal{A} \subset \mathbb{R}^n$ be closed. Suppose that V is a Lyapunov function candidate for $\mathcal{H}$ and there exist $\alpha_1, \alpha_2 \in \mathcal{K}_\infty$, and a continuous $\rho \in \mathcal{PD}$ such that (3.7a) and (3.7b) hold and*

$$V(g) - V(x) \;\leq\; 0 \quad \forall x \in D, \ g \in G(x) . \tag{3.9}$$

If, for each $r > 0$, there exists $\gamma_r \in \mathcal{K}_\infty$, $N_r \geq 0$ such that for every solution ϕ to $\mathcal{H}$, $|\phi(0,0)|_{\mathcal{A}} \in (0,r]$, $(t,j) \in \operatorname{dom}\phi$, $t + j \geq T$ imply $t \geq \gamma_r(T) - N_r$, then $\mathcal{A}$ is uniformly globally pre-asymptotically stable.

The next example illustrates the use of Proposition 3.27 and emphasizes that the condition on the time domain is needed only for establishing uniform convergence and thus is not needed for solutions that start in $\mathcal{A}$.

Example 3.28. (Illustrating Proposition 3.27) Let $A \in \mathbb{R}^{n \times n}$ be a Hurwitz matrix, let $f(x) = Ax$, let $g(x) = -x$, let $C = \mathbb{R}^n$ and let $D = \mathbb{R}^n_{\geq 0}$. Since A is Hurwitz, there exists a symmetric, positive definite matrix P such that $A^T P + PA = -I$. Define $V(x) = x^T P x$. Then (3.7a) holds with $\alpha_1(s) = \lambda_{min}(P)s^2$ and $\alpha_2(s) = \lambda_{max}(P)s^2$. Moreover, (3.7b) holds with $\rho(s) = s^2$. In addition, (3.9) holds since $V(g(x)) = (-x)^T P(-x) = x^T P x = V(x)$. Next consider a point $x \in D$ with $x^T x = 1$. Necessarily, all coordinates of x are nonnegative and at least one component of x is larger than or equal to $1/\sqrt{n}$. In turn it follows that at least one component of $g(x) = -x$ is less than or equal to $-1/\sqrt{n}$. In particular, $g(x) \notin D$ and $|g(x)|_D \geq 1/\sqrt{n}$. Thus, there exists a time $T_f > 0$ such that, for each $x \in D$ with $x^T x = 1$, a solution that jumps and then flows cannot reach the set D again until it flows for at least $T_f > 0$ units of time. Due to the nature of f, g, C, and D, scaling an initial condition by a positive constant results in solutions that are scaled by that same constant. Therefore, it follows that, except for solutions starting at the origin, each jump is followed by flowing for at least T_f units of time. In particular $(t, j) \in \operatorname{dom} \phi$ implies $j \leq 1 + t/T_f$. In turn, $(t, j) \in \operatorname{dom} \phi$ and $t + j \geq T$ imply $t \geq (T - 1)T_f/(T_f + 1)$. Therefore, Proposition 3.27 applies.

The next two results allow for the Lyapunov function candidate to increase. In the first one, the increases can be persistent but are compensated by strong and persistent decrease. In the second result, weaker decrease is assumed, but the increases are limited in duration.

Proposition 3.29. (Sufficient Lyapunov conditions: increase balanced by decrease) *Let $\mathcal{H} = (C, F, D, G)$ be a hybrid system and let $\mathcal{A} \subset \mathbb{R}^n$ be closed. Suppose that V is a Lyapunov function candidate for $\mathcal{H}$, there exist $\alpha_1, \alpha_2 \in \mathcal{K}_\infty$ such that (3.7a) holds, and*

$$
\begin{aligned}
\langle \nabla V(x), f \rangle &\leq \lambda_c V(x) \quad \forall x \in C, \ f \in F(x) \\
V(g) &\leq e^{\lambda_d} V(x) \quad \forall x \in D, \ g \in G(x).
\end{aligned}
\tag{3.10}
$$

If there exist $\gamma > 0$ and $M > 0$ such that, for each solution ϕ of $\mathcal{H}$, $(t, j) \in \operatorname{dom} \phi$ implies $\lambda_c t + \lambda_d j \leq M - \gamma(t + j)$ then $\mathcal{A}$ is uniformly globally pre-asymptotically stable.

PROOF. The proof is similar to that of Theorem 3.18. Just as inequalities (3.7b) and (3.7c) lead to (3.3), the inequalities (3.10) imply

$$
V(\phi(t, j)) \leq e^{-(\lambda_c t + \lambda_d j)} V(\phi(0, 0))
$$

for every $\phi \in \mathcal{S}_\mathcal{H}$. Then

$$
|\phi(t, j)|_\mathcal{A} \leq \alpha_1^{-1} \left(e^{M - \gamma(t+j)} \alpha_2 \left(|\phi(0, 0)|_\mathcal{A} \right) \right) \leq \alpha_1^{-1} \left(e^M \alpha_2 \left(|\phi(0, 0)|_\mathcal{A} \right) \right)
$$

for every $\phi \in \mathcal{S}_\mathcal{H}$, and uniform stability is established. The arguments for uniform pre-attractivity are the same as in Theorem 3.18, with (3.4) replaced by $T = 1 + (M - \ln \alpha_1(\delta) + \ln \alpha_2(r))/\gamma$. $\qquad\square$

Proposition 3.30. (Sufficient Lyapunov conditions: finite number of jumps or bounded time of flow) *Let $\mathcal{H}$ be a hybrid system and let $\mathcal{A} \subset \mathbb{R}^n$ be closed. Suppose that V is a Lyapunov function candidate for $\mathcal{H}$ and there exist $\alpha_1, \alpha_2 \in \mathcal{K}_\infty$ such that (3.7a) holds. If either of the assumptions (1) or (2) below holds, then $\mathcal{A}$ is uniformly globally pre-asymptotically stable.*

(1) There exist $\rho \in \mathcal{PD}$ and $\lambda \in \mathcal{K}_\infty$ such that (3.7b) holds,

$$V(g) \leq \lambda(V(x)) \qquad \forall x \in D, \ g \in G(x) \tag{3.11}$$

and there exist $\gamma \in \mathcal{K}$, $J > 0$ such that, for every solution ϕ to $\mathcal{H}$, $(t,j) \in$ dom ϕ implies $j \leq \gamma(|\phi(0,0)|_\mathcal{A}) + J$.

(2) There exist $\rho \in \mathcal{PD}$ and $\lambda \in \mathbb{R}$ such that (3.7c) holds,

$$\langle \nabla V(x), f \rangle \leq \lambda V(x) \qquad \forall x \in C, \ f \in F(x) \tag{3.12}$$

and, for every $r > 0$ there exists T_r such that, for every solution ϕ to $\mathcal{H}$ with $|\phi(0,0)|_\mathcal{A} \in (0,r]$, $(t,j) \in$ dom ϕ implies $t \leq T_r$.

PROOF. Only the case of assumption (1) is worked out; the other case is similar. Without loss of generality, suppose $\lambda(s) \geq s$ for all $s \geq 0$. Let $\tilde{\alpha} \in \mathcal{K}_\infty$ satisfy

$$\tilde{\alpha}(s) \geq \lambda^{k(s)}(\alpha_2(s))$$

where $k(s)$ is the smallest integer satisfying $k(s) \geq \gamma(s) + J$. For every $\phi \in \mathcal{S}_\mathcal{H}$ and $(t,j) \in$ dom ϕ,

$$V(\phi(t,j)) \leq \lambda^j \left(V(\phi(0,0)) \right) \leq \lambda^j (\alpha_2(|\phi(0,0)|_\mathcal{A})).$$

Using that $j \leq \gamma(|\phi(0,0)|_\mathcal{A}) + J$ and the lower bound on $\tilde{\alpha}$, it follows that $|\phi(t,j)|_\mathcal{A} \leq \alpha \left(|\phi(0,0)|_\mathcal{A} \right)$ where

$$\alpha(r) = \alpha_1^{-1} \left(\tilde{\alpha}(r) \right).$$

Since $\alpha \in \mathcal{K}_\infty$, this establishes uniform stability.

To establish uniform pre-attractivity, pick any $\varepsilon, r > 0$ and let $R = \alpha(r)$, $\delta = \alpha^{-1}(\varepsilon)$, and $m = \min \rho\left([\delta, R]\right)$. For convenience, set $J_r = \gamma(r) + J$. Let

$$T = J_r \left(\alpha_2(R) - \alpha_1(\delta) \right) / m + J_r.$$

It is left to show that, for every $\phi \in \mathcal{S}_\mathcal{H}$ with $|\phi(0,0)|_\mathcal{A} \leq r$ and length dom $\phi \geq T$, there exists $(t',j') \in$ dom ϕ with $t' + j' \leq T$ such that $|\phi(t',j')|_\mathcal{A} \leq \delta$. If this was false, then for some $\phi \in \mathcal{S}_\mathcal{H}$ with $|\phi(0,0)|_\mathcal{A} \leq r$ and length dom $\phi \geq T$, $|\phi(t,j)|_\mathcal{A} > \delta$ for all $(t,j) \in$ dom ϕ with $t + j \leq T$, and, for such (t,j), $\rho\left(|\phi(t,j)|_\mathcal{A}\right) \geq m$. Because length dom $\phi \geq T$ and the number of jumps of ϕ is bounded by J_r, there exists an interval of flow $[t',t''] \times \{j\} \subset$ dom ϕ with $t'' + j \leq T$ and $t'' - t' \geq (T - J_r)/J_r$. Since $|\phi(t',j)|_\mathcal{A} \leq R$,

$$V(\phi(t'',j)) \leq V(\phi(t',j)) - m(t'' - t') \leq \alpha_2\left(|\phi(t',j)|_\mathcal{A}\right) - m(T - J_r)/J_r \leq \alpha_1(\delta).$$

This contradicts $|\phi(t,j)|_\mathcal{A} > \delta$ for all $(t,j) \in$ dom ϕ with $t + j \leq T$, hence the proof is complete. $\qquad\square$

Example 3.31. (Uniting local and global controllers) Consider using Proposition 3.30 to establish stability for Example 3.23. Define $V(x) = W_q(z)$. The assumptions of Example 3.23 guarantee that there exist $\alpha_1, \alpha_2 \in \mathcal{K}_\infty$ and $\rho \in \mathcal{PD}$ such that (3.7a) and (3.7b) hold. Also, for each $x = (z, q) \in D$, i.e., $z \in D_q$, and $g \in G(x)$,

$$V(g) = W_{3-q}(z) \leq \alpha_2(|z|) \leq \alpha_2 \circ \alpha_1^{-1}(W_q(z)) = \alpha_2 \circ \alpha_1^{-1}(V(x)) \ .$$

Thus, (3.11) holds with $\lambda(s) = \alpha_2 \circ \alpha_1^{-1}(s)$. Finally, due to the construction of D_1 and D_2, the variable z cannot reach the set D_1 from the set D_2 so that no solution experiences more than two jumps. Therefore, Assumption (1) of Proposition 3.30 holds with $J = 2$ and $\gamma \in \mathcal{K}$ arbitrary.

3.4 STABILITY FROM CONTAINMENT

In this section, stability for a hybrid system is related to stability for its components. First are some straightforward observations.

Proposition 3.32. (UGpAS from containment) Let $C_2 \subset C_1 \subset \mathbb{R}^n$, $F_2(x) \subset F_1(x)$ for all $x \in C_2$, $D_2 \subset D_1 \subset \mathbb{R}^n$, $G_2(x) \subset G_1(x)$ for all $x \in D_2$. If the closed set $\mathcal{A} \subset \mathbb{R}^n$ is UGpAS for the hybrid system $\mathcal{H}_1 = (C_1, F_1, D_1, G_1)$ then $\mathcal{A}$ is UGpAS for the hybrid system $\mathcal{H}_2 = (C_2, F_2, D_2, G_2)$.

PROOF. The result follows directly from the fact, due to the assumptions on the data, each solution to $\mathcal{H}_2$ is a solution to $\mathcal{H}_1$. □

Corollary 3.33. (UGpAS implies UGpAS for flows and UGpAS for jumps) If the closed set $\mathcal{A} \subset \mathbb{R}^n$ is UGpAS for the hybrid system $\mathcal{H} = (C, F, D, G)$ then $\mathcal{A}$ is UGpAS for the system $\mathcal{H}_c = (C, F, \emptyset, \emptyset)$ and $\mathcal{A}$ is UGpAS for the system $\mathcal{H}_d = (\emptyset, \emptyset, D, G)$.

The converse of Corollary 3.33 does not hold in general. In particular, it is possible to construct a hybrid system for which the origin is UGpAS when the flows act alone and also when the jumps act alone, but not when the flows and jumps are combined together.

Example 3.34. (No converse to Corollary 3.33, I) On $\mathbb{R}$, consider the hybrid system with flow set $C = (-\infty, 1] \cup [2, \infty)$, flow map $f(x) = -x$ for all $x \in C$, jump set $D = [1, 2]$, and jump map $g(x) = 3x$ for all $x \in D$. For the system $\mathcal{H}_c = (C, f, \emptyset, \emptyset)$, the origin is UGpAS since each solution ϕ starting at a point in C has a domain of the form $[0, \bar{t}] \times \{0\}$, where $\bar{t} \in [0, \infty)$, or $[0, \infty) \times \{0\}$ and satisfies $\phi(t, 0) = \exp(-t)\phi(0, 0)$ for all t such that $(t, 0) \in \operatorname{dom} \phi$. Alternatively, this UGpAS follows from Proposition 3.32 by considering the system with $C = \mathbb{R}$, $f(x) = -x$, and empty D, for which the origin is UGpAS. For the system $\mathcal{H}_d = (\emptyset, \emptyset, D, g)$, the origin is UGpAS (although no solution is complete) since each solution ϕ starting at a point in D has a domain of the form $(0, 0) \cup (0, 1)$ and satisfies $\phi(0, 1) = 3\phi(0, 0)$. On the other hand, the origin is not UGpAS for the system $\mathcal{H} = (C, f, D, g)$ since the maximal solution starting at $x = 2$ is complete and satisfies $\phi(t, j) \geq 2$ for all $(t, j) \in \text{dom}\,\phi$.

Example 3.35. (No converse to Corollary 3.33, II) On $\mathbb{R}$, consider a hybrid system with flow set $C = [1, 2]$, flow map $f(x) = 1$ for all $x \in C$, jump set $D = [0, 1) \times [2, \infty)$, and jump map $g(x) = 0.75x$ for all $x \in D$. For the system $\mathcal{H}_d = (\emptyset, \emptyset, D, g)$, the origin is UGpAS since each solution starting at a point in D has a domain of the form $\{0\} \times \{0, \ldots, J\}$, where J is a positive integer, or $\{0\} \times \mathbb{N}$ and satisfies $\phi(0, j) = (0.75)^j \phi(0, 0)$ for all j such that $(0, j) \in \text{dom}\,\phi$. For the system $\mathcal{H}_c = (C, f, \emptyset, \emptyset)$, the set $\mathcal{A} = \{0\}$, that is, the origin, is UGpAS (although no solution is complete) since each solution ϕ starting at a point in C has a domain of the form $[0, \bar{t}] \times \{0\}$ where $\bar{t} \in [0, 1]$ and satisfies $\phi(t, j) = \phi(0, 0) + t$ for all $(t, j) \in \text{dom}\,\phi$. Figure 3.5 depicts the sets C and D as well as the flow map f. Therefore, uniform global stability follows from

$$|\phi(t, j)| = |\phi(0, 0) + t| \leq |\phi(0, 0)| + 1 \leq 2|\phi(0, 0)|,$$

which holds for all solutions ϕ and all $(t, j) \in \text{dom}\,\phi$. Uniform global pre-attractivity follows by taking any $T > 1$ and noting that there does not exist a solution ϕ to $\mathcal{H}$ and $(t, j) \in \text{dom}\,\phi$ such that $t + j \geq T$. In such a case, the condition for uniform global pre-attractivity holds trivially. On the other hand, for the system $\mathcal{H} = (C, f, D, g)$ the origin is not UGpAS since the maximal solution starting at $x = 2$ is complete and satisfies $\phi(t, j) \geq 3/2$ for all $(t, j) \in \text{dom}\,\phi$.

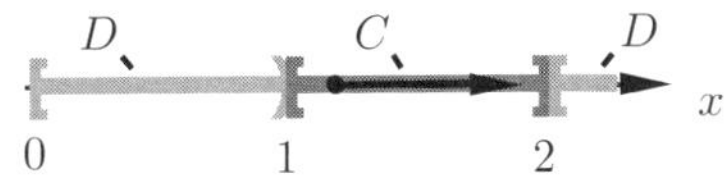

Figure 3.5: Sets C and D as well as the flow map f for Example 3.35.

It is in fact possible to combine an asymptotically stable differential equation, to which all solutions are complete, with an asymptotically stable difference equation, to which all solutions are complete, and end up with a hybrid system that is not asymptotically stable.

Example 3.36. (No converse to Corollary 3.33, III) On $\mathbb{R}^2$, consider a hybrid system with flow set $C = \mathbb{R}^2$, jump set $D = \mathbb{R}^2$, flow map

$$f(x) = \begin{pmatrix} -\varepsilon & 9 \\ -1 & -\varepsilon \end{pmatrix} x$$

where $\varepsilon > 0$, and jump map

$$g(x) = \lambda \begin{pmatrix} 0 & 1/3 \\ 3 & 0 \end{pmatrix} x$$

where $\lambda \in (0,1)$. Since the eigenvalues of the matrix defining the flow map have negative real part and the flow set is the entire space, the continuous-time system has the origin uniformly globally asymptotically stable with complete maximal solutions. Moreover, since the eigenvalues of the matrix defining the jump map have magnitude less than one and the jump set is the entire space, the discrete-time system has the origin uniformly globally asymptotically stable with complete maximal solutions. However, for ε sufficiently close to zero and λ sufficiently close to one, the origin of the hybrid system is not uniformly globally asymptotically stable. This fact can be established by first considering certain solutions for $\varepsilon = 0$ and $\lambda = 1$. Let $x(0,0) = (0,1)$ and consider a solution that flows until it reaches the x_1-axis, which will occur at the value $(3,0)$. Next let the solution jump once, to the value $(0,9)$. Now observe that with $\varepsilon > 0$ but small and $\lambda < 1$ but close to one, flowing from the x_2-axis at $(0,1)$ until reaching the x_1-axis and then jumping once will yield a solution that returns to the x_2-axis at a location above $(0,8)$. Figure 3.6 shows such a solution. More generally, since scaling the initial condition by a positive factor results in solutions that are scaled by that same factor for this system, a solution starting at $(0,c)$ with $c > 0$ can return to a point on the x_2-axis at a location above $(0,8c)$. This means that there are solutions that grow unbounded exponentially. In particular, the origin is not uniformly globally asymptotically stable.

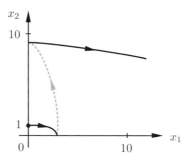

Figure 3.6: A solution to the hybrid system in Example 3.36.

The following theorem provides one situation where UGpAS for a system implies UGpAS for a "larger" system. Theorem 3.37 is useful for simplifying the analysis of systems that have jumps to points outside of $\overline{C} \cup D$.

Theorem 3.37. (Treating systems with "bad" jumps) Let $\mathcal{H} = (C, F, D, G)$ be a hybrid system and let $\mathcal{A} \subset \mathbb{R}^n$ be closed. The following statements are equivalent:

(a) The set $\mathcal{A}$ is uniformly globally pre-asymptotically stable for $\mathcal{H}$.

(b) Both of the following conditions hold:

(i) *there exists* $\gamma \in \mathcal{K}_\infty$ *such that*

$$|g|_\mathcal{A} \leq \gamma(|x|_\mathcal{A}) \qquad \forall x \in D, \; g \in G(x) \; .$$

(ii) *The set* $\mathcal{A}$ *is uniformly globally pre-asymptotically stable for* $\widetilde{\mathcal{H}} = (C, F, \widetilde{D}, \widetilde{G})$ *where* $\widetilde{G}(x) = G(x) \cap (\overline{C} \cup D)$ *for all* $x \in D$, *and* $\widetilde{D} \subset D$ *is the set of points where* $\widetilde{G}(x)$ *is nonempty.*

PROOF. That (a) implies (b) follows from Proposition 3.32.

To establish that (b) implies (a), note that the complete solutions of $\mathcal{H}$ are solutions of $\widetilde{\mathcal{H}}$ while the solutions of $\mathcal{H}$ that are not complete are solutions of $\widetilde{\mathcal{H}}$ perhaps with one extra jump, to a point outside of $\overline{C} \cup D$, appended. Let $\widetilde{\alpha}$ characterize uniform global stability of $\mathcal{A}$ for $\widetilde{\mathcal{H}}$. Then uniform global stability of $\mathcal{A}$ for $\mathcal{H}$ is established with the function

$$\alpha(s) = \max \left\{ \widetilde{\alpha}(s), \gamma(\widetilde{\alpha}(s)) \right\} \; .$$

In addition, using uniform global pre-attractivity for $\widetilde{\mathcal{H}}$, let $\varepsilon > 0$ and $r > 0$ be given and let $T > 0$ be such that, for any solution $\widetilde{\phi}$ to $\widetilde{\mathcal{H}}$ with $|\widetilde{\phi}(0,0)|_\mathcal{A} \leq r$, $(t, j) \in \operatorname{dom} \widetilde{\phi}$ and $t + j \geq T$ imply

$$|\widetilde{\phi}(t,j)|_\mathcal{A} \leq \min \left\{ \varepsilon, \gamma^{-1}(\varepsilon) \right\} \; .$$

It follows that, for any solution ϕ to $\mathcal{H}$ with $|\phi(0,0)|_\mathcal{A} \leq r$, $(t, j) \in \operatorname{dom} \phi$ and $t + j \geq T$, $|\phi(t,j)|_\mathcal{A} \leq \varepsilon$. In other words, $\mathcal{A}$ is uniformly globally pre-attractive for $\mathcal{H}$. Thus, item (a) holds. $\qquad\square$

Theorem 3.37 could have been used to establish uniform global pre-asymptotic stability of the origin for the system $\mathcal{H}_d = (\emptyset, \emptyset, D, g)$ in Example 3.34. Indeed, for that system, $g(x) = 3x$ so that the first item of condition (b) in Theorem 3.37 holds with $\gamma(s) = 3s$. Also, since $D = [1, 2]$, it follows that for all $x \in D$, $\widetilde{g}(x) = \{3x\} \cap D$ is the empty set. Thus, the second item of condition (b) in Theorem 3.37, namely uniform global pre-asymptotic stability of $\mathcal{A}$ for $\widetilde{\mathcal{H}}_d = (\emptyset, \emptyset, \emptyset, \emptyset)$, holds vacuously.

Theorem 3.37 can be applied iteratively to further simplify the analysis of systems with consecutive jumps that end at a point outside of $\overline{C} \cup D$. For example, consider the system $\mathcal{H}_d = (\emptyset, \emptyset, D, g)$ with $g(x) = 3x$ and $D = [1, 6]$. In this case, following the notation of Theorem 3.37, $\widetilde{g}(x) = 3x$ and $\widetilde{D} = [1, 2]$. Thus, $\widetilde{\mathcal{H}}_d$ is given by the discrete-time system of Example 3.34, which was analyzed using Theorem 3.37 in the preceding paragraph and was shown to have the origin globally pre-asymptotically stable. Applying Theorem 3.37 twice shows that the origin is UGpAS for $\mathcal{H}_d = (\emptyset, \emptyset, D, g)$ with $g(x) = 3x$ and $D = [1, 6]$.

3.5   EQUIVALENT CHARACTERIZATIONS

This section presents several equivalent characterizations of uniform global pre-asymptotic stability. The characterizations involve various uniform bounds on the solutions. The first characterization uses class-$\mathcal{KL}$ functions.

Definition 3.38. (Class-$\mathcal{KL}$ functions) *A function $\beta : \mathbb{R}_{\geq 0} \times \mathbb{R}_{\geq 0} \to \mathbb{R}_{\geq 0}$ is a class-$\mathcal{KL}$ function, also written $\beta \in \mathcal{KL}$, if it is nondecreasing in its first argument, nonincreasing in its second argument, $\lim_{r \to 0^+} \beta(r, s) = 0$ for each $s \in \mathbb{R}_{\geq 0}$, and $\lim_{s \to \infty} \beta(r, s) = 0$ for each $r \in \mathbb{R}_{\geq 0}$.*

Lemma 3.39. (Bounding class-$\mathcal{KL}$ functions by class-$\mathcal{K}_{\infty}$ functions) *For each $\beta \in \mathcal{KL}$ there exists $\alpha \in \mathcal{K}_{\infty}$ such that $\beta(r, s) \leq \alpha(r)$ for all $(r, s) \in \mathbb{R}_{\geq 0} \times \mathbb{R}_{\geq 0}$.*

PROOF. Since $\beta(r, s) \leq \beta(r, 0)$ for all $(r, s) \in \mathbb{R}_{\geq 0} \times \mathbb{R}_{\geq 0}$, it is enough to find a class-$\mathcal{K}_{\infty}$ function α such that $\beta(r, 0) \leq \alpha(r)$ for all $r \geq 0$. For each integer i, define $\beta_i = \beta(2^i, 0)$. The sequence β_i is nondecreasing. Define $\alpha_1(0) = 0$ and, for each integer i and all $r \in (2^{i-1}, 2^i]$, let

$$\alpha_1(r) = \beta_i + (\beta_{i+1} - \beta_i) \frac{r - 2^{i-1}}{2^i - 2^{i-1}} \ .$$

By construction, α_1 is continuous and nondecreasing. Then, for all $r \in (2^{i-1}, 2^i]$,

$$\beta(r, 0) \leq \beta(2^i, 0) = \beta_i \leq \alpha_1(r) \ .$$

The result holds by taking $\alpha(r) = \alpha_1(r) + r$ for all $r \geq 0$.   □

The first equivalent characterization of uniform global pre-asymptotic stability combines uniform global stability and uniform global pre-attractivity into a convenient, single bound on the solutions of the system.

Theorem 3.40. (Equivalence of UGpAS and a $\mathcal{KL}$ bound) *Let $\mathcal{H}$ be a hybrid system and $\mathcal{A} \subset \mathbb{R}^n$ be closed. The following statements are equivalent:*

(a) *The set $\mathcal{A}$ is uniformly globally pre-asymptotically stable for $\mathcal{H}$.*

(b) *There exists a $\mathcal{KL}$ function β such that any solution ϕ to $\mathcal{H}$ satisfies*

$$|\phi(t, j)|_{\mathcal{A}} \leq \beta(|\phi(0, 0)|_{\mathcal{A}}, t + j) \qquad \forall (t, j) \in \operatorname{dom} \phi \ .$$

PROOF. Suppose that (a) holds. Define a function $\beta_0 : \mathbb{R}_{\geq 0} \times \mathbb{R}_{\geq 0} \to [-\infty, \infty)$ by

$$\beta_0(r, s) = \sup \left\{ |\phi(t, j)|_{\mathcal{A}} : \phi \in \mathcal{S}_{\mathcal{H}}(\xi), \ |\xi|_{\mathcal{A}} \leq r, \ t + j \geq s \right\}.$$

The very definition implies that the bound

$$|\phi(t, j)|_{\mathcal{A}} \leq \beta_0(|\phi(0, 0)|_{\mathcal{A}}, t + j) \qquad \forall (t, j) \in \operatorname{dom} \phi$$

holds for all solutions ϕ to $\mathcal{H}$. (The bound is vacuously true when $\beta_0(r, s) = -\infty$, which only occurs when the supremum is taken over an empty set, i.e., when all solutions ϕ with $|\phi(0,0)|_{\mathcal{A}} \leq r$ are such that $\operatorname{length} \operatorname{dom} \phi < s$.) The definition further implies that $\beta_0(r, s)$ is nondecreasing in r and nonincreasing in s. One also has $\beta_0(r, s) \leq \alpha(r)$ for all $r, s \geq 0$, where α comes from the definition of uniform global stability of $\mathcal{A}$. Now define $\beta : \mathbb{R}_{\geq 0} \times \mathbb{R}_{\geq 0} \to \mathbb{R}_{\geq 0}$ by

$$\beta(r, s) = \max \{0, \beta_0(r, s)\} \, .$$

Then $\beta(r, s)$ is nondecreasing in r, nonincreasing in s, $\lim_{r \to 0^+} \beta(r, s) = 0$ for each $s \in \mathbb{R}_{\geq 0}$ since $\beta(r, s) \leq \alpha(r)$, and the bound in (b) holds for each solution ϕ. Finally, uniform global asymptotic pre-attractivity of $\mathcal{A}$ implies that $\lim_{s \to \infty} \beta(r, s) = 0$ for each $r \geq 0$. Hence, β is a class-$\mathcal{KL}$ function, and (b) is satisfied.

Suppose that (b) holds. Using Lemma 3.39, let $\alpha \in \mathcal{K}_\infty$ satisfy $\beta(r, s) \leq \alpha(r)$ for all $(r, s) \in \mathbb{R}_{\geq 0} \times \mathbb{R}_{\geq 0}$. Then

$$|\phi(t, j)|_{\mathcal{A}} \leq \beta(|\phi(0,0)|_{\mathcal{A}}, t + j) \leq \alpha(|\phi(0,0)|_{\mathcal{A}}) \, ,$$

which establishes uniform global stability. In addition, using the properties of $\mathcal{KL}$ functions, for each $r \geq 0$ and $\varepsilon > 0$ there exists $T > 0$ such that $\beta(r, s) \leq \varepsilon$ for all $s \geq T$. Then $|\phi(0,0)|_{\mathcal{A}} \leq r$, $(t, j) \in \operatorname{dom} \phi$, and $t + j \geq T$ imply

$$|\phi(t, j)|_{\mathcal{A}} \leq \beta(r, t + j) \leq \varepsilon \, ,$$

which establishes uniform global pre-attractivity. $\qquad\square$

The next characterization of uniform global pre-asymptotic stability establishes that, without loss of generality, the $\mathcal{KL}$ function in the previous characterization can be taken to be continuous. Moreover, it establishes that when the distance to the set $\mathcal{A}$ is viewed through an appropriate function, the convergence toward the attractor appears to be exponential convergence. The characterization relies on the following preliminary result, the Massera-Sontag Lemma.

Lemma 3.41. (Class $\mathcal{KL}$ and exponential decay) *For each class-$\mathcal{KL}$ function β and each $\lambda > 0$ there exist class-$\mathcal{K}_\infty$ functions α_1, α_2 such that, for all $r, s \in \mathbb{R}_{\geq 0}$,*

$$\alpha_1 (\beta(r, s)) \leq \alpha_2(r)e^{-\lambda s}.$$

PROOF. First, pick a class $\mathcal{K}_\infty$ function η and a function $\theta : \mathbb{R}_{\geq 0} \to \mathbb{R}$, decreasing to 0 as $s \to \infty$, such that

$$\beta(\eta(s), s) \leq \theta(s) \qquad \forall s \geq 0. \tag{3.13}$$

To do this, set $\varepsilon_0 = \beta(1, 0)$ and pick any decreasing to 0 sequence $\{\varepsilon_i\}_{i=1}^\infty$ with each $\varepsilon_i \in (0, \beta(1, 0))$. Set $s_0 = 0$ and pick an increasing to ∞ sequence of positive numbers $\{s_i\}_{i=1}^\infty$ such that $\beta(i, s_i) \leq \varepsilon_i$, $i = 1, 2, \dots$, which is possible since β is a class-$\mathcal{KL}$ function. Let η be a class-$\mathcal{K}_\infty$ function such that $\eta(s) \leq i$ for

$s \in [s_i, s_{i+1}]$, $i = 0, 1, \ldots$. Let $\theta : \mathbb{R}_{\geq 0} \to \mathbb{R}$ be a decreasing to 0 as $s \to \infty$ function such that $\theta(s) \geq \varepsilon_i$ for $s \in [s_i, s_{i+1}]$, $i = 0, 1, \ldots$. Then, whenever $s \in [s_i, s_{i+1}]$,

$$\beta(\eta(s), s) \leq \beta(i, s) \leq \beta(i, s_i) \leq \varepsilon_i \leq \theta(s),$$

and hence (3.13) holds.

Now pick a class $\mathcal{K}_\infty$-function α_1 such that $\alpha_1(u) \leq e^{-2\lambda\theta^{-1}(u)}$ for all $u \in (0, \theta(0)]$. Then for all $s \geq 0$, all $0 \leq r \leq \eta(s)$,

$$
\begin{aligned}
\alpha_1\left(\beta(r,s)\right) &\leq \sqrt{\alpha_1\left(\beta(r,0)\right)}\sqrt{\alpha_1\left(\beta(\eta(s),s)\right)} \leq \sqrt{\alpha_1\left(\beta(r,0)\right)}\sqrt{\alpha_1\left(\theta(s)\right)} \\
&\leq \sqrt{\alpha_1\left(\beta(r,0)\right)}\,e^{-\lambda s},
\end{aligned}
$$

while for all $r > \eta(s)$,

$$\alpha_1\left(\beta(r,s)\right) \leq \alpha_1\left(\beta(r,0)\right) \leq \alpha_1\left(\beta(r,0)\right)e^{\lambda\eta^{-1}(r)}e^{-\lambda s}.$$

Picking any class-$\mathcal{K}_\infty$ function α_2 such that

$$\alpha_2(r) \geq \max\left\{\sqrt{\alpha_1\left(\beta(r,0)\right)}, \alpha_1\left(\beta(r,0)\right)e^{\lambda\eta^{-1}(r)}\right\}$$

finishes the proof. □

Theorem 3.42. (Equivalent characterizations of UGpAS) *Let $\mathcal{H}$ be a hybrid system and $\mathcal{A} \subset \mathbb{R}^n$ be closed. The following statements are equivalent:*

(a) *The set $\mathcal{A}$ is uniformly globally pre-asymptotically stable for $\mathcal{H}$.*

(b) *For each $\lambda > 0$, there exist $\mathcal{K}_\infty$ functions α_1 and α_2 such that any solution ϕ to $\mathcal{H}$ satisfies*

$$\alpha_1\left(|\phi(t,j)|_\mathcal{A}\right) \leq \alpha_2\left(|\phi(0,0)|_\mathcal{A}\right)e^{-\lambda(t+j)} \qquad \forall(t,j) \in \operatorname{dom}\phi. \quad (3.14)$$

PROOF. Suppose that (a) holds. Let $\beta \in \mathcal{KL}$ satisfy condition (b) of Theorem 3.40. Let $\lambda > 0$ be given. Then, using Lemma 3.41, there exist class-$\mathcal{K}_\infty$ functions α_1 and α_2 such that

$$\alpha_1\left(|\phi(t,j)|_\mathcal{A}\right) \leq \alpha_1\left(\beta(|\phi(0,0)|_\mathcal{A}, t+j)\right) \leq \alpha_2(|\phi(0,0)|_\mathcal{A})e^{-\lambda(t+j)},$$

which establishes condition (b) of the theorem.

That (b) implies (a) follows from Theorem 3.40 by noting that the bound (3.14) can be written as $|\phi(t,j)|_\mathcal{A} \leq \alpha_1^{-1}\left(\alpha_2\left(|\phi(0,0)|_\mathcal{A}\right)e^{-\lambda(t+j)}\right)$, which establishes condition (b) of Theorem 3.40 with $\beta(r,s) = \alpha_1^{-1}\left(\alpha_2(r)e^{-\lambda s}\right)$. □

3.6   NOTES

Uniform global asymptotic stability as given in Definition 3.6 parallels the classical definition used for continuous-time systems, although completeness of maximal solutions is usually assumed in the classical definitions. A modern treatment of uniform global asymptotic stability for time-varying differential equations can be found in Khalil's textbook [62], and for certain differential inclusions and more general closed sets in Lin et al. [74]. Especially for time-varying systems, the literature has offered multiple definitions of uniform global asymptotic stability, which are compared and contrasted by Teel and Zaccarian [119]. The use of $\mathcal{K}$, $\mathcal{K}_\infty$, and $\mathcal{KL}$ functions in asymptotic stability characterizations dates back to at least Hahn's classical text [46] and, following [62], is prevalent in the recent nonlinear systems literature. An example in the spirit of Example 3.10 but where the set $\mathcal{A}$ is a single point was offered by Vinograd in [126]; the example also appears in [46] and Vidyasagar's text [125]. Example 3.11 is essentially the same as Example 4.18 of [62]. The proof of Theorem 3.40 is inspired by the proof of [74, Proposition 2.5]. Lemma 3.41 has connections to Massera's lemma (see Section 12 of [87] or [62, Lemma C.1]); it is based on Sontag's [113, Proposition 7] while the proof follows the one given by Teel and Praly in [118, Lemma 3]. The general concept of a Lyapunov function dates back to the 1892 thesis of Lyapunov [77] and has been a core component of nonlinear systems stability analysis since the middle of the last century. The next-to-last Lyapunov function for the bouncing ball in Example 3.19 was proposed by Cai et al. in [25] while the last Lyapunov function of that example is inspired by a construction due to Lamperski and Ames [67]. The average dwell-time condition used in Example 3.22 was introduced by Hespanha and Morse in [57] and the Lyapunov arguments in Example 3.22 are similar to arguments used by Liberzon [73] and Hespanha et al. [56]. Example 3.36 is based on a classical example of instability from the switched systems literature. Other stability results that exploit relaxed Lyapunov conditions but that have not been included here include the concept of "multiple Lyapunov functions" due to Branicky [19] and DeCarlo et al. [32], and Matrosov functions, introduced by Matrosov [89] and used more recently for continuous-time systems by Paden and Panja [96] and Loria et al. [76], for discrete-time systems by Nešić and Teel [94], and in the context of hybrid systems by Malisoff and Mazenc [82] and Sanfelice and Teel [107].

Chapter Four

Perturbations and generalized solutions

This chapter discusses the effect of state perturbations on solutions to a hybrid system. It is shown that state perturbations, of arbitrarily small size, can dramatically change the behavior of solutions. While such a phenomenon is also present in continuous-time and discrete-time dynamical systems, it is magnified in the hybrid setting, due to the flows and the jumps being constrained to the flow and the jump sets, respectively.

Perturbations affecting the whole state of a hybrid system are usually considered. The resulting behaviors are quite representative of what may occur if perturbations come from state measurement error in a hybrid feedback control system or from errors present in numerical simulation of hybrid systems. The case of hybrid feedback control is given some attention in this chapter and is revisited later. Throughout the chapter, effects of perturbations are related to the regularity properties of the data of hybrid systems.

4.1 DIFFERENTIAL AND DIFFERENCE EQUATIONS

Feedback control of a nonlinear system can require the application of a discontinuous feedback, and thus lead to the closed-loop system represented by a differential equation with a discontinuous right-hand side. (See Example 4.19 later in this chapter.) Such differential equations can be quite sensitive to state perturbations. For instance, the presence of two opposite values of the right-hand side of a differential equation, near an initial point, can lead to a solution chattering around such an initial point. The following example illustrates this phenomenon, and more general behaviors resulting from state perturbations.

Example 4.1. (Differential equation with discontinuous right-hand side) Consider the differential equation $\dot{z} = f(z)$ on $\mathbb{R}^2$, where

$$f(z) = \begin{bmatrix} 1 \\ 0 \end{bmatrix} \text{ if } z_2 \geq 0, \quad f(z) = \begin{bmatrix} -1 \\ 0 \end{bmatrix} \text{ if } z_2 < 0.$$

The unique solution from the origin is given by $z_1(t) = t$, $z_2(t) = 0$ for $t \in \mathbb{R}_{\geq 0}$. Now, given an arbitrarily small $\varepsilon > 0$, consider $e : \mathbb{R}_{\geq 0} \to \mathbb{R}^2$ given by $e_1(t) = 0$, $e_2(t) = \varepsilon \sin t$. The solution to $\dot{z}(t) = f(z(t) + e(t))$, in the first variable, is a see-saw function z_{1e} oscillating between 0 and π as shown in Figure 4.1. This solution is significantly different from the original, unperturbed, solution.

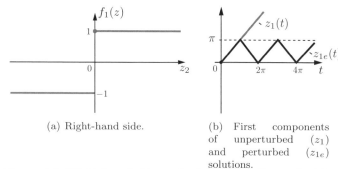

(a) Right-hand side.

(b) First components of unperturbed (z_1) and perturbed (z_{1e}) solutions.

Figure 4.1: Right-hand side and solutions to Example 4.1.

More generally, given an $\varepsilon > 0$ and any $\lambda \in (0,1)$, let $e : \mathbb{R}_{\geq 0} \to \mathbb{R}^2$ be given by $e_1(t) = 0$ for $t \in \mathbb{R}_{\geq 0}$ and by e_2 that is periodic, with period ε, and defined on $[0,\varepsilon)$ by $e_2(t) = \varepsilon$ for $t \in [0, \lambda\varepsilon)$, $e_2(t) = -\varepsilon$ for $t \in [\lambda\varepsilon, \varepsilon)$. The resulting solution, from $\xi = 0$, displays a "see-saw like" function in the first variable. The average rate of growth of that function, over $[0, k\varepsilon)$, where $k \in \mathbb{N}$, is $2\lambda - 1$. The limit (uniform on compact intervals, and in fact uniform), when $\varepsilon \to 0$, of such "see-saw like" functions is $z_1(t) = (2\lambda - 1)t$ for $t \in \mathbb{R}_{\geq 0}$. Together with $z_2(t) = 0$ for $t \in \mathbb{R}_{\geq 0}$, such a limit is in fact a solution to the differential inclusion $\dot{z} \in F(z)$, where

$$F(z) = \begin{bmatrix} 1 \\ 0 \end{bmatrix} \text{ if } z_2 > 0, \quad F(z) = \begin{bmatrix} -1 \\ 0 \end{bmatrix} \text{ if } z_2 < 0, \quad F(z) = \begin{bmatrix} [-1,1] \\ 0 \end{bmatrix} \text{ if } z_2 = 0.$$

In particular, at the points of discontinuity of f, the set-valued mapping F is given as the smallest convex set containing the values of f from nearby points.

The example suggests that the effect of state perturbations on solutions to a differential equation $\dot{z} = f(z)$ (with discontinuous f) may be reflected by solutions to a differential inclusion $\dot{z} \in F(z)$ with F being the "convex closure" of f. This observation is now made rigorous.

Definition 4.2. (Generalized solutions to differential equations) *Let $f : \mathbb{R}^n \to \mathbb{R}^n$ be a function and $z : [0,T] \to \mathbb{R}^n$ be an absolutely continuous function.*

- *The function z is a Hermes solution to $\dot{z} = f(z)$ if there exist a sequence of absolutely continuous functions $z_i : [0,T] \to \mathbb{R}^n$ and a sequence of measurable functions $e_i : [0,T] \to \mathbb{R}^n$ such that $\dot{z}_i(t) = f(z_i(t) + e_i(t))$ for almost all $t \in [0,T]$, the sequence $\{z_i\}_{i=1}^{\infty}$ converges uniformly to z on $[0,T]$, and the sequence $\{e_i\}_{i=1}^{\infty}$ converges uniformly to the zero function on $[0,T]$.*

- *The function z is a Krasovskii solution to $\dot{z} = f(z)$ if*

$$\dot{z}(t) \in F(z(t)) \text{ for almost all } t \in [0,T],$$

where, for each $\xi \in \mathbb{R}^n$,

$$F(\xi) = \bigcap_{\delta > 0} \overline{\mathrm{con}} f(\xi + \delta \mathbb{B}).$$

Above, $\overline{\mathrm{con}} f(\xi + \delta \mathbb{B})$ is the closed convex hull of the set $f(\xi + \delta \mathbb{B})$, in other words, the smallest closed convex set containing $f(\xi + \delta \mathbb{B})$.

In Example 4.1, it was argued that $z_1(t) = (2\lambda - 1)t$, $z_2(t) = 0$ for $t \in \mathbb{R}_{\geq 0}$ describes a Hermes solution z to the differential equation discussed in that example. It was also noted that this Hermes solution is a Krasovskii solution. In fact, every Hermes solution is a Krasovskii solution, and furthermore, the converse statement is also true.

Theorem 4.3. (Hermes and Krasovskii solutions for differential equations) *Let $f : \mathbb{R}^n \to \mathbb{R}^n$ be a locally bounded function. Then, an absolutely continuous $z : [0, T] \to \mathbb{R}^n$ is a Hermes solution to $\dot{z} = f(z)$ if and only if it is a Krasovskii solution to $\dot{z} = f(z)$.*

One direction, that Krasovskii solutions to differential equations are Hermes solutions, is a direct consequence of Theorem 4.21, which is shown later in the course of proving Theorem 4.17 – a hybrid version of Theorem 4.3. Theorem 4.17 is a generalization of Theorem 4.3, but seeing this requires minor technicalities relating different types of convergence to be worked out.

Corresponding definitions of Hermes and Krasovskii solutions, and a result paralleling Theorem 4.3, can be given for discrete-time dynamical systems given by difference equations of the type $z^+ = g(z)$. The discrete-time case is, however, immediately captured by the far more general case of hybrid systems, which is treated in the upcoming Section 4.3. For illustration purposes, a discrete-time example involving state perturbations is included.

Example 4.4. (Difference equation with discontinuous right-hand side) Consider the discrete-time system $z^+ = g(z)$ on $\mathbb{R}$, where g is given as in Figure 4.2(a). For every initial point $\xi \in \mathbb{R}$, solutions from it converge to zero in a finite number of jumps. In fact, the number of jumps it takes for the solution from ξ to reach zero is equal ξ if $\xi \in \mathbb{Z}$ and floor$(\xi) + 1$ otherwise, where floor(ξ) is the greatest integer less than or equal to $|\xi|$. Given any $\varepsilon > 0$, consider $e : \mathbb{R} \to \mathbb{R}$ given by $e(j) := \varepsilon$ for each $j \in \mathbb{N}$. The solution to $z^+ = g(z + e)$ from $\xi' \in \mathbb{R}$ is given by $z(0) = \xi'$, $z(j) = \mathrm{floor}(\xi' + \varepsilon)$ for $j = 1, 2, \ldots$. When $\varepsilon \to 0$, such solutions converge uniformly to $z(0) = \xi'$, $z(j) = \xi'$ for $j = 1, 2, \ldots$. This limiting solution is a solution to the difference inclusion $z^+ \in G(z)$ where G is given as in Figure 4.2(b). At the points of discontinuity, the set-valued map G is given by all of the values of g at nearby points.

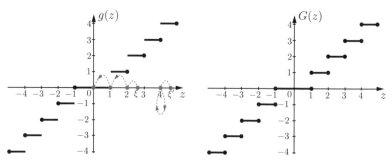

(a) Right-hand side g, solution without state perturbation from ξ, and with state perturbation from ξ'.

(b) Right-hand side for set-valued discrete-time system.

Figure 4.2: Right-hand sides and solutions to Example 4.4.

4.2   SYSTEMS WITH STATE PERTURBATIONS

The hybrid system $\mathcal{H} = (C, F, D, G)$ with a state perturbation e is denoted by $\mathcal{H}_e$, and following (1.1), is written in the suggestive form:

$$\begin{cases} x + e \in C & \dot{x} \in F(x+e) \\ x + e \in D & x^+ \in G(x+e). \end{cases} \tag{4.1}$$

Before formally defining solutions to $\mathcal{H}_e$, a class of admissible state perturbation is specified.

Definition 4.5. (Admissible state perturbation) *A mapping e is an admissible state perturbation if $\mathrm{dom}\, e$ is a hybrid time domain and the function $t \to e(t, j)$ is measurable on $\mathrm{dom}\, e \cap (\mathbb{R}_{\geq 0} \times \{j\})$ for each $j \in \mathbb{N}$.*

In several instances, the naturally arising state perturbations in a hybrid system depend on time $t \in \mathbb{R}_{\geq 0}$ only, and not on j. Then, every measurable signal $e' : \mathbb{R}_{\geq 0} \to \mathbb{R}^n$ corresponding to the state perturbation can be considered as given on an arbitrary hybrid time domain E, by setting

$$e(t, j) := e'(t) \qquad (t, j) \in \mathrm{dom}\, e := E . \tag{4.2}$$

In most cases, e is given on the hybrid time domain of the solution to the system it affects.

A formal definition of solutions to $\mathcal{H}_e$ with admissible state perturbation e follows.

Definition 4.6. (Solution to a hybrid system with state perturbation) *A hybrid arc ϕ is a solution to the hybrid system $\mathcal{H}_e$ with admissible state perturbation e if $\mathrm{dom}\, \phi = \mathrm{dom}\, e$, $\phi(0,0) + e(0,0) \in \overline{C} \cup D$, and*

(S1$_e$) for all $j \in \mathbb{N}$ such that I^j has nonempty interior, where $I^j \times \{j\} :=$ dom $\phi \cap ([0, +\infty) \times \{j\})$,

$$\phi(t, j) + e(t, j) \in C \quad \text{for all } t \in \text{int } I^j,$$
$$\dot{\phi}(t, j) \in F(\phi(t, j) + e(t, j)) \quad \text{for almost all } t \in I^j; \tag{4.3}$$

(S2$_e$) for all $(t, j) \in \text{dom } \phi$ such that $(t, j + 1) \in \text{dom } \phi$,

$$\phi(t, j) + e(t, j) \in D, \quad \phi(t, j + 1) \in G(\phi(t, j) + e(t, j)). \tag{4.4}$$

With some abuse of terminology, it is sometimes said that a measurable function $e' : \mathbb{R}_{\geq 0} \to \mathbb{R}^n$ leads to a solution ϕ to $\mathcal{H}_e$ if ϕ with e given by $e(t, j) = e'(t)$ is a solution to $\mathcal{H}_e$ in the sense of Definition 4.6. Similarly, the statement that $e' : \mathbb{R}_{\geq 0} \to \mathbb{R}^n$ leads to nonexistence of solutions to $\mathcal{H}_e$ (possibly, with a specified initial point) means that there are no solutions ϕ to $\mathcal{H}_e$ with e given by $e(t, j) = e'(t)$ for $(t, j) \in \text{dom } \phi$.

The behavior of solutions to $\mathcal{H}_e$ can be dramatically different from the behavior of solutions to $\mathcal{H}$. Even when the flow map and the jump map are continuous functions, in which case state perturbations do not affect the flow and the jump maps significantly, the flow set or the jump set not being closed can lead to $\mathcal{H}$ being very sensitive to state perturbations. One simple consequence of this is that state perturbations can lead to solutions that "miss" the jump set. This means that it is possible that solutions that jump in the absence of state perturbations can flow forever when perturbations are present. Such phenomenon may be undesired in, say, hybrid feedback control, as stabilization may rely on certain variables jumping.

Example 4.7. (Solutions miss the jump set) Consider the hybrid system $\mathcal{H}$ on $\mathbb{R}^2$ with jump set D being the line of slope -1 passing through $(0, \sqrt{2})$, as in Figure 4.3, flow set $C = \mathbb{R}^2 \setminus D$, jump map $g(x) = 0$ for all $x \in D$, and flow map $f(x) = [1 \ 1]^T / \sqrt{2}$ for all $x \in C$. The unique solution to $\mathcal{H}$ from $\xi = 0$ is "periodic": $\phi(t, j) = [t - j \ t - j]^T$ for $(t, j) \in \text{dom } \phi = \bigcup_{j \in \mathbb{N}} ([j, j+1] \times \{j\})$. (Note that the distance from the origin to D is unitary.) The unique solution to $\mathcal{H}_e$ from $\xi = 0$ with $e(t, 0) = 0$ if $t \neq 1$ and $e(1, 0) \neq 0$ is continuous: $\phi_e(t, 0) = (t, t)$ for $(t, 0) \in \text{dom } \phi_e = \mathbb{R}_{\geq 0} \times \{0\}$, and so, quite different from ϕ; in fact, different from all solutions to $\mathcal{H}$ from initial points close to 0.

A more extreme, while somewhat exotic, example is as follows:

Example 4.8. (From few to many solutions) Let C and D be two dense subsets of $\mathbb{R}$ such that $C \cap D = \emptyset$, fix $c \neq 0$ and let $f(x) = c$ for all $x \in \mathbb{R}$, and let $g : \mathbb{R} \to \mathbb{R} \setminus D$ be an arbitrary function. Flow is impossible for $\mathcal{H}$: the only continuous functions $z : [0, \varepsilon) \to \mathbb{R}$, $\varepsilon > 0$, that satisfy $z(t) \in C$ for all $t \in (0, \varepsilon)$ are constant, but f is not 0. The only maximal solutions to $\mathcal{H}$ are then hybrid arcs that jump once, from $\phi(0, 0) \in D$ to $\phi(0, 1) \in g(\phi(0, 0)) \notin D$. Note that ϕ is nontrivial. With a state perturbation, there exist solutions with any a priori

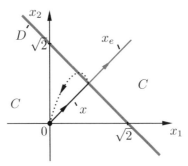

Figure 4.3: Flow and jump set for the hybrid system in Example 4.7. Solution ϕ without measurement noise is periodic while solution ϕ_e with measurement noise escapes to infinity.

chosen hybrid time domains. For example, a continuous hybrid arc ϕ_e from ξ, given by $\phi_e(t,0) = \xi + ct$, is a solution to $\mathcal{H}_e$ if one considers a perturbation such that $ct + e(t,0) \in C$ for all $t \in \mathbb{R}_{\geq 0}$. Such perturbation can be arbitrarily small and Lipschitz in t.

With the effect of state perturbations so dramatic (for hybrid systems with data missing some regularity, for example, with the flow sets or the jump sets not closed), one should not expect asymptotic stability in a hybrid system to be robust. This issue is illustrated by the next example.

Example 4.9. (Asymptotic stability without robustness) Consider a hybrid system with $f(x) = -x$ for all $x \in C := (-\infty, 1]$, and $g(x) = 1$ for all $x \in D := (1, \infty)$. Solutions starting from C (in particular, the unique solution starting from $\xi = 1$) converge to 0 exponentially. Solutions from D jump to $1 \in C$ instantly, and then converge to 0 exponentially. It can be easily verified that the system is globally uniformly asymptotically stable. However, the unique solution to $\mathcal{H}_e$ from $\xi = 1$ with constant and equal to $\varepsilon > 0$ perturbation is given by $\phi_e(0, j) = 1$ for all $j \in \mathbb{N}$. Figure 4.4 illustrates this behavior. Thus, an arbitrarily small state perturbation leads to a large qualitative change in the behavior of solutions. In particular, perturbed solutions do not converge to a small neighborhood of the origin.

Example 4.8 showed how perturbations can lead to the existence of a plethora of solutions that do not resemble any of the solutions to the nominal hybrid system. A different phenomenon is also possible: for some hybrid systems, even with quite regular data, certain state perturbations can lead to nonexistence of solutions. This is certainly possible for initial conditions ξ that are on the boundary of $C \cup D$, independently of regularity of the data (in fact, C, D can be closed in $\mathbb{R}^n$, f, g continuous, and the necessary conditions for existence of solutions, as in Proposition 2.10, can be assumed). Indeed, for such ξ, there

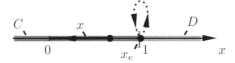

Figure 4.4: The effect of state perturbations in Example 4.9. ϕ is a solution to $\mathcal{H}$ and ϕ_e is a solution to $\mathcal{H}_e$ with admissible measurement noise e.

exists an arbitrarily small $\Delta \in \mathbb{R}^n$ such that $\xi + \Delta \notin \overline{C \cup D}$. Then, there are no nontrivial solutions to $\mathcal{H}_e$ with $e(t) = \Delta$, $t \geq 0$. In fact, when ξ is in the intersection of the boundaries of C and of D (but is possibly in the interior of $C \cup D$), existence may still be problematic for some perturbations. For such ξ, there exists (arbitrarily small) Δ_1 such that $\xi + \Delta_1 \notin D$ and Δ_2 such that $\xi + \Delta_2 \notin \overline{C}$. Taking noise defined by $e(0) = \Delta_1$, $e(t) = \Delta_2$ for $t > 0$, results in no solutions to $\mathcal{H}_e$ from ξ. Indeed, a solution to $\mathcal{H}_e$ from ξ would either satisfy $\phi(0,0) + \Delta_1 \in D$, which does not hold, or $\phi(t,0) + \Delta_2 \in C$ for small enough t, which also does not hold (this follows since $\phi(0,0) + \Delta_2 \notin \overline{C}$, the complement of $\overline{C}$ is open, and $\phi(t,0)$ is close to $\phi(0,0)$ for small t).

Proposition 4.10. (Basic existence with state perturbations) *Consider a hybrid system $\mathcal{H} = (C, F, D, G)$. Let $\xi \in \overline{C} \cup D$. If there exists $\delta > 0$ such that either $\xi + \delta \mathbb{B} \subset D$ or*

(VCe) *$\xi + \delta \mathbb{B} \subset C$ and for every measurable $e' : \mathbb{R}_{\geq 0} \to \mathbb{R}^n$ with $e'(t) \in \delta \mathbb{B}$ for all $t \in \mathbb{R}_{\geq 0}$ there exists $\varepsilon > 0$ and an absolutely continuous $z_{e'} : [0, \varepsilon] \to \mathbb{R}^n$ such that $\dot{z}_{e'}(t) \in F(z_{e'}(t) + e'(t))$ for almost all $t \in [0, \varepsilon]$,*

then there exists a nontrivial solution ϕ to $\mathcal{H}_e$ with $\phi(0,0) = \xi$ for every admissible state perturbation e with $e(t,j) \in \delta \mathbb{B}$ for all $(t,j) \in \operatorname{dom} e$.

Limits of solutions to hybrid systems, as the perturbations vanish, are discussed in the next section.

4.3   GENERALIZED SOLUTIONS

The previous sections illustrated that the effect of state perturbations on solutions to a hybrid system can be quite significant. In this section, the effect of (arbitrarily small) state perturbations is related to an operation that regularizes the data of the hybrid system. Briefly speaking, an appropriately understood limit of a sequence of solutions to a hybrid system, generated with state perturbation decreasing in magnitude, turns out to be a solution to the regularized hybrid system. Conversely, every solution to the regularized system can be approximated, with arbitrary precision, with solutions to the original system generated with state perturbations.

To make these statements precise, a way to measure whether two hybrid arcs are close to one another is needed. Certainly, because two different hybrid arcs need not have the same domains, relying on the uniform metric is impossible. The ε-closeness of hybrid arcs, defined below, is related to the Hausdorff distance between the graphs of the arcs, as Figure 4.5 shows. It will be used only when compact hybrid arcs are involved. Further details, and a generalization of the ε-closeness to a concept better suited for dealing with hybrid arcs that are not compact are in Chapter 5.

Definition 4.11. (ε-closeness of hybrid arcs) *Given $\varepsilon > 0$, two hybrid arcs ϕ_1 and ϕ_2 are ε-close if*

(a) *for all $(t,j) \in \mathrm{dom}\,\phi_1$ there exists s such that $(s,j) \in \mathrm{dom}\,\phi_2$, $|t - s| < \varepsilon$, and*

$$|\phi_1(t,j) - \phi_2(s,j)| < \varepsilon,$$

(b) *for all $(t,j) \in \mathrm{dom}\,\phi_2$ there exists s such that $(s,j) \in \mathrm{dom}\,\phi_1$, $|t - s| < \varepsilon$, and*

$$|\phi_2(t,j) - \phi_1(s,j)| < \varepsilon.$$

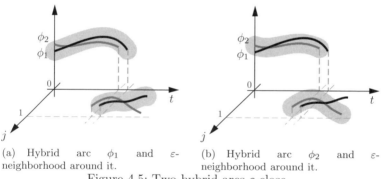

(a) Hybrid arc ϕ_1 and ε-neighborhood around it. (b) Hybrid arc ϕ_2 and ε-neighborhood around it.

Figure 4.5: Two hybrid arcs ε-close.

Accounting for state perturbations, or regularizing the data of a hybrid system, leads to two concepts of generalized solutions to a hybrid system. The terminology used for these concepts, Hermes solutions and Krasovskii solutions, is borrowed from what was established for differential equations; recall Definition 4.2. For convenience, the suggestive description of the hybrid system $\mathcal{H} = (C, F, D, G)$ is recalled:

$$\begin{cases} x \in C & \dot{x} \in F(x) \\ x \in D & x^+ \in G(x) \end{cases}$$

along with the description of $\mathcal{H}_e$ that represents $\mathcal{H} = (C, F, D, G)$ when a state perturbation e is present:

$$\begin{cases} x + e \in C & \dot{x} \in F(x + e) \\ x + e \in D & x^+ \in G(x + e). \end{cases}$$

Definition 4.12. (Hermes solutions to hybrid systems) *A compact hybrid arc ϕ is a compact Hermes solution to $\mathcal{H}$ if there exist a sequence $\{\phi_i\}_{i=1}^{\infty}$ of compact hybrid arcs and a sequence $\{e_i\}_{i=1}^{\infty}$ of admissible state perturbations such that*

- *ϕ_i is a solution to $\mathcal{H}_e$ with state perturbation e_i for every $i \in \mathbb{N}$;*

- *for every $\varepsilon > 0$ there exists i_0 such that for all $i > i_0$, ϕ_i and ϕ are ε-close;*

- *the sequence of $\sup_{(t,j) \in \operatorname{dom} e_i} |e_i(t,j)|$ converges to 0.*

A hybrid arc ϕ is a Hermes solution to $\mathcal{H}$ if the restriction of ϕ to any compact hybrid time domain that is a subset of $\operatorname{dom} \phi$ is a compact Hermes solution to $\mathcal{H}$.

Definition 4.13. (Krasovskii solutions to hybrid systems) *A hybrid arc ϕ is a Krasovskii solution to $\mathcal{H} = (C, F, D, G)$ if ϕ is a solution to the regularized hybrid system*

$$\begin{cases} x \in \widehat{C} & \dot{x} \in \widehat{F}(x) \\ x \in \widehat{D} & x^+ \in \widehat{G}(x) \end{cases} \tag{4.5}$$

where $\widehat{C} := \overline{C}$, $\widehat{D} := \overline{D}$, and

$$\forall x \in \widehat{C} \qquad \widehat{F}(x) := \bigcap_{\delta > 0} \overline{\operatorname{con}} F((x + \delta \mathbb{B}) \cap C),$$

$$\forall x \in \widehat{D} \qquad \widehat{G}(x) := \bigcap_{\delta > 0} \overline{G((x + \delta \mathbb{B}) \cap D)}.$$

For the hybrid system $\mathcal{H}$ in Example 4.7, the regularized hybrid system $\widehat{\mathcal{H}}$ has the same jump set, flow map, and jump map as the original $\mathcal{H}$. The only change is in the flow set, with $\widehat{C} = \mathbb{R}$. This leads to nonuniqueness of solutions to $\widehat{\mathcal{H}}$ from $\xi = 0$. One solution is the solution already present for $\mathcal{H}$ (and unique for $\mathcal{H}$). Another solution, that reflects what was exhibited by $\mathcal{H}$ under perturbations, is continuous, and flows through D (while remaining in C). Of course, other solutions are also present.

For the system in Example 4.8, regularizing the data has a dramatic effect: both $\widehat{C}$ and $\widehat{D}$ are equal to $\mathbb{R}$. Solutions to $\widehat{\mathcal{H}}$ abound and can have any a priori chosen domains. This reflects what was seen under state perturbations for $\mathcal{H}$.

Regarding Example 4.9, the only difference between $\mathcal{H}$ and $\widehat{\mathcal{H}}$ is the jump set, which, for $\widehat{\mathcal{H}}$, is given by $\widehat{D} = \overline{D} = [1, \infty)$. Solutions from $\xi = 1$ are no longer unique: the solution that (exponentially) converges to the origin is still present while the discrete solution that is always equal to 1 is new. The new solution is the Krasovskii solution to $\mathcal{H}$ that does not converge to the origin as discussed in Example 4.9.

Example 4.14. (Bouncing Ball — Krasovskii regularization) The Krasovskii regularization of the Bouncing Ball system from Example 1.1 with

$$f(x) = \begin{cases} \begin{pmatrix} x_2 \\ -\gamma \end{pmatrix} & x_1 > 0, \text{ or } x_1 = 0 \text{ and } x_2 > 0 \\ \begin{pmatrix} 0 \\ 0 \end{pmatrix} & x = 0 \end{cases}$$

is given by the following data:

$$\widehat{C} = \{ x \in \mathbb{R}^2 : x_1 \geq 0 \} \qquad \widehat{F}(x) = \begin{cases} \begin{pmatrix} x_2 \\ -\gamma \end{pmatrix} & \text{if} \quad x \in \widehat{C}, x \neq 0, \\ \begin{pmatrix} 0 \\ [0, -\gamma] \end{pmatrix} & \text{if} \quad x = 0, \end{cases}$$

$$\widehat{D} = \{ x \in \mathbb{R}^2 : x_1 = 0, x_2 \leq 0 \} \qquad \widehat{G}(x) = \begin{pmatrix} 0 \\ -\lambda x_2 \end{pmatrix}.$$

The regularization admits only one new solution, in comparison to the original model in Example 1.1. This solution is the discrete solution with $\mathrm{dom}\,\phi = \{0\} \times \mathbb{N}$ given by $\phi(0, j) = 0$ for all $j \in \mathbb{N}$. It is not a solution to the original model with a state perturbation: if $x + e \in D$ then $g(x + e) \neq 0$. However, it is a Hermes solution to the original model. Considering perturbations is not needed to see this fact, since ϕ can be approximated by solutions to the original model starting from initial conditions near 0. In fact, for every $\varepsilon > 0$ there exists $\delta > 0$ such that the maximal and hence complete solution ψ to the original model from $x_1 = \delta$, $x_2 = 0$ is ε-close to ϕ. This is a stronger property than that required in Definition 4.12. To find a needed δ explicitly, note that it is sufficient that $\sup_t \mathrm{dom}\,\psi < \varepsilon$ and $|\psi(t, j)| < \varepsilon$ for all $(t, j) \in \mathrm{dom}\,\psi$. The first inequality is guaranteed if $\frac{1+\lambda}{1-\lambda} \sqrt{\frac{2\delta}{\gamma}} < \varepsilon$; recall (2.4). The second is guaranteed if $\sqrt{\delta^2 + 2\gamma\delta} < \varepsilon$, since $|\psi_1(t, j)| \leq \delta$ and $|\psi_2(t, j)| \leq \sqrt{2\gamma\delta}$ for all $(t, j) \in \mathrm{dom}\,\psi$.

Example 4.15. (Flashing fireflies — Krasovskii regularization) Consider the hybrid system describing internal clocks of a group of fireflies, described in Example 1.2. Then

$$\widehat{C} = [0, 1]^n \qquad \widehat{F}(x) = (f_1(x_1), f_2(x_2), \dots, f_n(x_n))^T$$
$$\widehat{D} = D \qquad \widehat{G}(x) = (\widehat{g}_1(x_1), \widehat{g}_2(x_2), \dots, \widehat{g}_n(x_n))^T$$

where

$$\widehat{g}_i(x_i) = \begin{cases} (1+\varepsilon)x_i, & \text{when } (1+\varepsilon)x_i < 1, \\ 0, & \text{when } (1+\varepsilon)x_i > 1, \\ \{0,1\}, & \text{when } (1+\varepsilon)x_i = 1. \end{cases}$$

The regularization affects only the flow set and the jump map. Closing the flow set, in this case, does not affect the solutions. Regularization of the jump map does introduce new solutions.

Example 4.16. ("Uniformly nonuniform attractivity" explained) Let g be the function from Example 3.12. Then

$$\widehat{g}(x) = \{g(x), 0, x\} \quad \forall x \in \mathbb{R}.$$

For the discontinuous function g, the system $x^+ = g(x)$ had 0 asymptotically stable and attractive, but with the number of jumps it takes solutions to reach 0 arbitrarily large over any open subset of $\mathbb{R}$. For $x^+ = \widehat{g}(x)$, there are constant solutions from every initial point.

Theorem 4.17. (Hermes and Krasovskii solutions to hybrid systems) *Suppose that F and G are locally bounded. Then a hybrid arc ϕ is a Hermes solution to $\mathcal{H}$ if and only if it is a Krasovskii solution to $\mathcal{H}$.*

That Hermes solutions are Krasovskii solutions is concluded in Proposition 6.32. This fact reflects a more general property that for every hybrid system that possesses the regularity that the Krasovskii regularization possesses, limits of convergent sequences of solutions under vanishing perturbations are solutions. That Krasovskii solutions are Hermes is concluded in Corollary 4.23. The proof includes a continuous-time result about approximation of Krasovskii solutions to constrained differential inclusions with solutions under perturbations, in Theorem 4.21, and an inductive extension of it to hybrid systems, in Corollary 4.22.

The following example illustrates a situation where a hybrid system with data coinciding with the data of its regularized version can be easily and intuitively derived.

Example 4.18. (Robust zero-crossing detection) The effect of state perturbations illustrated in Example 4.7 can arise in general when decisions are made at a surface or "thin" jump set D. Consider the problem of counting the number of times that the trajectories of the planar system

$$\dot{x} = \begin{pmatrix} x_2 \\ -x_1 \end{pmatrix}$$

cross the x_1-axis. Suppose that every time that the x_1 component of the solution hits zero, a counter is incremented. The resulting system is hybrid and can be written as $\mathcal{H}$ with state given by $\begin{pmatrix} x \\ p \end{pmatrix}$, where $p \in \mathbb{N}$, flow map $f(x,p) = \begin{pmatrix} x_2 \\ -x_1 \\ 0 \end{pmatrix}$, jump map $g(x,p) = \begin{pmatrix} x \\ p+1 \end{pmatrix}$, jump set $D = \{x \in \mathbb{R}^2 : x_2 = 0\} \times \mathbb{N}$, and flow set given by $C := (\mathbb{R}^2 \setminus \{x \in \mathbb{R}^2 : x_2 = 0\}) \times \mathbb{N}$. Following the discussion in Example 4.7, the presence of arbitrarily small state perturbations can cause the hybrid system to miss jumps, and consequently, miss crossings of the x_1 axis. In fact, for instance, in numerical simulations, state perturbations appear due to the approximation of the value of the system state. However, it is likely that such pathology is not actually observed since typically commercial numerical simulators include a zero-crossing detection algorithm in their integration schemes. These algorithms usually include a memory variable that keeps track of the side of the decision boundary in which the state is located. The hybrid system above with a zero-crossing detection algorithm results in the following hybrid system in $\mathbb{R}^4$:

$$ C := \left\{(x,p,q) \in \mathbb{R}^2 \times \mathbb{N} \times \{-1,1\} : x_2 q \geq 0\right\}, \quad f(x,p,q) := \left(\begin{pmatrix} x_2 \\ -x_1 \\ 0 \\ 0 \end{pmatrix}\right) $$

$$ D := \cup_{q \in \{-1,1\}} (D_q \times \mathbb{N} \times \{q\}), \quad G(x,p,q) := \begin{pmatrix} x \\ p+1 \\ -\mathrm{sign}(x_1) \end{pmatrix}, $$

where $D_1 := \{x \in \mathbb{R}^2 : x_1 \geq 0, x_2 = 0\}$ and $D_{-1} := \{x \in \mathbb{R}^2 : x_1 \leq 0, x_2 = 0\}$. It follows that this hybrid system and its regularization coincide. As a consequence, state perturbations do not affect the number of crosses of the x_1-axis. Figure 4.6 illustrates solutions to $\mathcal{H}$ and this hybrid system with state perturbations.

4.4 MEASUREMENT NOISE IN FEEDBACK CONTROL

In general feedback control systems, hybrid or not, *measurement noise* enters the system not as a state perturbation affecting every occurrence of the state in the equations of motion, but only through feedback. More specifically, given a general nonlinear control system $\dot{z} = \varphi(z,u)$ and a feedback mapping $u = \kappa(z)$ (certain control objectives require the use of discontinuous feedback κ), measurement noise enters the closed loop $\dot{z} = \varphi(z,\kappa(z))$ through the feedback, leading to differential equations like $\dot{z} = \varphi(z,\kappa(z+e))$. When the feedback is hybrid, or when the control system is hybrid to begin with, the measurement error can also interplay with the flow sets, jump sets, and the jump map. Some of this is illustrated in the following example.

Example 4.19. (Robust stabilization and measurement noise) Consider a simple control system $\dot{z} = u$ with $z \in \mathbb{R}$, $u \in [-1,1]$. The goal is to robustly stabilize via feedback the set consisting of two points given by $\mathcal{A} = \{0,6\}$. Let $sat : \mathbb{R} \to [-1,1]$ be the standard saturation function, that is

$sat(u) = -1$ if $u < -1$, $sat(u) = u$ if $u \in [-1,1]$, $sat(u) = 1$ if $u > 1$.

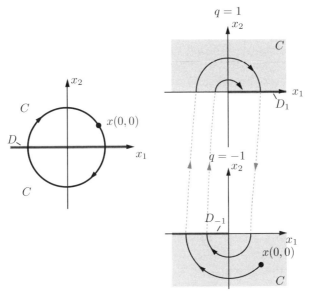

(a) Solutions with state perturbations may miss counts when state perturbations are present.

(b) Solutions with robust zero-crossing detection: flows are not possible any longer once the jump set is crossed, which forces q to jump and register a count.

Figure 4.6: Solutions to zero-crossing detection systems in Example 4.18.

A nonhybrid feedback that results in asymptotic stability of $\mathcal{A}$ for the closed loop $\dot{z} = k(z)$ and that closed-loop system are described below:

$$\dot{z} = k(z) := \begin{cases} -sat(z) & \text{if } z \leq 3, \\ -sat(z-6) & \text{if } z > 3 \ . \end{cases}$$

Note that this k is discontinuous at $z = 3$, and because of this, the resulting asymptotic stability is not robust to measurement noise. Indeed, for arbitrarily small $\varepsilon > 0$, the unique solution from 3 to $\dot{z} = k(z+e)$ with $e(t) := \varepsilon \cos(\pi t/\varepsilon)$ is a see-saw function oscillating between $3-\varepsilon$ and $3+\varepsilon$ with period 2ε. The uniform limit of such see-saw functions is a constant function $z(t) = 3$ for all $t \in \mathbb{R}_{\geq 0}$. Of course, the nonrobustness of asymptotic stability can be detected by looking for Krasovskii solutions to $\dot{z} = k(z)$ (recall Definition 4.2 and Theorem 4.3). These are the solutions to the differential inclusion $\dot{z} \in K(z)$, where K differs from k only at the point of discontinuity of k, that is, $K(3) = [-1, 1]$. Obviously, 3 is an equilibrium of $\dot{z} \in K(z)$. (And so the constant function $z(t) = 3$ is a Krasovskii solution to $\dot{z} = f(z)$, and equivalently, a Hermes solution, by Theorem 4.3.

That the constant function is a Hermes solution can be seen directly, as it is the uniform limit, as $\varepsilon \searrow 0$, of the see-saw functions constructed above.)

Asymptotic stability of $\mathcal{A}$ that is robust to measurement noise can be accomplished by hybrid feedback. One possible approach is based on hysteresis, with the hybrid feedback involving an additional logic variable $q \in \{1, 2\}$. To make an analogy with the nonhybrid feedback analyzed above, the logic variable, in a sense, keeps track of whether $z \leq 3$ or $z > 3$, and will prohibit switching between these two instances too often. More specifically, the hybrid feedback sets $u = -sat(z)$ if the logic variable q equals 1 and $z \leq 4$, and sets $u = -sat(z - 6)$ if $q = 2$ and $z \geq 2$. If neither of the conditions are met, the logic variable q is toggled. In closed loop, this leads to a hybrid system in $\mathbb{R}^2$ with the variable $x = \begin{pmatrix} q \\ z \end{pmatrix}$, the flow and jump sets

$$C = (-\infty, 4] \times \{1\} \cup [2, \infty) \times \{2\}, \quad D = (4, \infty) \times \{1\} \cup (-\infty, 2) \times \{2\},$$

and the flow and jump maps

$$\dot{z} = f(z, q) := \begin{cases} -sat(z) & \text{if } (z, 1) \in C, \\ -sat(z - 6) & \text{if } (z, 2) \in C, \end{cases} \qquad q^+ = g(q) = 3 - q.$$

The logic variable q remains constant during flow, and the state z remains constant during jumps. Figure 4.7 depicts the flow and jump sets and two solutions. It is easy to verify that solutions $\phi = (z, q)$ to $\mathcal{H}$ jump at most once, the maximal ones are complete, and the set $\mathcal{A} \times \{1, 2\}$ is globally uniformly asymptotically stable. The solutions are also unique for every initial condition. The uniqueness would no longer be true if the jump set D was replaced by its closure. Indeed, then a solution from $(4, 1)$ could only flow (with z converging to 0) or jump first (with q changing from 1 to 2) and then flow (with z converging to 6). Similar nonuniqueness would also occur from $(2, 2)$. Still, such nonuniqueness does not affect asymptotic stability.

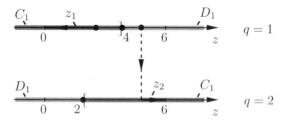

Figure 4.7: Sets $C := (C_1 \times \{1\}) \cup (C_2 \times \{2\})$, $D := (D_1 \times \{1\}) \cup (D_2 \times \{2\})$ for the robust feedback controller in Example 4.19.

For robustness analysis of $\mathcal{H}$, it is natural to only consider measurement noise affecting z and not q. This suggests considering the following system:

$$\begin{cases} \dot{z} = f(z + e) & (z + e, q) \in C \\ q^+ - g(q) & (z + e, q) \in D \,. \end{cases} \tag{4.6}$$

It can be verified that if $e : \mathbb{R}_{\geq 0} \to [-\varepsilon, \varepsilon]$ with $\varepsilon < 2$, solutions to (4.6) converge to the set $([-\varepsilon, \varepsilon] \cup [6 - \varepsilon, 6 + \varepsilon]) \times \{1, 2\}$ and a bound as required by global asymptotic stability also exists. (Solutions to (4.6) may either only flow, instantly jump from the initial state and then only flow – these two behaviors are the same as for the system without noise – or flow for at most ε amount of time before jumping and then flowing forever.) Considering arbitrarily small ε suggests that practical stability of $\mathcal{A} \times \{1, 2\}$ is preserved. Certainly, the behavior that appeared when measurement noise affected the nonhybrid feedback – a constant solution remaining at the initial state 3 – is not possible for the proposed hybrid feedback.

However, the very existence of solutions to (4.6) may still be an issue, as alluded to in Proposition 4.10. Indeed, for the initial condition $(4, 1)$ and measurement noise $e(0) = -\varepsilon$, $e(t) = \varepsilon$ for $t > 0$, there are no solutions to (4.6). Following Proposition 4.10, robust existence can be guaranteed by altering the data so that the flow set and the jump set overlap. This can be done without affecting the asymptotic stability and its robustness. For example, it is sufficient to alter the jump set to be

$$C = ((-\infty, 5] \times \{1\}) \cup ([1, \infty) \times \{2\}) \,.$$

This introduces nonuniqueness, even in the absence of measurement noise: from every point in $(4, 5] \times \{1\} \cup [1, 2) \times \{2\}$ there exists a solution that only flows and a solution that jumps first. However, asymptotic stability is preserved, it is robust to measurement noise as discussed above, and existence of solutions to (4.6) is guaranteed for $e : \mathbb{R}_{\geq 0} \to [-\varepsilon, \varepsilon]$ with $\varepsilon < 1$.

In general, application of hybrid feedback to a nonlinear control system $\dot{z} = \varphi(z, u)$ can lead, in the closed loop, to a hybrid system of the kind

$$\begin{cases} x \in C & \dot{x} \in F\left(x, \kappa_c(x)\right) \\ x \in D & x^+ \in G\left(x, \kappa_d(x)\right) \end{cases} \tag{4.7}$$

with the state x including the original state z and other variables as well, for example, a discrete variable q as it was the case in Example 4.19, or a timer variable, as it is the case in sample-and-hold control. For such a system, it may be natural to consider F and G quite regular, but allow discontinuous "feedbacks" κ_c, κ_d. Assuming that the measurement error enters (4.7) through the mappings κ_c, κ_d and also affects the flow and the jump sets, leads to the system

$$\begin{cases} x + e \in C & \dot{x} \in F\left(x, \kappa_c(x + e)\right) \\ x + e \in D & x^+ \in G\left(x, \kappa_d(x + e)\right) \,. \end{cases} \tag{4.8}$$

Two notions of generalized solutions to (4.7) above can be considered, paralleling the notions of Hermes solutions and Krasovskii solutions in Definitions 4.12 and 4.13. *Control Hermes solutions* can be defined as limits of sequences of solutions to (4.8) generated with sequences of measurement noise vanishing in the limit. Control Krasovskii solutions can be defined like Krasovskii solutions were defined in Definition 4.13. (While a seemingly different approach would be to consider $\dot{\phi} \in \cap_{\delta > 0} \overline{\mathrm{con}} F(\phi, \kappa_c((\phi + \delta\mathbb{B}) \cap C))$ and a similarly defined jump equation, mild regularity — including continuity — of F and just local boundedness — but not continuity — of κ_c make such an approach lead to Krasovskii solutions.) Equivalences between control-Hermes and Krasovskii solutions to (4.7), similar to what is stated in Theorem 4.17, can be shown.

4.5 KRASOVSKII SOLUTIONS ARE HERMES SOLUTIONS

This section is devoted to proving that Krasovskii solutions to a hybrid system are Hermes solutions. One consequence of what is proven below, Corollary 4.24, is that arbitrary switching between differential equations can be captured by considering a differential inclusion.

Lemma 4.20. (ε-piecewise affine approximation) *Let* $z : [a, b] \to \mathbb{R}^n$ *be a Lipschitz continuous arc. Then, for every* $\varepsilon > 0$, *there exists a Lipschitz continuous arc* $w : [a, b] \to \mathbb{R}^n$ *such that*

$$|z(t) - w(t)| \le \varepsilon \quad \text{for all } t \in [a, b]$$

and that is piecewise affine in the following sense: there exists a finite family $\mathcal{I}$ *of mutually disjoint intervals* $(\alpha, \beta) \subset [a, b]$ *with* $\beta - \alpha \le \varepsilon$, *such that* $\bigcup_{(\alpha, \beta) \in \mathcal{I}} [\alpha, \beta] = [a, b]$, *and such that for each member of* $\mathcal{I}$, *there exists* $\eta \in [\alpha, \beta]$, $\eta \ne a$, $\eta \ne b$ *such that*

$$\dot{w}(t) = \dot{z}(\eta) \quad \text{for all } t \in (\alpha, \beta).$$

PROOF. Pick $\varepsilon > 0$. Let $\varepsilon_1 = \varepsilon/3((b - a) + 1)$ and $\varepsilon_2 = \varepsilon/(6(M + 1))$. Let $M > 0$ be a Lipschitz constant for z on $[a, b]$. As z is Lipschitz continuous, the set $N \subset [a, b]$ of all points of nondifferentiability of z satisfies $\mu(N) = 0$ (here and in what follows, μ stands for the one-dimensional Lebesgue measure). At each $\eta \in (a, b) \setminus N$, z is differentiable and $|\dot{z}(t)| \le M$. For each such η pick an interval (c, d) with $\gamma = (c + d)/2$, $d - c \le \varepsilon_2$ such that

$$|z(t) - z(\eta) - \dot{z}(\eta)(t - \eta)| \le \varepsilon_1(t - \eta) \quad \text{for all } t \in [c, d]. \quad (4.9)$$

Let $\mathcal{J}$ stand for the family of all these intervals. The union of all intervals in $\mathcal{J}$ is open, and so is a union of countably many disjoint open intervals $(a_i, b_i) \subset (a, b)$, $i = 1, 2, \ldots$. It also contains $[a, b] \setminus N$, and so is of full measure in $[a, b]$. Thus, there exist $I > 0$ such that $\mu \left(\bigcup_{i=1}^{I} (a_i, b_i) \right) \ge 1 - \epsilon_2/2$.

Fix $i \in \{1, 2, \ldots, I\}$. Pick a compact interval $K_i \subset (a_i, b_i)$ with $\mu\left((a_i, b_i) \setminus K_i\right) \leq \varepsilon_2/2^{i+1}$. Pick finitely many intervals $(c_j, d_j) \subset (a_i, b_i)$, $j = 1, 2, \ldots, J$ from the family $\mathcal{J}$ so that their union covers K_i. Without loss of generality, it can be assumed that no (c_j, d_j) is a subset of $(c_{j'}, d_{j'})$, $j \neq j'$ and that $\eta_1 < \eta_2 < \cdots < \eta_J$. (Recall that $\eta_j = (c_j + d_j)/2$.) Then in fact $c_1 < c_2 < \cdots < c_J$, $d_1 < d_2 < \cdots < d_J$, and hence $c_{j+1} < d_j$, $j = 1, 2, \ldots, J-1$ (otherwise, $d_j < c_{j+1}$ and the intervals do not cover K_i). Now, let $\alpha_1 = c_1$, $\beta_1 = \min\{d_1, \eta_2\}$. Note that $(\alpha_1, \beta_1) \subset (c_1, d_1)$ as $\beta_1 \leq d_1$ and that $\eta_1 \in (\alpha_1, \beta_1)$. For $j = 2, 3, \ldots, J-1$, let $\alpha_j = \beta_{j-1}$, $\beta_j = \min\{d_j, \eta_{j+1}\}$. Note that $(\alpha_j, \beta_j) \subset (c_j, d_j)$, as $c_j < \alpha_j = \min\{d_{j-1}, \eta_j\}$ and $\beta_j \leq d_j$. Note also that $\eta_j \in [\alpha_j, \beta_j]$. Indeed, $\eta_j < \min\{d_j, \eta_{j+1}\} = \beta_j$ while either $\alpha_j = \beta_{j-1} = \eta_j$ or $\alpha_j = \beta_{j-1} = d_{j-1} < \eta_j$. Finally, let $\alpha_J = \beta_{J-1}$ and $\beta_J = d_J$, and note that $(\alpha_J, \beta_J) \subset (c_J, d_J)$ and $\eta_J \in [\alpha_J, \beta_J]$. This yields J disjoint intervals $(\alpha_1, \beta_1), (\alpha_2, \beta_2), \ldots, (\alpha_J, \beta_J)$ whose union covers $K_i \setminus \{\alpha_2, \alpha_3, \ldots, \alpha_J\}$ and so $\mu\left((a_i, b_i) \setminus \bigcup_{j=1}^{J} \mu(\alpha_j, \beta_j)\right) \leq \varepsilon_2/2^{i+1}$. In what follows, let $J^i = J$ and $(\alpha_j^i, \beta_j^i) = (\alpha_j, \beta_j)$, $j = 1, 2, \ldots, J$.

Let $\mathcal{I}^0$ consist of all (mutually disjoint) intervals (α_j^i, β_j^i), $i = 1, 2, \ldots, I$, $j = 1, 2, \ldots, J^i$. Let (α_k^0, β_k^0), $k = 1, 2, \ldots, K$ be all of the intervals in $\mathcal{I}^0$, ordered by $\alpha_1^0 < \alpha_2^0 < \cdots < \alpha_K^0$. Let $\mathcal{I}$ consist of intervals (α_k, β_k), $k = 1, 2, \ldots, K$ where

$$\alpha_1 = a, \ \beta_1 = \alpha_2 = \beta_1^0, \ \beta_2 = \alpha_3 = \beta_2^0, \ \ldots, \beta_K = b,$$

so that $\eta_k \in [\alpha_k^0, \beta_k^0] \subset [\alpha_k, \beta_k]$ and $\bigcup_{k=1}^{K} [\alpha_k, \beta_k] = [a, b]$. Note that since $\mu\left((a, b) \setminus \bigcup_{i=1}^{I} K_i\right) \leq \varepsilon_2$, $\mu\left((a, b) \setminus \bigcup_{k=1}^{K} (\alpha_k^0, \beta_k^0)\right) \leq \varepsilon_2$. Thus, for each $k = 1, 2, \ldots, K$, $\beta_k - \alpha_k \leq \beta_k^0 - \alpha_k^0 + \varepsilon_2 \leq \varepsilon_2 + \varepsilon_2 < \varepsilon$.

Consider the measurable and integrable function $v : [a, b] \to \mathbb{R}^n$ given by $v(t) = \dot{z}(\eta_k)$ if $t \in (\alpha_k, \beta_k)$ for some $k \in \{1, 2, \ldots, K\}$ and $v(t) = 0$ otherwise. Define $w : [a, b] \to \mathbb{R}^n$ by

$$w(t) = z(a) + \int_a^t v(s)\, ds \quad \text{for all } t \in [a, b].$$

Then w is absolutely continuous and in fact Lipschitz continuous with constant K. Let $S = \bigcup_{k=1}^{K} (\alpha_k^0, \beta_k^0)$. Then for each $t \in [a, b]$,

$$
\begin{aligned}
|z(t) - w(t)| &= \left| \int_0^t \dot{z}(t) - \dot{w}(t)\, dt \right| \\
&\leq \left| \int_{[a,t] \cap S} \dot{z}(t) - \dot{w}(t)\, dt \right| + \left| \int_{[a,t] \setminus S} \dot{z}(t) - \dot{w}(t)\, dt \right|,
\end{aligned}
$$

where

$$\left| \int_{[a,t] \setminus S} \dot{z}(t) - \dot{w}(t)\, dt \right| \leq \int_{[a,t] \setminus S} |\dot{z}(t) - \dot{w}(t)|\, dt \leq 2M\varepsilon_2 < \varepsilon/3$$

because $\mu((a,b)\setminus S) \leq \varepsilon_2$ while, for $T = ([a,t]\cap S)\setminus\bigcup_{\beta_k<t}(\alpha_k,\beta_k)$ with measure $\mu(T) \leq \varepsilon_2$,

$$\left|\int_{[a,t]\cap S} \dot{z}(t) - \dot{w}(t)\,dt\right| \leq \sum_{\beta_k<t}\int_{\alpha_k}^{\beta_k}|\dot{z}(t)-\dot{w}|\,dt + \int_T |\dot{z}(t)-\dot{w}|\,dt$$

$$\leq \sum_{\beta_k<t}|z(\beta_k)-z(\alpha_k)-\dot{w}(\eta_k)(\beta_k-\alpha_k)| + 2M\varepsilon_2$$

$$\leq \sum_{\beta_k<t}\varepsilon_1(\beta_k-\alpha_k) + 2M\varepsilon_2$$

$$\leq \varepsilon_1(b-a) + 2M\varepsilon_2 < \varepsilon/3 + \varepsilon/3.$$

Consequently, $|z(t)-w(t)| < \epsilon$ for all $t \in [a,b]$. □

Theorem 4.21. (Approximation of continuous-time constrained Krasovskii solutions) *Let $z : [a,b] \to \mathbb{R}^n$ be an absolutely continuous arc that satisfies*

$$z(t) \in \widehat{C} \quad \text{for all } t \in (a,b), \qquad \dot{z}(t) \in \widehat{F}(z(t)) \quad \text{for almost all } t \in [a,b]. \quad (4.10)$$

Suppose that F is locally bounded. Then, for any $\varepsilon > 0$ and $\psi_a \in C$ such that $|\psi_a - z(a)| < \varepsilon$, there exists a Lipschitz and piecewise linear $\psi : [a,b] \to \mathbb{R}^n$ and a measurable $e : [a,b] \to \mathbb{R}^n$ such that $\sup_{t\in[a,b]}|e(t)| \leq \varepsilon$,

$$\psi(a) = \psi_a,$$
$$\psi(t) + e(t) \in C \quad \text{for all } t \in [a,b],$$
$$\dot{\psi}(t) \in F(\psi(t)+e(t)) \quad \text{for almost all } t \in [a,b],$$

and $|\psi(t)-z(t)| \leq \varepsilon$ for all $t \in [a,b]$.

PROOF. Fix $\varepsilon > 0$, pick $\psi_0 \in C$ so that $|\psi_0 - z(0)| < \varepsilon$, and pick $\varepsilon' \in (0, \varepsilon - |\psi_0 - z(0)|)$. Let $L > 0$ be a bound on F on the ε' neighborhood of z, i.e., the set $\{x \in \mathbb{R}^n : \exists t \in [a,b]\ |x - z(t)| \leq \varepsilon'\}$. (In particular, L is a Lipschitz constant for z.) Let $\varepsilon_w = \frac{\varepsilon'}{8(1+L)}$ and use Lemma 4.20 to obtain a ε_w-piecewise affine approximation $w : [a,b] \to \mathbb{R}^n$ with

$$|z(t) - w(t)| \leq \varepsilon_w < \varepsilon'/8$$

and $\beta - \alpha \leq \varepsilon_w$ for each $(\alpha,\beta) \in \mathcal{I}$. Then, in particular, $|z(t) - z(\alpha)| \leq \varepsilon'/8$ for all $t \in (\alpha,\beta)$ and all intervals $(\alpha,\beta) \in \mathcal{I}$, where γ are as described in Lemma 4.20. For each $(\alpha,\beta) \in \mathcal{I}$ and the associated γ, by the definition of $\widehat{F}$ and since $\dot{z}(\gamma) \in \widehat{F}(z(\gamma))$, there exist points $z_j^\gamma \in C$ and $v_j^\gamma \in F(z_j^\gamma)$, $j = 1,2,\ldots,n+1$, so that $|z_j^\gamma - z(\gamma)| \leq \varepsilon'/8$ and constants $\lambda_j^\gamma \geq 0$, $j = 1,2,\ldots,n+1$, so that $\sum_{j=1}^{n+1}\lambda_j^\gamma = 1$ and

$$\left|\left(\sum_{j=1}^{n+1}\lambda_j^\gamma v_j^\gamma\right) - \dot{z}(\gamma)\right| \leq \frac{\varepsilon'}{8(b-a+1)}.$$

Divide (α, β) into $n+1$ subintervals $(\delta_{j-1}^{\gamma}, \delta_j^{\gamma})$ with $\delta_0^{\gamma} = \alpha$, $\delta_{n+1}^{\gamma} = \beta$ of lengths $\lambda_j^{\alpha}(\beta - \alpha)$. Now define $v : [a, b] \to \mathbb{R}^n$ by $v(t) = v_j^{\gamma}$ if $t \in (\delta_{j-1}^{\gamma}, \delta_j^{\gamma})$, $v(t) = 0$ otherwise. Note that $v \in L_1[a, b]$, and consequently, the function $\psi : [a, b] \to \mathbb{R}^n$ defined by

$$\psi(t) = \psi_a + \int_a^t v(\tau)\, d\tau$$

is absolutely continuous. Furthermore, $\dot{\psi}(t) = v(t)$ for almost all $t \in [a, b]$ (in fact, whenever $t \in (\delta_{j-1}^{\gamma}, \delta_j^{\gamma}), j = 1, 2, \ldots, n+1$), and for every (α, β) in $\mathcal{I}$,

$$\int_\alpha^\beta \dot{\psi}(t)\, dt = \sum_{j=1}^{n+1} \int_{\gamma_{j-1}^{\alpha}}^{\gamma_j^{\alpha}} v(t)\, dt = \sum_{j=1}^{n+1} \lambda_j^{\alpha}(\beta - \alpha) v_j^{\gamma} = (\beta - \alpha) \sum_{j=1}^{n+1} \lambda_j^{\alpha} v_j^{\gamma}$$

and thus

$$\left| \int_\alpha^\beta \dot{\psi}(t)\, dt - \dot{w}(\gamma)(\beta - \alpha) \right| \le (\beta - \alpha) \left| \sum_{j=1}^{n+1} \lambda_j^{\alpha} v_j^{\gamma} - \dot{z}(\gamma) \right| \le \frac{\varepsilon'(\beta - \alpha)}{8(b - a + 1)}.$$

Also, define $e : [a, b] \to \mathbb{R}^n$ as follows: set

$$e(t) = z_j^{\alpha} - \psi(t) \quad \text{if} \quad t \in (\delta_{j-1}^{\gamma}, \delta_j^{\gamma})$$

and note that this defines e for almost all $t \in [a, b]$, and that for those t's, we have $\dot{\psi}(t) \in F(\psi(t) + e(t))$ as well as $\psi(t) + e(t) \in C$. For each t at which $e(t)$ has not been defined yet, one can find $c_t \in C$ such that $|c_t - z(t)| < \varepsilon'/4$ and then set $e(t) = c - \psi(t)$. Then $\psi(t) + e(t) \in C$ for all $t \in [a, b]$. Also, $\psi + e$ is constant on each $(\delta_{j-1}^{\gamma}, \delta_j^{\gamma})$, as so $\psi + e$ is piecewise constant on $[a, b]$. The functions ψ and e have the other desired properties.

For each initial point α of some interval $(\alpha, \beta) \in \mathcal{I}$

$$
\begin{aligned}
|\psi(\alpha) - w(\alpha)| &\le |\psi(a) - w(a)| + \left| \int_a^\alpha (\dot{\psi}(t) - \dot{w}(t))\, dt \right| \\
&\le |\psi_a - w(a)| + \sum_{(\alpha', \beta') \in \mathcal{I}, \beta' \le \alpha} \left| \int_{\alpha'}^{\beta'} \dot{\psi}(t)\, dt - \dot{w}(\gamma)(\beta' - \alpha') \right| \\
&\le |\psi_a - z(a)| + |z(a) - w(a)| + \sum_{(\alpha', \beta') \in \mathcal{I}, \beta' \le \alpha} \frac{\varepsilon'(\beta' - \alpha')}{8(b - a + 1)} \\
&\le |\psi_0 - z(0)| + \varepsilon'/8 + \frac{\varepsilon'}{8(b - a + 1)} \sum_{(\alpha', \beta') \in \mathcal{I}, \beta' \le \alpha} (\beta' - \alpha') \\
&= |\psi_a - z(a)| + \varepsilon'/8 + \frac{\varepsilon'}{8(b - a + 1)}(\alpha - a) \\
&\le |\psi_a - z(a)| + \varepsilon'/4
\end{aligned}
$$

where the sum above is over all intervals $(\alpha', \beta') \in \mathcal{I}$ with $\beta' \le \alpha$, so that in particular, $\sum(\beta' - \alpha') = \alpha - a < b - a$. Furthermore, for each interval (α, β)

and every $\tau \in (\alpha, \beta)$,

$$\left| \int_\alpha^\tau \dot{\psi}(t) - \dot{w}(t)\, dt \right| \leq 2L(\tau - \alpha) \leq 2L(\beta - \alpha) \leq \varepsilon'/4$$

since both $\dot{\psi}$ and $\dot{w}$ are bounded by L. Consequently, for all $t \in [a, b]$,

$$|\psi(t) - w(t)| \leq |\psi_0 - z(0)| + \varepsilon'/2$$

and hence

$$|\psi(t) - z(t)| \leq |\psi_0 - z(0)| + 5\varepsilon'/8 < \varepsilon$$

for all $t \in [a, b]$.

For the error bound, one obtains, for t in one of the intervals $(\delta_{j-1}^\gamma, \delta_j^\gamma)$,

$$\begin{aligned} |e(t)| &\leq |z_j^\alpha - z(\alpha)| + |z(\alpha) - z(t)| + |z(t) - \psi(t)| \\ &\leq \varepsilon'/8 + \varepsilon'/8 + |\psi_0 - z(0)| + 5\varepsilon'/8 \\ &= |\psi_0 - z(0)| + 7\varepsilon'/8 < \varepsilon. \end{aligned}$$

For the remaining t's,

$$|e(t)| \leq |c_t - z(t)| + |z(t) - \psi(t)| \leq \varepsilon'/4 + |\psi_0 - z(0)| + 5\varepsilon'/8 < \varepsilon$$

by the choice of c_t and the previously established bound on $|z(t) - \psi(t)|$. Thus $|e(t)| \leq \varepsilon$ for all $t \in [a, b]$. $\qquad\square$

Corollary 4.22. (Approximation of Krasovskii solutions) *Suppose that F is locally bounded and let ϕ be a compact Krasovskii solution to $\mathcal{H} = (C, F, D, G)$. For any $\varepsilon > 0$ there exists a hybrid arc y with $\mathrm{dom}\, y = \mathrm{dom}\, \phi$ and an admissible state perturbation e such that*

$$\sup_{(t,j)\in\mathrm{dom}\,\phi} |\phi(t,j) - y(t,j)| \leq \varepsilon, \qquad \sup_{(t,j)\in\mathrm{dom}\,e} |e(t,j)| \leq \varepsilon \qquad (4.11)$$

and that y is a solution to $\mathcal{H}_e$ with perturbation e (recall Definition 4.6).

PROOF. The proof uses induction on the number of jumps for ϕ. Let $\mathrm{dom}\, \phi = \bigcup_{j=0}^J ([t_j, t_{j+1}] \times \{j\})$, and fix $\varepsilon > 0$.

If $J = 0$ and $0 = t_0 < t_1$, then the conclusion is exactly Theorem 4.21. If $J = 0$ and $0 = t_0 = t_1$, then the meaning of ϕ being a Krasovskii solution reduces to $\phi(0,0) \in \widehat{C} \cup \widehat{D} = \overline{C} \cup \overline{D}$, and it is sufficient to find $e(0,0)$ with $|e(0,0)| \leq \varepsilon$ and $\phi(0,0) + e(0,0) \in C \cup D$.

Now consider $J > 0$ and let ϕ' be a compact Krasovskii solution to $\mathcal{H}$ given as a truncation of ϕ: $\phi'(t,j) = \phi(t,j)$ for $(t,j) \in \mathrm{dom}\, \phi' := \bigcup_{j=0}^{J-1}([t_j, t_{j+1}] \times \{j\})$. Suppose that the conclusions of the corollary hold for ϕ': there exists a hybrid arc y' with $\mathrm{dom}\, y = \mathrm{dom}\, \phi'$ and an admissible state perturbation e' with

$$\sup_{(t,j)\in\mathrm{dom}\,\phi'} |\phi'(t,j) - y'(t,j)| \leq \varepsilon/2, \qquad \sup_{(t,j)\in\mathrm{dom}\,e'} |e'(t,j)| \leq \varepsilon/2,$$

and such that y' is a solution to $\mathcal{H}_e$ with e replaced by e'. As ϕ is a Krasovskii solution to $\mathcal{H}$, $\phi(t_J, J-1) \in \widehat{D}$ and $\phi(t_J, J) \in \widehat{G}(\phi(t_J, J-1))$. Thus, there exists $u \in D$ with $|u - \phi(t_J, J-1)| \leq \varepsilon/2$ and $v \in G(u)$ with $|v - \phi(t_J, J)| \leq \varepsilon/2$. If $t_J < t_{J+1}$, rely on Theorem 4.21 to obtain $y'' : [t_J, t_{J+1}] \times \{j\} \rightarrow \mathbb{R}^n$ and $e'' : [t_J, t_{J+1}] \times \{j\} \rightarrow \mathbb{R}^n$ with

$$\sup_{t \in [t_J, t_{J+1}]} |\phi''(t, J) - y''(t, J)| \leq \varepsilon/2, \qquad \sup_{t \in [t_J, t_{J+1}]} |e''(t, J)| \leq \varepsilon/2,$$

and such that $y''(t_J, J) = v$, $y''(t, J) + e''(t, J) \in C$ for all $t \in [t_J, t_{J+1}]$, $\dot{y}''(t, J) \in F(y''(t, J) + e''(t, J))$ for almost all $t \in [t_J, t_{J+1}]$. If $t_J = t_{J+1}$, let $y''(t_J, J) = v$. Consider a hybrid arc y, with $\operatorname{dom} y = \operatorname{dom} \phi$, given by

$$\begin{aligned}
y(t, j) &= y'(t, j) \text{ if } j \leq J - 1, t < t_J, \\
y(t, J) &= y''(t, J) \text{ if } t \in [t_J, t_{J+1}],
\end{aligned}$$

and an admissible state perturbation e, with $\operatorname{dom} e = \operatorname{dom} y = \operatorname{dom} \phi$, given by

$$\begin{aligned}
e(t, j) &= e'(t, j) \text{ if } j \leq J - 1, t < t_J, \\
e(t_J, J - 1) &= u - y'(t_J, J - 1), \\
e(t, J) &= e''(t, J) \text{ if } t \in [t_J, t_{J+1}].
\end{aligned}$$

Then y is a solution to $\mathcal{H}_e$ and the inequalities (4.11) hold. $\square$

Corollary 4.23. (Krasovskii solutions are Hermes solutions) *Suppose that F and G are locally bounded. If ϕ is a Krasovskii solution to $\mathcal{H}$ then it is a Hermes solution to $\mathcal{H}$.*

Another consequence of Theorem 4.21, of slightly different nature than those discussed above, states that solutions to a switching system $\dot{z} = f_s(z)$ under arbitrary switching signals are dense in the set of solutions to the differential inclusion generated by the functions f_s. In contrast to Section 2.4, where switching between finitely many functions was discussed, the result below allows far more general sets of functions. Of course, every finite set can be identified with a subset of a sphere, even in two dimensions.

Corollary 4.24. (Switching and differential inclusions) *Let S be a nonempty subset of the unit sphere in $\mathbb{R}^m$; for each $s \in S$, let $f_s : \mathbb{R}^n \rightarrow \mathbb{R}^n$ be functions such that $f_s(x)$ is locally Lipschitz continuous in $x \in \mathbb{R}^n$ uniformly in $s \in S$. Define a (possibly set-valued) mapping $\Phi : \mathbb{R}^n \rightrightarrows \mathbb{R}^n$ at each $x \in \mathbb{R}^n$ by*

$$\Phi(x) = \overline{\operatorname{con}} \bigcup_{s \in S} f_s(x).$$

Suppose that Φ is locally bounded and $z : [a, b] \rightarrow \mathbb{R}^n$ is a solution to $\dot{z} \in \Phi(z)$. Pick $\varepsilon > 0$. Then there exists a piecewise constant function $s : [a, b] \rightarrow S$, and an absolutely continuous $\psi : [a, b] \rightarrow \mathbb{R}^n$, such that

$$\dot{\psi}(t) = f_{s(t)}(\psi(t))$$

for almost all $t \in [a, b]$ and

$$|\psi(t) - z(t)| \leq \varepsilon$$

for all $t \in [a, b]$.

PROOF. Pick $\varepsilon > 0$ and and let K be a Lipschitz constant for all f_s, $s \in S$ on the ε neighborhood of z. Fix $s_0 \in S$. Consider $F : \mathbb{R}^{n+m} \mapsto \mathbb{R}^{n+m}$ defined by

$$F\begin{pmatrix} x \\ u \end{pmatrix} = \begin{pmatrix} f_s(x) \\ 0 \end{pmatrix} \quad \text{if } u = \lambda s \text{ for some } \lambda > 0, \qquad F\begin{pmatrix} x \\ 0 \end{pmatrix} = \begin{pmatrix} f_{s_0}(x) \\ 0 \end{pmatrix}.$$

Note that, for each $x \in \mathbb{R}^n$,

$$\widehat{F}\begin{pmatrix} x \\ 0 \end{pmatrix} = \begin{pmatrix} \Phi(x) \\ 0 \end{pmatrix},$$

and $t \mapsto (z(t), 0)$ is a solution to the differential inclusion

$$\begin{pmatrix} \dot{z} \\ \dot{u} \end{pmatrix} \in \widehat{F}\begin{pmatrix} z \\ u \end{pmatrix}.$$

Consider $\varepsilon' = \varepsilon e^{-K(b-a)}$. Theorem 4.21 implies that there exist an absolutely continuous $\psi' : [a, b] \to \mathbb{R}^n$ with $\psi'(a) = z(a)$ and measurable $e_1 : [a, b] \to \mathbb{R}^n$, $e_2 : [a, b] \to \mathbb{R}^m$ such that $\psi' + e_1$ and e_2 are piecewise constant (this property is shown in the proof of Theorem 4.21), $\sup_{t \in [a,b]} |e_1(t)| \leq \varepsilon'$, $\sup_{t \in [a,b]} |e_2(t)| \leq \varepsilon'$,

$$\begin{pmatrix} \dot{\psi}'(t) \\ 0 \end{pmatrix} \in F\begin{pmatrix} \psi'(t) + e_1(t) \\ e_2(t) \end{pmatrix} = \begin{pmatrix} f_{s(t)}(\psi'(t) + e_1(t)) \\ 0 \end{pmatrix}$$

where $s : [a, b] \to S$ is such that $s(t) = \lambda e_2(t)$ for some $\lambda > 0$ or $s(t) = s_0$ if $e_2(t) = 0$, and $|\psi'(t) - z(t)| \leq \varepsilon'$ for all $t \in [a, b]$. Now, let $\psi : [a, b] \to \mathbb{R}^n$ be the unique function such that

$$\dot{\psi}(t) = f_{s(t)}(\psi(t)) \quad \text{for almost all } t \in [a, b], \qquad \psi(a) = z(a).$$

Note that $|\dot{\psi}(t) - \dot{\psi}'(t)| \leq K|\psi(t) - \psi'(t)| + K\varepsilon'$ for almost all $t \in [a, b]$ and hence $|\psi(t) - \psi'(t)| \leq \varepsilon'(e^{K(b-a)} - 1)$ for all $t \in [a, b]$. Hence $|\psi(t) - z(t)| \leq \varepsilon$ and the proof is finished. □

4.6 NOTES

A classical reference for differential equations with discontinuous right-hand sides is the book by Filippov [35]. In Definition 4.2, the concept of a Hermes solution to a differential equation comes from Hermes [52], and the concept of a Krasovskii solution, in Definition 4.2, comes from Krasovskii [64]. A concept of a generalized solution to a differential equation $\dot{z} = f(z)$ proposed by Filippov in [34] ignores the behavior of f on sets of measure zero, and hence is problematic for hybrid systems, or even for constrained differential equations. Further discussion of the various notions of generalized solutions to differential equations can be found in Filippov [34], [52], [64], and Hájek [47]. The result contained in Theorem 4.3 was noted in [52] and expanded upon in [47]. An alternative proof, included here as Lemma 4.20 and Theorem 4.21, is given in Sanfelice et al. [105].

Chapter Five

Preliminaries from set-valued analysis

Further developments in the theory of hybrid systems, for example, making rigorous the concept of convergence of sequences of hybrid arcs or considering perturbations of the flow and the jump sets, can be conveniently carried out with the help of some concepts from set-valued analysis. This chapter includes the necessary background. Section 5.1 presents the concept of convergence of sets. Section 5.2 deals with set-valued mappings and their continuity properties. Section 5.3 specializes some of the concepts, such as graphical convergence, to hybrid arcs and provides further details in such a setting. Finally, Section 5.4 discusses differential inclusions. The presentation of the set-valued analysis material is by no means complete. The results and examples are selected with the developments of the following chapters in mind. Proofs of many background results are omitted, while some are included for illustration purposes.

5.1 SET CONVERGENCE

The notion of convergence for a sequence of sets generalizes the notion of convergence of a sequence of points. The formal definitions are as follows.

Definition 5.1. (Set convergence) *Let $\{S_i\}_{i=1}^{\infty}$ be a sequence of sets in $\mathbb{R}^n$.*

- *The inner limit of the sequence $\{S_i\}_{i=1}^{\infty}$, denoted $\liminf_{i \to \infty} S_i$, is the set of all $x \in \mathbb{R}^n$ for which there exist points $x_i \in S_i$, $i = 1, 2, \ldots$, such that $\lim_{i \to \infty} x_i = x$;*

- *The outer limit of the sequence $\{S_i\}_{i=1}^{\infty}$, denoted $\limsup_{i \to \infty} S_i$, is the set of all $x \in \mathbb{R}^n$ for which there exists a subsequence $\{S_{i_k}\}_{k=1}^{\infty}$ of $\{S_i\}_{i=1}^{\infty}$ and points $x_k \in S_{i_k}$, $k = 1, 2, \ldots$, such that $\lim_{k \to \infty} x_k = x$.*

When the inner limit and the outer limit of the sequence $\{S_i\}_{i=1}^{\infty}$ are equal, the sequence $\{S_i\}_{i=1}^{\infty}$ is convergent, and its limit is given by

$$\lim_{i \to \infty} S_i = \liminf_{i \to \infty} S_i = \limsup_{i \to \infty} S_i.$$

If each element of the sequence $\{S_i\}_{i=1}^{\infty}$ consists of a single point, i.e., $S_i = \{x_i\}$, then the outer limit of the sequence is the set of all accumulation points of the sequence $\{x_i\}_{i=1}^{\infty}$. The inner limit is nonempty if and only if the sequence $\{x_i\}_{i=1}^{\infty}$ is convergent, and then $\{\lim_{i \to \infty} x_i\} = \liminf_{i \to \infty} S_i = \lim_{i \to \infty} S_i$.

A simple example of convergence of sets that are not singletons can be given in terms of balls in $\mathbb{R}^n$ of radius r_i and centered at 0, so $S_i = r_i\mathbb{B}$. Then the ball $r\mathbb{B}$ is the inner limit $\liminf_{i\to\infty} S_i$ if $r = \liminf_{i\to\infty} r_i$; it is the outer limit $\limsup_{i\to\infty} S_i$ if $r = \limsup_{i\to\infty} r_i$; and thus it is the limit $\lim_{i\to\infty} S_i$ if $r = \lim_{i\to\infty} r_i$. A similar example can be given in terms of a sequence of intervals in $\mathbb{R}$. Another example is provided by "rotating rays." In $\mathbb{R}^2$, consider the sets $S_i = \{x \in \mathbb{R}^2 : x_2 = a_i x_1\}$. Then the sequence $\{S_i\}_{i=1}^{\infty}$ converges when the sequence $\{a_i\}_{i=1}^{\infty}$ converges or is divergent to ∞ or $-\infty$. In the former case, $\lim_{i\to\infty} S_i = \{x \in \mathbb{R}^2 : x_2 = \lim_{i\to\infty} a_i x_1\}$. In the latter, the limit is $\{x \in \mathbb{R}^2 : x_1 = 0\}$. Another example of interest is the sequence of sets S_i given by the graph of the function $t \mapsto t^i$ on $[0,1]$, that is, $S_i := \{(t, t^i) : t \in [0,1]\}$. This sequence is convergent and has limit $S = ([0,1] \times \{0\}) \cup (\{1\} \times [0,1])$. Figure 5.1 depicts S_i and S.

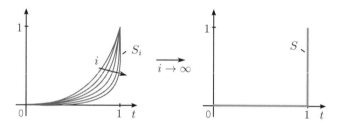

Figure 5.1: A sequence of sets S_i converging to the set S given by the reflected L shape.

Example 5.2. (Reachable set for a differential equation) Let $K \subset \mathbb{R}^n$ be a compact set and $f : \mathbb{R}^n \to \mathbb{R}^n$ be a continuous function such that solutions to the differential equation $\dot{z} = f(z)$ are unique. For $\tau \geq 0$, let S_τ be the reachable set from K in time τ, that is, the set of all $x \in \mathbb{R}^n$ such that there exists a solution $z : [0, \tau] \to \mathbb{R}^n$ to $\dot{z} = f(z)$ such that $z(0) \in K$ and $z(\tau) = x$. Consider a sequence $\{\tau_i\}_{i=1}^{\infty}$ such that $\lim_{i\to\infty} \tau_i = \tau$ for some $\tau \geq 0$.

One has $S_\tau \subset \liminf_{i\to\infty} S_{\tau_i}$. To see this, pick any $x \in S_\tau$. Then $x = z(\tau)$ for some solution $z : [0, \tau] \to \mathbb{R}^n$ to $\dot{z} = f(z)$ such that $z(0) \in K$. Since f is continuous, the solution z can be extended to a solution $z : [0, \tau + \varepsilon] \to \mathbb{R}$ for some $\varepsilon > 0$. Then, for large enough i, $\tau_i \in [0, \tau + \varepsilon]$ and $z(\tau_i) \in S_{\tau_i}$. Since z is continuous, $\lim_{i\to\infty} z(\tau_i) = z(\tau)$. Hence $z(\tau) = x \in \liminf_{i\to\infty} S_{\tau_i}$, which is what needed to be shown. Note that the conclusion is true whether the reachable sets under discussion are bounded or not.

If, additionally, every maximal solution to the differential equation starting in K is complete, then $\limsup_{i\to\infty} S_{\tau_i} \subset S_\tau$. Indeed, pick $x_i \in S_{\tau_i}$ such that $\lim_{i\to\infty} x_i = x$ for some $x \in \mathbb{R}^n$. For each $i = 1, 2, \ldots$, let $z_i : [0, \tau_i] \to \mathbb{R}^n$ be the maximal, hence complete, solution to $\dot{z} = f(z)$ with $z_i(\tau_i) = x_i$ and $z_i(0) \in K$. Since K is compact, without loss of generality one may assume that $\lim_{i\to\infty} z_i(0)$ exists. Denote it by x_0 and note that $x_0 \in K$. Then, continuous dependence of the solution on initial conditions and parameters implies that $z(\tau) = \lim_{i\to\infty} z_i(\tau_i)$, where z is the solution from x_0. Thus $\lim_{i\to\infty} x_i \in S_\tau$, which is what needed to be shown. Consequently, under the additional completeness assumption, $\limsup_{i\to\infty} S_{\tau_i} \subset S_\tau \subset \liminf_{i\to\infty} S_{\tau_i}$ which means that $\lim_{i\to\infty} S_{\tau_i} = S_\tau$.

Some basic properties of set limits are

- the inner limit, the outer limit, and, if it exists, the limit of a sequence $\{S_i\}_{i=1}^\infty$ are closed;

- a sequence $\{S_i\}_{i=1}^\infty$ either *escapes to the horizon*, in the sense that for each compact $K \subset \mathbb{R}^n$ there exists i_K such that for all $i > i_K$, $S_i \cap K = \emptyset$, or it has a nonempty outer limit;

- if the sequence $\{S_i\}_{i=1}^\infty$ is monotone, in the sense that either $S_i \subset S_{i+1}$ for all $i \in \mathbb{N}$ or $S_i \supset S_{i+1}$ for all $i \in \mathbb{N}$, and the sequence $\{S_i\}_{i=1}^\infty$ does not escape to the horizon, then the limit exists.

Example 5.3. (Convergence of hybrid time domains)   Let $\{E_i\}_{i=1}^\infty$ be a convergent sequence of hybrid time domains. Then $E = \lim_{i\to\infty} E_i$ is a hybrid time domain and $\text{length}(E) = \lim_{i\to\infty} \text{length}(E_i)$.

Indeed, directly from the definition of set convergence, $\lim_{i\to\infty} E_i = E$ if and only if for all $J \in \mathbb{N}$, $E_i^J := E_i \cap (\mathbb{R}_{\geq 0} \times \{J\})$ converge to $E^J := E \cap (\mathbb{R}_{\geq 0} \times \{J\})$. Because set limits are always closed, each E^J is a closed interval (possibly empty, consisting of one point, or unbounded to the right). If E^{J+1} is nonempty, then so is E^J. Indeed, the right endpoints of E_i^J agree with the left endpoints of E_i^{J+1}, and the latter converge to the left endpoint of E^{J+1}. This is enough to conclude that E is a hybrid time domain.

Because for each $(t,j) \in E$ there exist $(t_i, j_i) \in E_i$ with $(t_i, j_i) \to (t, j)$, $\text{length}(E) \leq \liminf_{i\to\infty} \text{length}(E_i)$. This is enough to conclude that

$$\text{length}(E) = \lim_{i\to\infty} \text{length}(E_i)$$

if $\text{length}(E) = \infty$. If $\text{length}(E) < \infty$, for each $\varepsilon > 0$, then the sets

$$E_i \cap \{(t,j) : \text{length}(E) + \varepsilon \leq t + j \leq \text{length}(E) + \varepsilon + 1\}$$

must be empty for large enough i. Indeed, otherwise $\lim_{i\to\infty} E_i = \limsup_{i\to\infty} E_i$ would include a point (t,j) with $t + j \geq \text{length}(E) + \varepsilon$, which is impossible. Then $\text{length}(E) \geq \limsup_{i\to\infty} \text{length}(E_i)$ and this finishes the argument.

The example of "rotating rays" suggests that a sequence $\{S_i\}_{i=1}^\infty$ can converge to a limit S even if the Hausdorff distance between S_i and S is always infinite. Indeed, the Hausdorff distance between two sets S' and S'' in $\mathbb{R}^n$ is the infimum of all $d \geq 0$ such that $S' \subset S'' + d\mathbb{B}$, $S'' \subset S' + d\mathbb{B}$. For S_i, S in the "rotating rays" example, for no $\varepsilon > 0$ and no $i \in \mathbb{N}$ it holds that $S_i \subset S + \varepsilon\mathbb{B}$, and the Hausdorff distance between S_i and S is always infinite. However, uniformity in set convergence can be concluded if one considers truncations of sets S_i and S.

Theorem 5.4. (Uniformity in set convergence) *For a sequence of sets $\{S_i\}_{i=1}^{\infty}$ in $\mathbb{R}^n$ and a closed set $S \subset \mathbb{R}^n$,*

(a) $S \subset \liminf_{i \to \infty} S_i$ *if and only if for every $r > 0$ and $\varepsilon > 0$ there exists i_0 such that for all $i > i_0$,*
$$S \cap r\mathbb{B} \subset S_i + \varepsilon\mathbb{B};$$

(b) $S \supset \limsup_{i \to \infty} S_i$ *if and only if for every $r > 0$ and $\varepsilon > 0$ there exists i_0 such that for all $i > i_0$,*
$$S_i \cap r\mathbb{B} \subset S + \varepsilon\mathbb{B}.$$

Consequently, $S = \lim_{i \to \infty} S_i$ if and only if for every $r > 0$ and $\varepsilon > 0$ there exists i_0 such that for all $i > i_0$, both inclusions hold.

In particular, when $S \subset \mathbb{R}^n$ is closed and a sequence of sets $\{S_i\}_{i=1}^{\infty}$ in $\mathbb{R}^n$ is uniformly bounded, $\lim_{i \to \infty} S_i = S$ if and only if the sequence of Hausdorff distances between S_i and S converges to 0.

Example 5.5. (Sets under measurement error) Let $\{\rho_i\}_{i=1}^{\infty}$ be a sequence of continuous functions $\rho_i : \mathbb{R}^n \to \mathbb{R}_{\geq 0}$ that converges locally uniformly to a function identically equal to 0 on $\mathbb{R}^n$. For example, one could consider $\rho_i(x) = \delta_i \rho(x)$ for a sequence $\{\delta_i\}_{i=1}^{\infty}$ of numbers $\delta_i \in (0, 1)$ with $\delta_i \to 0$ and a continuous $\rho : \mathbb{R}^n \to \mathbb{R}_{\geq 0}$; such sequences will be considered in later chapters. Let $S \subset \mathbb{R}^n$ be a closed set. Consider

$$S_i = \{x \in \mathbb{R}^n \ : \ x + \rho_i(x)\mathbb{B} \cap S \neq \emptyset\}. \tag{5.1}$$

Then each of the sets S_i is closed and $\lim_{i \to \infty} S_i = S$. Figure 5.2 depicts the sequence S_i and the set S. Regarding the latter equation, obviously $S \subset \liminf_{i \to \infty} S_i$ since $S \subset S_i$ for each $i = 1, 2, \ldots$. To see that $S \supset \limsup_{i \to \infty} S_i$, rely on Theorem 5.4 (b). Pick $r > 0$, $\varepsilon > 0$. For each i large enough so that $\rho_i(r\mathbb{B}) \leq \varepsilon$, it holds that $x + \rho_i(x)\mathbb{B} \subset x + \varepsilon\mathbb{B}$ for each $x \in r\mathbb{B}$. Thus, for such i's, $S_i \cap r\mathbb{B} \subset S + \varepsilon\mathbb{B}$.

Set convergence need not behave well under linear transformations. For example, take a special case of the "rotating rays": for $i \in \mathbb{N}$, let $S_i = \{x \in \mathbb{R}^2 \ : \ x_2 = x_1/i\}$ so that the sequence $\{S_i\}_{i=1}^{\infty}$ converges to $S = \mathbb{R} \times \{0\}$. Consider the orthogonal projection of $\mathbb{R}^2$ onto $\{0\} \times \mathbb{R}$, that is, the linear mapping given by $A = \begin{bmatrix} 0 & 0 \\ 0 & 1 \end{bmatrix}$. Then $AS_i = \{0\} \times \mathbb{R}$ for each $i \in \mathbb{N}$ while $AS = 0 \in \mathbb{R}^2$. In particular, $A \lim_{i \to \infty} S_i \neq \lim_{i \to \infty} AS_i$. A similar phenomenon can occur when each set S_i is bounded, but the sequence $\{S_i\}_{i=1}^{\infty}$ is not uniformly bounded.

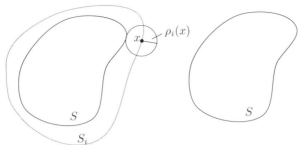

Figure 5.2: Set under state-dependent measurement noise as in (5.1).

Lemma 5.6. (Set convergence and linear mappings) *Let $\{S_i\}_{i=1}^{\infty}$ be a uniformly bounded sequence of sets in $\mathbb{R}^n$ that is convergent. Let $A : \mathbb{R}^n \to \mathbb{R}^m$ be a linear mapping. Then*

$$\lim_{i\to\infty} AS_i = A \lim_{i\to\infty} S_i.$$

A classical and fundamental result in analysis, the Bolzano-Weierstrass Theorem, concludes that any bounded sequence of points in $\mathbb{R}^n$ contains a convergent subsequence. In fact, it is sufficient to assume that the sequence does not "diverge to infinity," because then it contains a bounded subsequence. A similar result is true for sequences of sets.

Theorem 5.7. (Compactness in set convergence) *Any sequence $\{S_i\}_{i=1}^{\infty}$ of sets $S_i \subset \mathbb{R}^n$ either escapes to the horizon or it has a subsequence converging to a nonempty set.*

5.2   SET-VALUED MAPPINGS

The initial discussion of set-valued mappings and the definition of a domain of a set-valued mapping were given in Chapter 2. Further basic terminology is as follows.

Definition 5.8. (Domain, range, graph) *Given a set-valued mapping $M : \mathbb{R}^m \rightrightarrows \mathbb{R}^n$,*

- *the range of M is the set*

$$\text{rge}\, M = \{y \in \mathbb{R}^n \ : \ \exists x \in \mathbb{R}^m \text{ such that } y \in M(x)\};$$

- *the graph of M is the set*

$$\text{gph}\, M = \{(x, y) \in \mathbb{R}^m \times \mathbb{R}^n \ : \ y \in M(x)\}.$$

Given a set $S \subset \mathbb{R}^m$, a set-valued mapping $M : S \rightrightarrows \mathbb{R}^n$ can be trivially extended to a mapping (with some abuse of notation) $M : \mathbb{R}^m \rightrightarrows \mathbb{R}^n$, by setting

$M(x) = \emptyset$ for $x \notin S$. The domain, range, graph, and other concepts defined for mappings from $\mathbb{R}^m$ to $\mathbb{R}^n$, when applied to a mapping from $S \subset \mathbb{R}^m$, should be understood as the domain, range, graph, etc. of such trivial extension.

A set-valued mapping M is fully determined by its graph, in the sense that $M(x) = \{y : (x,y) \in \operatorname{gph} M\}$. A mapping M is empty valued, single valued, or multivalued at x if $M(x)$ is empty, a singleton, or a set containing more than one element, respectively. Every function defined on a set S is a set-valued mapping that is single valued at each point of S.

Definition 5.9. (Outer semicontinuity) *A set-valued mapping* $M : \mathbb{R}^m \rightrightarrows \mathbb{R}^n$ *is outer semicontinuous* (osc) *at* $x \in \mathbb{R}^m$ *if for every sequence of points* x_i *convergent to* x *and any convergent sequence of points* $y_i \in M(x_i)$, *one has* $y \in M(x)$, *where* $\lim_{i\to\infty} y_i = y$. *The mapping* M *is outer semicontinuous if it is outer semicontinuous at each* $x \in \mathbb{R}^m$. *Given a set* $S \subset \mathbb{R}^m$, $M : \mathbb{R}^m \rightrightarrows \mathbb{R}^n$ *is outer semicontinuous relative to* S *if the set-valued mapping from* $\mathbb{R}^n$ *to* $\mathbb{R}^m$ *defined by* $M(x)$ *for* $x \in S$ *and* $\emptyset$ *for* $x \notin S$ *is outer semicontinuous at each* $x \in S$.

Outer semicontinuity of $M : \mathbb{R}^m \rightrightarrows \mathbb{R}^n$ relative to $S \subset \mathbb{R}^m$ just means that for each $x \in S$, each sequence of points $x_i \in S_i$ convergent to x, and each sequence of points $y_i \in M(x_i)$ convergent to y, $y \in M(x)$. Inner semicontinuity and continuity of a set-valued mapping are not defined here, as they will not play a significant role in the analysis that follows.

In terms of convergence of sets, a mapping $M : \mathbb{R}^m \rightrightarrows \mathbb{R}^n$ is osc at $x \in \mathbb{R}^m$ if, for every sequence $\{x_i\}_{i=1}^\infty$ with $\lim_{i\to\infty} x_i = x$, it satisfies $\limsup_{i\to\infty} M(x_i) \subset M(x)$. This can be verified directly from the definitions.

A basic example of an outer semicontinuous set-valued mapping is provided by continuous functions. That is, if $M : \mathbb{R}^m \to \mathbb{R}^n$ is a function, it is outer semicontinuous (as a set-valued mapping) at each point where it is continuous (as a function). However, if $M : \mathbb{R}^m \rightrightarrows \mathbb{R}^n$ is single valued on a neighborhood of $x \in \mathbb{R}^m$ and outer semicontinuous at x, it may still fail to be continuous. For example, take $M : \mathbb{R} \rightrightarrows \mathbb{R}$ given by $M(x) = 0$ for $x \le 0$, $M(x) = 1/x$ for $x > 0$. This mapping is osc at $x = 0$ but not continuous there. Continuity can be guaranteed with a local boundedness assumption.

Obviously, outer semicontinuity of M implies that values of M are closed, but the reverse implication is not valid. However, outer semicontinuity follows from the graph of a mapping being closed.

Lemma 5.10. (Osc and closed graph) *A set-valued mapping* $M : \mathbb{R}^m \rightrightarrows \mathbb{R}^n$ *is outer semicontinuous if and only if* $\operatorname{gph} M$ *is closed. More generally, given a set* $S \subset \mathbb{R}^m$, *a set-valued mapping* $M : \mathbb{R}^m \rightrightarrows \mathbb{R}^n$ *is outer semicontinuous relative to* S *if and only if the set* $\{(x,y) \in \mathbb{R}^m \times \mathbb{R}^n : x \in S, \ y \in M(x)\}$ *is relatively closed in* $S \times \mathbb{R}^n$.

Thus, for example, the function $f : \mathbb{R} \to \mathbb{R}$ given by $f(x) = -1$ for $x < 0$, $f(x) = 1$ for $x \ge 0$ is not outer semicontinuous (as a set-valued mapping), but the closely related $g : \mathbb{R} \rightrightarrows \mathbb{R}$ given by $g(x) = -1$ for $x < 0$, $g(0) = \{-1, 1\}$,

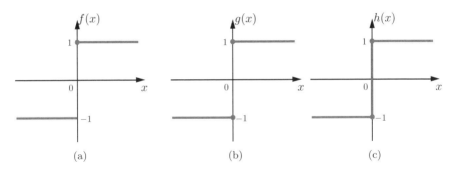

Figure 5.3: Mappings: (a) not outer semicontinuous, (b) outer semicontinuous, (c) outer semicontinuous and convex valued.

$g(x) = 1$ for $x > 0$ is outer semicontinuous, and so is $h : \mathbb{R} \rightrightarrows \mathbb{R}$ given by $h(x) = -1$ for $x < 0$, $h(0) = [-1, 1]$, $h(x) = 1$ for $x > 0$. (Note that gph g is the closure of gph f, while h can also be obtained from f by closing the graph and then by convexification of the values; see Figure 5.3. Lemma 5.16 sheds more light on this.) Similarly, the set-valued mapping $M : \mathbb{R} \rightrightarrows \mathbb{R}$ given by $M(x) = 0$ for $x \leq 0$, $M(x) = [-1, 1]$ for $x > 0$ is not outer semicontinuous (at $x = 0$) but changing the definition to $M(x) = 0$ for $x < 0$, $M(x) = [-1, 1]$ for $x \geq 0$ does yield an osc M.

Example 5.11. (Optimal solutions in parametric optimization) Let $\phi : \mathbb{R}^m \times \mathbb{R}^n \to \mathbb{R}$ be a continuous function and $K \subset \mathbb{R}^n$ be a nonempty compact set. Then, the (set-valued) mapping $M : \mathbb{R}^m \to \mathbb{R}^n$ given at each $x \in \mathbb{R}^m$ by

$$M(x) = \left\{ y \in K \ : \ \phi(x, y) = \min_{z \in K} \phi(x, z) \right\}$$

is outer semicontinuous (and nonempty valued). If furthermore $y \mapsto \phi(x, y)$ is a convex function for every x and K is a convex set, then $M(x)$ is convex for all $x \in \mathbb{R}^m$.

Another example of set-valued mappings, that naturally arise in optimization but also in the analysis of differential equations or inclusions whose solutions are constrained to a set — see Section 5.4 — is provided by tangent cones. Tangent cones also play an important role in approximation of hybrid systems, in the spirit of linearization, as presented in Chapter 9.

Definition 5.12. (Tangent cone) *The tangent cone to a set $S \subset \mathbb{R}^n$ at a point $x \in \mathbb{R}^n$, denoted $T_S(x)$, is the set of all vectors $w \in \mathbb{R}^n$ for which there exist $x_i \in S$, $\tau_i > 0$ with $x_i \to x$, $\tau_i \searrow 0$, and*

$$w = \lim_{i \to \infty} \frac{x_i - x}{\tau_i}.$$

The tangent cone is sometimes called the contingent cone, or the Bouligand tangent cone. For each not isolated point $x \in \overline{S}$, i.e., when $x \in \overline{S} \setminus \{x\}$, $T_S(x)$ is nonempty, closed, and contains a nonzero element. Then, automatically, $T_S(x)$ is unbounded because it is a cone. For each $x \notin \overline{S}$, $T_S(x) = \emptyset$. When S is a smooth surface, the tangent cone T_S amounts to the concept of a tangent space.

Figure 5.4 depicts a flow map F and the tangent cone to a flow set C at different points in the set. The flow map is single valued at x^1, x^3 and set valued at x^2. At points in the interior of C, such as x^1, the tangent cone is the entire space. At x^1 and x^2, the intersection between the flow map and the tangent cone is nonempty, whereas at x^3 this intersection is empty.

Some more interesting examples are given below.

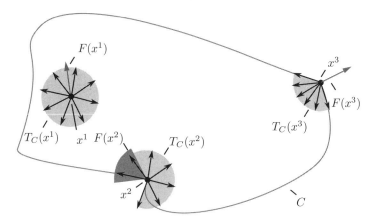

Figure 5.4: A flow map F and the tangent cone to a set C represented at several points $x \in C$.

Example 5.13. (Tangent cones) Let $S = [a, b] \subset \mathbb{R}$. Then

$$
T_S(x) = \begin{cases} (-\infty, 0] & \text{if} \quad x = a, \\ (-\infty, \infty) & \text{if} \quad x \in (a, b), \\ [0, \infty) & \text{if} \quad x = b. \end{cases}
$$

For boxes in $\mathbb{R}^n$, i.e., sets given by products of intervals

$$
S = [a_1, b_1] \times [a_2, b_2] \times \cdots \times [a_n, b_n],
$$

the tangent cone $T_S(x)$ can be found coordinate-wise:

$$
T_S(x) = T_{[a_1, b_1]}(x_1) \times T_{[a_2, b_2]}(x_2) \times \cdots \times T_{[a_n, b_n]}(x_n).
$$

Some further examples are

$$
S = \left\{ x \in \mathbb{R}^2 : x_1 \geq 0, \, x_2 \leq \sqrt{x_1} \right\}, \qquad T_S(0) = \left\{ w \in \mathbb{R}^2 : w_1 \geq 0 \right\}.
$$

$$S = \left\{ x \in \mathbb{R}^2 \ : \ x_1 \geq 0,\, 0 \leq x_2 \leq \sin x_1 \right\}, \quad T_S(0) = \left\{ w \in \mathbb{R}^2 \ : \ w_1 \geq 0,\, w_2 \leq w_1 \right\}.$$

In general, the tangent cone mapping T_S is not outer semicontinuous relative to $\overline{S}$, as evidenced by the first formula in Example 5.13.

Definition 5.14. (Local boundedness) *A set-valued mapping* $M : \mathbb{R}^m \rightrightarrows \mathbb{R}^n$ *is* locally bounded *at* $x \in \mathbb{R}^m$ *if there exists a neighborhood* U_x *of* x *such that* $M(U_x) \subset \mathbb{R}^n$ *is bounded. The mapping* M *is* locally bounded *if it is locally bounded at each* $x \in \mathbb{R}^m$. *Given a set* $S \subset \mathbb{R}^m$, *the mapping* M *is* locally bounded relative to S *if the set-valued mapping from* $\mathbb{R}^m$ *to* $\mathbb{R}^n$ *defined by* $M(x)$ *for* $x \in S$ *and* $\emptyset$ *for* $x \notin S$ *is locally bounded at each* $x \in S$.

It was mentioned before that convergence of sets need not agree with convergence of Hausdorff distances, unless uniform boundedness assumption is present. Similarly, outer semicontinuity of a set-valued mapping need not agree with a property often referred to as *upper semicontinuity*. The following lemma makes the relationship between the two concepts precise.

Lemma 5.15. (Osc vs. upper semicontinuity) *Let* $M : \mathbb{R}^m \rightrightarrows \mathbb{R}^n$ *be a set-valued mapping. Consider* $x \in \mathbb{R}^m$ *such that* $M(x)$ *is closed. If* M *is upper semicontinuous at* x, *that is, for every* $\varepsilon > 0$ *there exists* $\delta > 0$ *such that* $x' \in x + \delta\mathbb{B}$ *implies* $M(x') \subset M(x) + \varepsilon\mathbb{B}$, *then* M *is outer semicontinuous at* x. *If* M *is locally bounded at* x, *then the reverse implication is true.*

PROOF. If a sequence $\{x_i\}_{i=1}^{\infty}$ converges to x and a sequence $\{y_i\}_{i=1}^{\infty}$ of $y_i \in M(x_i)$ is convergent, then the ε-δ condition implies that $\lim_{i \to \infty} y_i \in M(x) + \varepsilon\mathbb{B}$ for every $\varepsilon > 0$. Since $M(x)$ is closed, this implies that $\lim_{i \to \infty} y_i \in M(x)$, and hence M is outer semicontinuous at x. Now assume that M is outer semicontinuous and locally bounded at x. Suppose that for some $\varepsilon > 0$ and any $\delta > 0$, there exists $x' \in x + \delta\mathbb{B}$ such that $M(x') \not\subset M(x) + \varepsilon\mathbb{B}$. In particular, there exists a sequence $\{x_i\}_{i=1}^{\infty}$ with $\lim_{i \to \infty} x_i = x$ and $M(x_i) \not\subset M(x) + \varepsilon\mathbb{B}$. Since M is locally bounded at x, there exists $r > 0$ such that, for all large enough i, $M(x_i) \subset r\mathbb{B}$. Then (b) of Theorem 5.4 implies that $\limsup_{i \to \infty} M(x_i) \not\subset M(x)$, which is a contradiction. $\qquad\square$

The following result will imply outer semicontinuity and local boundedness of the regularizations of the flow map and the jump map of a hybrid system used in the definition of Krasovskii solutions; recall Definition 4.13.

Lemma 5.16. (Osc regularization) *Let* $M : \mathbb{R}^m \rightrightarrows \mathbb{R}^n$ *be any set-valued mapping. Define* $M_1, M_2 : \mathbb{R}^m \rightrightarrows \mathbb{R}^n$ *by setting, for each* $x \in \mathbb{R}^m$,

$$M_1(x) = \bigcap_{\delta > 0} \overline{M(x + \delta\mathbb{B})}, \qquad M_2(x) = \bigcap_{\delta > 0} \overline{\mathrm{con}}\, M(x \mid \delta\mathbb{B}).$$

Then both M_1, M_2 *are outer semicontinuous. If additionally* M *is locally bounded, then so are* M_1 *and* M_2.

PROOF. Let $x_i \to x \in \mathbb{R}^m$, $y_i \in M_k(x_i)$, $y_i \to y$, where $k = 1$ or $k = 2$. It needs to be shown that $y \in M_k(x)$. Take $k = 1$. For all $i \in \mathbb{N}$, by the definition of $M_1(x_i)$, there exists $x_i' \in x_i + 1/i\mathbb{B}$ and $y_i' \in M(x_i')$ with $y_i \in y_i' + 1/i\mathbb{B}$. Then $x_i' \to x$, $y_i' \to y$, which by the definition of $M_1(x)$ yields $y \in M_1(x)$. Now take $k = 2$. For all $i \in \mathbb{N}$, by the definition of $M_2(x_i)$, there exist $\lambda_i^l \geq 0$, $x_i^l \in x_i + 1/i\mathbb{B}$, and $y_i^l \in M(x_i^l)$ for $l = 1, 2, \ldots, m+1$ with $\sum_{l=1}^{m+1} \lambda_i^l = 1$ and $\sum_{l=1}^{m+1} \lambda_i^l y_i^l \in y_i + 1/i\mathbb{B}$. For any arbitrarily small $\delta > 0$, there exists i_0 such that for all $i > i_0$ such that $x_i^l \in x + \delta\mathbb{B}$ and so $y_i \in \mathrm{con}M(x + \delta\mathbb{B}) + 1/i\mathbb{B} \subset \overline{\mathrm{con}}M(x + \delta\mathbb{B}) + 1/i\mathbb{B}$. As $y_i \to y$, $y \in \overline{\mathrm{con}}M(x + \delta\mathbb{B})$. Since this holds for any arbitrarily small $\delta > 0$, $y \in M_2(x)$. Local boundedness is straightforward. □

Lemma 5.17. (Inflations of set-valued mappings) *Let $M : \mathbb{R}^m \rightrightarrows \mathbb{R}^n$ be a set-valued mapping, and $\rho : \mathbb{R}^m \to \mathbb{R}_{\geq 0}$ be a continuous function. Define $M_1, M_2 : \mathbb{R}^m \rightrightarrows \mathbb{R}^n$ at each $x \in \mathbb{R}^m$ by*

$$M_1(x) = M(x + \rho(x)\mathbb{B}), \qquad M_2(x) = \overline{\mathrm{con}}M(x + \rho(x)\mathbb{B}).$$

If M is outer semicontinuous and locally bounded, then so are M_1, M_2. If furthermore $m = n$, then $M_3 : \mathbb{R}^m \rightrightarrows \mathbb{R}^n$ defined at each $x \in \mathbb{R}^m$ by

$$M_3(x) = \bigcup_{u \in M(x)} u + \rho(u)\mathbb{B}$$

is outer semicontinuous and locally bounded.

PROOF. Continuity of ρ implies that for each compact $K \subset \mathbb{R}^m$ there exists a compact $K' \subset \mathbb{R}^n$ such that $x + \rho(x)\mathbb{B} \subset K'$ for all $x \in K$. This in turn implies that M_1, M_2 are locally bounded. When $m = n$, local boundedness of M yields that for each compact $K'' \subset \mathbb{R}^n$ there exists a compact $K \subset \mathbb{R}^n$ such that $M(x) \subset K$ for all $x \in K''$. Hence M_3 is locally bounded.

To prove that M_k, $k = 1, 2, 3$ is osc, consider $x \in \mathbb{R}^m$, a sequence $x_i \to x$, a convergent sequence of $y_i \in M_k(x_i)$ with $y_i \to y$. For the case of $k = 2$, it needs to be shown that $y \in M_2(x)$. From the definition of M_2, $y_i \in \sum_{l=1}^{n+1} \lambda_i^l u_i^l + 1/i\mathbb{B}$, where $u_i^l \in F(w_i^l)$ for some $w_i^l \in x_i + \rho(x_i)\mathbb{B}$, $\lambda_i^l \in [0, 1]$ for all $l = 1, 2, \ldots, n+1$, and $\sum_{l=1}^{n+1} \lambda_i^l = 1$ for all $i \in \mathbb{N}$. For each $l = 1, 2, \ldots, n+1$, the sequences of w_i^l's, u_i^l's, and λ_i^l's are bounded. Without relabeling, pass to a subsequence of x_i's such that the corresponding sequences of w_i^l's, u_i^l's, and λ_i^l's are convergent. Then $u^l := \lim_{i \to \infty} u_i^l \in F(w_l)$ where $w_l := \lim_{i \to \infty} w_i^l \in x + \rho(x)\mathbb{B}$, and $y = \sum_{l=1}^{n+1} (\lim_{i \to \infty} \lambda_i^l) u^l \in \mathrm{con}F(x + \rho(x)\mathbb{B})$. Thus $y \in M_2(x)$, and M_2 is osc.

To prove that M_1 is osc, the arguments above can be repeated with $\lambda_i^1 = 1$ and $\lambda_i^k = 0$, $l = 2, 3, \ldots, n+1$. For M_3, note that $y_i = u_i + v_i$ for $u_i \in M(x_i)$ and $v_i \in \rho(x_i)\mathbb{B}$, $i \in \mathbb{N}$. Passing to a subsequence of x_i's for which the corresponding u_i and v_i converge finishes the argument. □

Definition 5.18. (Graphical convergence) *A sequence $\{M_i\}_{i=1}^\infty$ of set-valued mappings $M_i : \mathbb{R}^m \rightrightarrows \mathbb{R}^n$ converges graphically if the sequence of sets*

$\{\operatorname{gph} M_i\}_{i=1}^{\infty}$ *converges in the sense of set convergence (Definition 5.1). The* *graphical limit of a graphically convergent sequence* $\{M_i\}_{i=1}^{\infty}$ *is the mapping* $M : \mathbb{R}^m \rightrightarrows \mathbb{R}^n$ *such that* $\operatorname{gph} M = \lim_{i \to \infty} \operatorname{gph} M_i$.

Even for sequences of continuous functions, differences between graphical convergence and the classical notions of pointwise or uniform convergence are visible. For example, consider $\phi_i : \mathbb{R} \to \mathbb{R}$ given by $\phi_i(x) = |x|^i$. Pointwise, ϕ_i converge to 0 on $(-1, 1)$, to 1 at $x = -1$ and $x = 1$, and to ∞ otherwise. The convergence is uniform on each compact subset of $(-1, 1)$, but not on $[-1, 1]$. Graphically, the sequence of ϕ_i's converges to a set-valued mapping $\phi : \mathbb{R} \rightrightarrows \mathbb{R}$ given by $\phi(x) = 0$ if $x \in (-1, 1)$, $\phi(-1) = \phi(1) = [0, \infty)$, and $\phi(x) = \emptyset$ otherwise. If discontinuous functions are considered, the graphical limit of even a constant sequence $\phi_i = \phi$ for some discontinuous ϕ may be set valued. Indeed, the graph of the limit is the closure of $\operatorname{gph} \phi$.

Example 5.19. (Limits of domains and ranges) Let $\{M_i\}_{i=1}^{\infty}$ be a sequence of set-valued mappings $M_i : \mathbb{R}^m \rightrightarrows \mathbb{R}^n$ that is graphically convergent and set $M = \operatorname{gph-lim}_{i\to\infty} M_i$. Then $\operatorname{dom} M \subset \lim_{i\to\infty} \operatorname{dom} M_i$ and $\operatorname{rge} M \subset \lim_{i\to\infty} \operatorname{rge} M_i$. If $\{M_i\}_{i=1}^{\infty}$ is locally eventually bounded, in the sense that for any compact set $K \subset \mathbb{R}^m$ there exists i_0 and a compact $K' \subset \mathbb{R}^n$ such that $M_i(K) \subset K'$ for all $i > i_0$, then in fact

$$\operatorname{dom} M = \lim_{i \to \infty} \operatorname{dom} M_i. \tag{5.2}$$

Let P be the projection of $\mathbb{R}^m \times \mathbb{R}^n$ onto $\mathbb{R}^m$, so that $\operatorname{dom} M_i = P(\operatorname{gph} M_i)$, $\operatorname{dom} M = P(\operatorname{gph} M)$. Directly from the definitions,

$$P(\lim_{i\to\infty} \operatorname{gph} M_i) = P(\liminf_{i\to\infty} \operatorname{gph} M_i) \subset \liminf_{i\to\infty} P(\operatorname{gph} M_i).$$

The inclusion for ranges is shown similarly. Now suppose that $\{M_i\}_{i=1}^{\infty}$ is locally eventually bounded. Fix any $(t, j) \in \limsup_{i\to\infty} P(\operatorname{gph} M_i)$ and let $(t_{i_k}, j_{i_k}) \in \operatorname{dom} M_{i_k}$ converge to (t, j). As M_i's are locally eventually bounded, there exists a subsequence (which we do not relabel) such that $M_{i_k}(t_{i_k}, j_{i_k})$ converges. The limit is an element of $M(t, j)$, which, in particular, implies that $(t, j) \in P(\operatorname{gph} M)$. Consequently,

$$\limsup_{i\to\infty} P(\operatorname{gph} M_i) \subset P(\operatorname{gph} M) = P(\lim_{i\to\infty} \operatorname{gph} M_i).$$

This, and the previously displayed inclusion, implies (5.2).

5.3   GRAPHICAL CONVERGENCE OF HYBRID ARCS

This section specializes some of the definitions and results, given previously for general set-valued mappings, to hybrid arcs.

Definition 5.20. (Graph of a hybrid arc) *The graph of a hybrid arc* ϕ : dom $\phi \to \mathbb{R}^n$ *is a set in* $\mathbb{R}^{n+2}$ *given by*

$$\mathrm{gph}\,\phi = \{(t, j, x) \,:\, (t, j) \in \mathrm{dom}\,\phi, \ x = \phi(t, j)\}.$$

Definition 5.21. (Graphical convergence of hybrid arcs) *A sequence* $\{\phi_i\}_{i=1}^{\infty}$ *of hybrid arcs* ϕ_i : dom $\phi_i \to \mathbb{R}^n$ *converges graphically if the sequence of sets* $\{\mathrm{gph}\,\phi_i\}_{i=1}^{\infty}$ *converges in the sense of set convergence (Definition 5.1). The graphical limit of a graphically convergent sequence* $\{\phi_i\}_{i=1}^{\infty}$ *is the mapping* $M : \mathbb{R}^2 \rightrightarrows \mathbb{R}^n$ *such that* gph $M = \lim_{i \to \infty} \mathrm{gph}\,\phi_i$.

Directly from the definition, and from the structure of domains of hybrid arcs, it can be concluded that a sequence $\{\phi_i\}_{i=1}^{\infty}$ of hybrid arcs converges graphically if and only if, for each $j \in \mathbb{N}$, the sequence of mappings $\phi_i(\cdot, j)$ converges graphically.

Definition 4.11 proposed one way to measure how close two hybrid arcs are. This was the concept of ϵ-closeness, closely related to the Haussdorff distance between the graphs of the hybrid arcs. The following example illustrates it.

Example 5.22. (Closeness of hybrid arcs) Consider a hybrid system $\mathcal{H}$ on $\mathbb{R}$ with $C = [0, 9]$, $D = \{4\}$, $F(x) = 2\sqrt{x}$, and $G(4) = 1$. Given an initial point $1 + \varepsilon$, for ε in $[-1, 1]$, the hybrid arc ϕ_ε is a maximal solution to $\mathcal{H}$, where

$$\phi_\varepsilon(t, 0) = (t + \sqrt{1 + \varepsilon})^2 \ \text{ for } \ t \in [0, t_1], \quad \phi_\varepsilon(t, 1) = (t - t_1 + 1)^2 \ \text{ for } \ t \in [t_1, t_2],$$

with $t_1 = 2 - \sqrt{1 + \varepsilon}$, $t_2 = t_1 + 2$. For such ϕ_ε,

$$\mathrm{dom}\,\phi_\varepsilon = \left[0, 2 - \sqrt{1 + \varepsilon}\right] \times \{0\} \cup \left[2 - \sqrt{1 + \varepsilon}, 4 - \sqrt{1 + \varepsilon}\right] \times \{1\}.$$

Note that ϕ_ε is not the unique maximal solution from ξ_ε; in fact there are solutions that never jump, or complete solutions that jump infinitely many times. (Of course, there is also nonuniqueness to solutions to $\dot{z} = F(z)$ for the initial point 0.) The solution ϕ_0, with initial point 1, is

$$\phi_0(t, 0) = (t + 1)^2 \ \text{ for } \ t \in [0, 1], \quad \phi_0(t, 1) = t^2 \ \text{ for } \ t \in [1, 3].$$

Consider $\varepsilon > 0$. It can be verified that

- for each $(t, j) \in \mathrm{dom}\,\phi_\varepsilon$, one can take $(s, j) \in \mathrm{dom}\,\phi_0$ with $s = t + (\sqrt{1 + \varepsilon} - 1)$ and then $\phi_\varepsilon(t, j) = \phi_0(s, j)$;

- for each $(t, 0) \in \mathrm{dom}\,\phi_0$ with $t < \sqrt{1 + \varepsilon} - 1$ one can take $(0, 0) \in \mathrm{dom}\,\phi_\varepsilon$ in which case $|\phi_0(t, 0) - \phi_\varepsilon(0, 0)| \leq |\phi_0(0, 0) - \phi_\varepsilon(0, 0)| = \varepsilon$, while for each $(t, j) \in \mathrm{dom}\,\phi_0$ with $t \geq \sqrt{1 + \varepsilon} - 1$ one can take $(s, j) \in \mathrm{dom}\,\phi_\varepsilon$ with $s = t - (\sqrt{1 + \varepsilon} - 1)$ and then $\phi_0(t, j) = \phi_\varepsilon(s, j)$.

In particular, since $\sqrt{1 + \varepsilon} - 1 \leq \varepsilon$, the arcs ϕ_0 and ϕ_ε are ε-close.

In Section 3.6 and in Example 5.22, ε-closeness was used for hybrid arcs with compact domains. When hybrid arcs with unbounded domains are considered, the concept of ε-closeness may turn out to be too restrictive. A similar issue is already apparent for solutions of differential equations. The arcs $z_1(t) = \xi_1 e^t$ and $z_2(t) = \xi_2 e^t$ on $[0, \infty)$, which are the solutions to $\dot{z} = z$ from ξ_1 and ξ_2, do not satisfy $|z_1(t) - z_2(t)| < \varepsilon$ for all $t \geq 0$ unless, of course, $\xi_1 = \xi_2$, independently of the size of ε.

In other words, it is possible for a sequence of hybrid arcs ϕ_i to converge graphically to a hybrid arc ϕ even if the arcs ϕ_i and ϕ are not ε-close for any ε and any i. Consequently, the following refinement of ε-closeness will be used; see Figure 5.5.

Definition 5.23. ((τ, ε)-closeness of hybrid arcs) *Given $\tau, \varepsilon > 0$, two hybrid arcs ϕ_1 and ϕ_2 are (τ, ε)-close if*

(a) *for all $(t, j) \in \operatorname{dom} \phi_1$ with $t + j \leq \tau$ there exists s such that $(s, j) \in \operatorname{dom} \phi_2$, $|t - s| < \varepsilon$, and*
$$|\phi_1(t, j) - \phi_2(s, j)| < \varepsilon;$$

(b) *for all $(t, j) \in \operatorname{dom} \phi_2$ with $t + j \leq \tau$ there exists s such that $(s, j) \in \operatorname{dom} \phi_1$, $|t - s| < \varepsilon$, and*
$$|\phi_2(t, j) - \phi_1(s, j)| < \varepsilon.$$

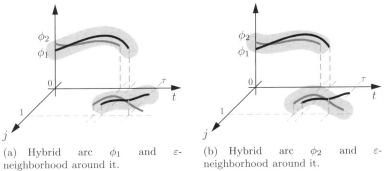

(a) Hybrid arc ϕ_1 and ε-neighborhood around it.

(b) Hybrid arc ϕ_2 and ε-neighborhood around it.

Figure 5.5: Two hybrid arcs (τ, ε)-close.

A slightly different way to measure how close two hybrid arcs are is the following: given $T \geq 0$, $J \geq 0$, and $\varepsilon > 0$, two hybrid arcs ϕ_1 and ϕ_2 are (T, J, ε)-close if

(a) for all $(t, j) \in \operatorname{dom} \phi_1$ with $t \leq T$, $j \leq J$, there exists s such that $(s, j) \in \operatorname{dom} \phi_2$, $|t - s| < \varepsilon$, and $|\phi_1(t, j) - \phi_2(s, j)| < \varepsilon$;

(b) for all $(t, j) \in \operatorname{dom} \phi_2$ with $t \leq T$, $j \leq J$, there exists s such that $(s, j) \in \operatorname{dom} \phi_1$, $|t - s| < \varepsilon$, and $|\phi_2(t, j) - \phi_1(s, j)| < \epsilon$.

Obviously, if ϕ_1, ϕ_2 are (τ, ε)-close then they are (T, J, ε)-close if $T + J \leq \tau$. Similarly, if ϕ_1, ϕ_2 are (T, J, ε)-close then they are (τ, ε)-close if $\tau \leq T$, $\tau \leq J$. Thus, given a sequence $\{\phi_i\}_{i=1}^{\infty}$ of hybrid arc and a hybrid arc ϕ, the following two statements are equivalent:

- for each $\tau \geq 0$, $\varepsilon > 0$ there exists $i_0 > 0$ such that for all $i > i_0$, ϕ_i and ϕ are (τ, ε)-close;

- for each $T, J \geq 0$, $\varepsilon > 0$ there exists $i_0 > 0$ such that for all $i > i_0$, ϕ_i and ϕ are (T, J, ε)-close.

Definition 5.24. (Local eventual boundedness) *A sequence of hybrid arcs* $\{\phi_i\}_{i=1}^{\infty}$ *is locally eventually bounded if for any $m > 0$, there exists $i_0 > 0$ and a compact set $K \subset \mathbb{R}^n$ such that for all $i > i_0$, all $(t, j) \in \operatorname{dom} \phi_i$ with $t + j < m$, $\phi_i(t, j) \in K$.*

Theorem 5.25. (Graphical convergence and closeness of hybrid arcs) *Let* $\{\phi_i\}_{i=1}^{\infty}$ *be a sequence of hybrid arcs* $\phi_i : \operatorname{dom} \phi_i \to \mathbb{R}^n$ *and let* $\phi : \operatorname{dom} \phi \to \mathbb{R}^n$ *be a hybrid arc. Then the condition*

(a) *for each $\tau \geq 0$, $\varepsilon > 0$ there exists $i_0 > 0$ such that for all $i > i_0$, ϕ_i and ϕ are (τ, ε)-close*

implies that

(b) *the sequence* $\{\phi_i\}_{i=1}^{\infty}$ *converges graphically to* ϕ.

If the sequence $\{\phi_i\}_{i=1}^{\infty}$ *is locally eventually bounded, then also (b) implies (a).*

PROOF. If ϕ_i and ϕ are $\left(r, \varepsilon/\sqrt{2}\right)$-close, then

$$\operatorname{gph} \phi \cap r\mathbb{B} \subset \operatorname{gph} \phi_i + \varepsilon\mathbb{B}, \qquad \operatorname{gph} \phi_i \cap r\mathbb{B} \subset \operatorname{gph} \phi + \varepsilon\mathbb{B}. \qquad (5.3)$$

This property and Theorem 5.4 show that (a) implies (b). To show the other implication, fix $\tau \geq 0$ and $\varepsilon > 0$. If $\{\phi_i\}_{i=1}^{\infty}$ is locally eventually bounded, there exist $r' > 0$ and $i' > 0$ such that $\phi(t, j) \in r'\mathbb{B}$ for all $(t, j) \in \operatorname{dom} \phi$ and $\phi_i(t, j) \in r'\mathbb{B}$ for all $(t, j) \in \operatorname{dom} \phi$, as long as $t + j < \tau$ and $i > i'$. Let $r = \sqrt{r'^2 + \tau^2}$, and using Theorem 5.4 pick $i_0 > i'$ such that (5.3) holds for all $i > i_0$. Then, for all such i's, ϕ_i and ϕ are (τ, ε)-close. $\qquad \square$

This in particular guarantees that given a sequence $\{\varepsilon_i\}_{i=1}^{\infty}$ with $\varepsilon_i \geq 0$, $\varepsilon_i \to 0$, the sequence of hybrid arcs ϕ_{ε_i} from Example 5.22 converges graphically to ϕ_0. (One could easily show that this is the case for any sequence of $\varepsilon_i \to 0$.) Describing the convergence of ϕ_{ε_i} to ϕ_0 with a classical notion, such as pointwise or uniform convergence, is problematic though, even if one abandons hybrid time domains and relies on parameterization of solutions by t only. For example, suppose that z_1 represents a piecewise continuous function, continuous to the right, and having left limits, that corresponds to ϕ_0 of Example 5.22. That is,

$$z_0(t) = (t + 1)^2 \text{ for } t \in [0, 1), \quad z_0(t, 1) = t^2 \text{ for } t \in [1, 3].$$

For $\varepsilon \in (-1, 1)$, define z_ε similarly:

$$z_\varepsilon(t) = \begin{cases} (t + \sqrt{1+\varepsilon})^2 & t \in [0, 2 - \sqrt{1+\varepsilon}) \\ (t + \sqrt{1+\varepsilon} - 1)^2 & t \in [2 - \sqrt{1+\varepsilon}, 4 - \sqrt{1+\varepsilon}]. \end{cases}$$

Then, if $\varepsilon > 0$, $\varepsilon \searrow 0$, z_ε do converge pointwise, on $[0, 3]$, to z_0. However, if $\varepsilon < 0$, $\varepsilon \nearrow 0$, then z_ε do converge pointwise on $[0, 3]$, but to a function that is piecewise continuous, continuous to the left, and having right limits. Such function differs from z_0 at $t = 1$. In particular, the family z_ε does not have a pointwise limit as $\varepsilon \to 0$. Regarding uniform convergence, even for $\varepsilon > 0$ one has $\lim_{\varepsilon \searrow 0, \varepsilon \neq 0} (z_0(1) - z_\varepsilon(1)) = 3$. Thus, in such a case, z_ε do not converge uniformly to the pointwise limit.

5.4   DIFFERENTIAL INCLUSIONS

Motivation for considering differential inclusions, from the control engineering angle, was discussed in Chapter 4. This section collects some basic facts about solutions to differential inclusions.

The following assumption will be used in this section:

($\star$)   the set $C \subset \mathbb{R}^n$ is closed, the set-valued mapping $F : \mathbb{R}^n \rightrightarrows \mathbb{R}^n$ is outer semicontinuous and locally bounded relative to C, and $F(x)$ is nonempty and convex for all $x \in C$.

In particular, the assumption covers the case of a closed set C and a continuous function $f : C \to \mathbb{R}^n$. Even for this simple setting, results such as Lemma 5.26 are interesting.

Consider the constrained differential inclusion

$$\dot{z}(t) \in F(z(t)) \text{ and } z(t) \in C \quad \text{for almost all } t \in I, \tag{5.4}$$

where I is some interval. A simple necessary condition and a sufficient condition for existence of solution to (5.4) involve tangent cones to the set C, as presented in Section 5.2.

Lemma 5.26. (Viability) *Suppose that* ($\star$) *holds.*

(a) *If* $z : I \to \mathbb{R}^n$, *with* $I = [0, T]$ *for some* $T > 0$, *is a solution to* (5.4), *then*

$$F(z(0)) \cap T_C(z(0)) \neq \emptyset.$$

(b) *Given* $\xi \in C$, *if there exists a neighborhood* U *of* ξ *such that for all* $x \in U \cap C$,

$$F(x) \cap T_C(x) \neq \emptyset,$$

then there exists $T > 0$ *and a solution* $z : I := [0, T] \to \mathbb{R}^n$ *to* (5.4) *with* $z(0) = \xi$.

The necessary condition in (a) of Lemma 5.26 follows directly from the definition of the tangent cone, in the case when $\dot{z}(0)$ exists. Indeed, let $z : I \to \mathbb{R}^n$, with $I = [0, T]$ for some $T > 0$, be a solution to (5.4), and suppose that $\dot{z}(0)$ exists. Then, for any sequence $\{t_i\}_{i=1}^\infty$ with $t_i > 0$ and $\lim_{i \to \infty} t_i = 0$, $\dot{z}(0) = \lim_{i \to \infty} (z(t_i) - z(0))/t_i$. But $z(t_i) \in C$ and Definition 5.12 yields $\lim_{i \to \infty} (z(t_i) - z(0))/t_i \in T_C(z(0))$. The general case, and the sufficient argument, require more involved arguments.

The following assumption will now be used, in the discussion of convergence of solutions to constrained differential inclusions under state perturbations:

$(\star\star)$ the function $\rho : \mathbb{R}^n \to \mathbb{R}_{\geq 0}$ is continuous; $\{\delta_i\}_{i=1}^\infty$ is a decreasing sequence of numbers in $(0, 1)$ with $\lim_{i \to \infty} \delta_i = 0$; for $i = 1, 2, \ldots, I_i \subset \mathbb{R}$ is an interval and $z_i : I_i \to \mathbb{R}^n$ is a locally absolutely continuous function satisfying

$$\dot{z}_i(t) \in F_i(z_i(t)) \quad \text{and} \quad z(t) \in C_i \quad \text{for almost all } t \in I_i \qquad (5.5)$$

where

$$C_i = \{x \in \mathbb{R}^n : x + \delta_i \rho(x)\mathbb{B} \cap C \neq \emptyset\}$$

and, for all $x \in C_i$,

$$F_i(x) = \text{con}F\left((x + \delta_i \rho(x)\mathbb{B}) \cap C\right) + \delta_i \rho(x)\mathbb{B}.$$

Lemma 5.27. (Uniform convergence of solutions to differential inclusions) Under assumptions $(\star)$, $(\star\star)$, suppose that $I_i = I$ for $i = 1, 2, \ldots$, where $I = [a, b]$, $a, b \in \mathbb{R}$, $a < b$, and the sequence $\{z_i\}_{i=1}^\infty$ converges uniformly to a function $z : I \to \mathbb{R}^n$. Then z is absolutely continuous and satisfies (5.4).

Usually, results such as Lemma 5.27 do not assume uniform convergence, but some weaker condition ensuring boundedness of the sequence of points $z_i(a)$, and conclude the existence of a uniformly convergent subsequence of $\{z_i\}_{i=1}^\infty$, the limit of which is a solution to (5.4). In this book, only the weaker formulation is needed. Extraction of convergent subsequences is handled through Theorem 5.7, which guarantees the existence of a graphically convergent subsequence of $\{z_i\}_{i=1}^\infty$. Translating graphical convergence to uniform convergence is then possible through the following results.

Lemma 5.28. (From graphical to uniform convergence) Under assumptions $(\star)$, $(\star\star)$, suppose that the sequence $\{z_i\}_{i=1}^\infty$ is locally eventually bounded, in the sense that for each compact interval $T \in \mathbb{R}$ there exists a compact set $K \subset \mathbb{R}^n$ and $i^* > 0$ such that for all $i > i^*$, all $t \in I_i \cap T$, $z_i(t) \in K$, and converges graphically, I_i are closed to the left, and $a_i = \min I_i$ converge to $a \in \mathbb{R}$. Then, the graphical limit z of the sequence $\{z_i\}_{i=1}^\infty$ has $\text{dom } z = \lim_{i \to \infty} I_i$, is single valued and locally Lipschitz continuous, and the sequence $\{z_i\}_{i=1}^\infty$ converges to z uniformly on every compact subset of $\text{int dom } z$.

PROOF. Let $I = \text{dom } z$. Since the sequence $\{z_i\}_{i=1}^\infty$ is locally eventually bounded, Example 5.19 implies that $I = \lim_{i \to \infty} \text{dom } z_i = \lim_{i \to \infty} I_i$. Note that I is a closed interval that starts at a.

Now note that the sequence $\{z_i\}_{i=1}^{\infty}$ is locally eventually uniformly Lipschitz, in the sense that for any compact interval $T \subset \mathbb{R}$ there exists $L > 0$ and $i_0 > 0$ such that, for all $i > i_0$ and $t', t'' \in I_i \cap T$, $|z_i(t') - z_i(t'')| < L|t' - t''|$. Indeed, local eventual boundedness of $\{z_i\}_{i=1}^{\infty}$ implies that there exist a compact set $K \subset \mathbb{R}^n$ and $i_0 > 0$ such that $z_i(t) \in K$ for all i_0 and $t \in I_i \cap T$. Since $C_i \subset C_1$, $F_i(x) \subset F_1(x)$ for all $x \in C_i$, and, by Lemma 5.17, F_1 is locally bounded relative to C_1, there exists a compact set $K' \subset \mathbb{R}^n$ such that, for all $i > i_0$ and $x \in C_i \cap K$, one has $F_i(x) \subset K'$. Then, for all $i > i_0$ and almost all $t \in I_i \cap T$, $\dot{z}_i(t) \in K'$. This implies the needed Lipschitz condition.

Take an arbitrarily large compact interval $T \subset \mathbb{R}$ and a compact interval $T' \subset \mathbb{R}$ containing T in its interior. Let $i_0 > 0$ and L be such that, for all $i > i_0$ and $t', t'' \in I_i \cap T'$, $|z_i(t') - z_i(t'')| < L|t' - t''|$. Take any $t', t'' \in T$ and any $y' \in z(t')$, $y'' \in z(t'')$ — at this stage it has not been shown yet that z is single valued. There exist sequences $t_i' \to t'$, $t_i'' \to t''$ with $t_i', t_i'' \in I_i \cap T'$ such that $z_i(t_i') \to y'$, $z_i(t_i'') \to y''$. Then, for any $\varepsilon > 0$, the following holds for all large enough i's:

$$|y' - y''| \le |z_i(t_i') - z_i(t_i'')| + \varepsilon/2 \le L|t_i' - t_i''| + \varepsilon/2 \le L|t' - t''| + \varepsilon.$$

Since $\varepsilon > 0$ is arbitrary, $|y' - y''| \le L|t' - t''|$, which, by first considering $t' = t''$ to see that $y' = y''$ and hence z is single valued on T, implies that z is Lipschitz continuous on T with constant L.

Now suppose that there is a compact interval $T \subset \operatorname{int} \operatorname{dom} z$ on which the convergence of z_i to z is not uniform: for some $\varepsilon > 0$ there exist, subject to passing to a subsequence, points $t_i \in T$ such that $|z_i(t_i) - z(t_i)| > \varepsilon$. Subject to passing to a subsequence again, it can be assumed that $t_i \to t$ for some $t \in T$, and since $\{z_i\}_{i=1}^{\infty}$ is eventually bounded on T, that $z_i(t_i) \to y$ for some $y \in \mathbb{R}^n$. Graphical convergence of $\{z_i\}_{i=1}^{\infty}$ to z implies that $y \in z(t)$ and hence $y = z(t)$, because z is single valued. This is a contradiction with $|z_i(t_i) - z(t_i)| > \varepsilon$. $\qquad \square$

Theorem 5.29. (Graphical convergence of solutions to differential inclusions) *Under assumptions* $(\star)$, $(\star\star)$, *suppose that the sequence* $\{z_i\}_{i=1}^{\infty}$ *converges graphically,* I_i *are closed to the left, and* $a_i = \min I_i$ *converge to* $a \in \mathbb{R}$.

(a) *Suppose that the sequence* $\{z_i\}_{i=1}^{\infty}$ *is locally eventually bounded. Then* $z = \operatorname{gph-lim}_{i\to\infty} z_i$ *has* $\operatorname{dom} z = \lim_{i\to\infty} I_i$, z *is locally absolutely continuous and satisfies* (5.4).

(b) *Suppose that* $z_i(a_i)$ *converge to some* $\xi \in \mathbb{R}^n$ *but the sequence* $\{z_i\}_{i=1}^{\infty}$ *is not uniformly bounded. Then*

   — *there exists the smallest* $t^* > a$ *for which any sequence of times* $t_i \in I_i$ *such that* $\lim_{i\to\infty} t_i = t^*$ *leads to* $\lim_{i\to\infty} |z_i(t_i)| = \infty$, *and*

   — *the mapping* $z = (\operatorname{gph-lim}_{i\to\infty} z_i)\big|_{t\in[a,t^*)}$ *is locally absolutely continuous, it satisfies* $\dot{z}(t) \in F(z(t))$ *for almost all* $t \in [a, t^*)$, *and* $\lim_{t\to t^*} |z(t, j^*)| = \infty$.

PROOF. By Lemma 5.28, dom $z = \lim_{i\to\infty} I_i$, z is single valued and locally Lipschitz on $I := \operatorname{dom} z$. In particular, it is locally absolutely continuous. On each compact subinterval of int I, z_i converge to z uniformly, and Lemma 5.27 implies that z satisfies (5.4). Consequently, z satisfies (5.4) on I.

We now show (b). As the sequence $\{z_i\}_{i=1}^\infty$ is not uniformly bounded, there exists at least one $t \geq a$ for which there exists a subsequence $\{z_{i_k}\}_{k=1}^\infty$ of $\{z_i\}_{i=1}^\infty$ and a sequence $t_k \in I_k$, $k = 1, 2, \ldots$ such that $\lim_{k\to\infty} t_k = t$ and $\lim_{k\to\infty} \omega(z_{i_k}(t_k)) = \infty$. Let T be the set of all such t's and t^* be the infimum of T. Suppose that, for this t^*, there exists a subsequence $\{z_{i_k}\}_{k=1}^\infty$ of $\{z_i\}_{i=1}^\infty$ and a sequence of times $t_k \in I_k$, $k = 1, 2, \ldots$ such that $\lim_{k\to\infty} t_k = t^*$, and a compact set $K \subset \mathbb{R}^n$ such that $z_{i_k}(t_k) \in K$ for $k = 1, 2, \ldots$. Then gph-$\limsup_{k\to\infty} z_{i_k}(t^*) \cap K \neq \emptyset$, and since the sequence $\{z_i\}_{i=1}^\infty$ is graphically convergent, there exist $t_i \in I_i$ with $\lim_{i\to\infty} t_i = t^*$ and $\lim_{i\to\infty} z_i(t_i) \in K$. Let $K' \subset \mathbb{R}^n$ be any compact set with $K \subset \operatorname{int} K'$. Then, for all large enough i, $z_i(t_i) \in K'$. For any compact $K'' \subset \mathbb{R}^n$ with $K' \subset \operatorname{int} K''$, there exists $\tau \in (0,1)$ such that $z_i(t) \in K''$ for all $t \in I_i \cap [t_i - \tau, t_i + \tau]$, $i = 1, 2, \ldots$. Then $T \cap (t^* - \tau, t^* + \tau) = \emptyset$, which contradicts t^* being the infimum of T. Thus for any sequence of times $t_i \in I_i$, $i = 1, 2, \ldots$ such that $\lim_{k\to\infty} t_k = t^*$, $\lim_{i\to\infty} |z_i(t_i)| = \infty$, which also shows that $t^* \in T$ and $t^* > a$.

Now, to show that $z = (\text{gph-}\lim_{i\to\infty} z_i)|_{t<t^*}$ is locally absolutely continuous on $[a, t^*)$ and satisfies $\dot{z}(t) \in F(z(t))$ for almost all $t \in [a, t^*)$, it is enough to show these statements for $z|_{t\in[a,t']}$ on each interval $[a, t']$ with $t' \in [a, t^*)$. To this end, note that the sequence $\{z_i|_{t\in[a,t']}\}_{i=1}^\infty$ is eventually bounded, by the choice of t^*. Furthermore, its graphical limit is exactly $z|_{t\in[a,t']}$ – this easily follows from z being a pointwise limit of z_i's for each $t \in (a, t^*)$. Thus, statement (a) applies. It remains to verify that $\lim_{t\to t^*} \omega(z(t, j^*)) = \infty$. $\square$

Example 5.30. (Reachable set for a differential inclusion) Recall that Example 5.2 showed how a reachable set, in time τ, for a differential equation with a regular right-hand side, depends on τ. Similar results are true for the constrained differential inclusion (5.4). Let $K \subset \mathbb{R}^n$ be a compact set, and, for $\tau \geq 0$, let S_τ be the set of all $x \in \mathbb{R}^n$ such that there exists a solution $z : [0, \tau] \to \mathbb{R}^n$ to (5.4) such that $z(0) \in K$ and $z(\tau) = x$. Consider a sequence $\{\tau_i\}_{i=1}^\infty$ such that $\lim_{i\to\infty} \tau_i = \tau$ for some $\tau \geq 0$.

If assumption $(\star)$ holds and, for every $z_0 \in C$, condition (b) of Lemma 5.26 holds, then $S_\tau \subset \liminf_{i\to\infty} S_{\tau_i}$. The argument showing this is virtually identical to the one given in Example 5.2, with (b) of Lemma 5.26 ensuring the existence of a needed solution.

If, additionally, every maximal solution to (5.4) starting in K is complete, as it is automatically the case if C is compact, then $\limsup_{i\to\infty} S_{\tau_i} \subset S_\tau$. In other words, S_τ depends continuously on τ. The argument showing this is, again, virtually identical to the one given in Example 5.2, with Theorem 5.7 providing a graphically convergent subsequence of the sequence $\{z_i\}_{i=1}^\infty$ and (a) of Theorem 5.29 providing the needed solution.

5.5 NOTES

The presentation of the set-valued analysis concepts, and the terminology, follows the book [100] by Rockafellar and Wets. Another reference for this topic is the book [7] by Aubin and Frankowska. References for set-valued analysis in the setting of dynamical systems, in particular differential inclusions, include books by Aubin [4] and Aubin and Cellina [6]. Further references for applications of set-valued analysis include Clarke [26], which focuses on optimal control theory and Clarke et al. [28], where general control and stabilization problems are covered as well.

Definition 5.1 follows [100, Definition 4.1]. Theorem 5.4 is [100, Theorem 4.10]. Theorem 5.7 is [100, Theorem 4.18]. Definition 5.8 follows the early definitions of [100, Chapter 5]. Definition 5.9 follows [100, Definition 5.4]. A global version of Lemma 5.10 is contained in [100, Theorem 5.7]. Definition 5.12 follows [100, Definition 6.1]. Definition 5.14 follows [100, Definition 5.14]. Lemma 5.15 is included in [100, Proposition 5.12]. Definition 5.18 is a part of [100, Definition 5.32]. Statement (a), respectively (b), in Lemma 5.26 is [4, Proposition 3.4.1], respectively [4, Proposition 3.4.2]. For Lemma 5.27, see [26, Theorem 3.1.7]. While the perturbation used in [26, Theorem 3.1.7] differs from that in Lemma 5.27, the arguments in the proof of [26, Theorem 3.1.7] do apply, with $h(t, s, p) = \max \{p \cdot f \ : \ f \in \mathrm{con} F\left((s + \delta_i \rho(s)\mathbb{B}) \cap C\right)\}$. Furthermore, [26, Theorem 3.1.7] does not assume uniform convergence but does conclude the existence of a uniformly convergent subsequence. In this book, the extraction of convergent subsequences is handled via Theorem 5.7.

Chapter Six

Well-posed hybrid systems and their properties

In a classical setting, for example in differential equations or in optimization, a well-posed problem is often defined as one in which a solution exists, is unique, and depends continuously on parameters. For hybrid dynamical systems, insisting on uniqueness of solutions and on their continuous dependence on initial conditions (and possibly, on perturbations) is very restrictive and, as it turns out, not necessary to develop a reasonable stability theory. In fact, stability theory results such as converse smooth Lyapunov results and invariance principles are possible for a quite general class of hybrid systems. In what follows, *nominally well-posed hybrid systems* and *well-posed hybrid systems* are defined to be those hybrid systems, vaguely speaking, for which graphical limits of graphically convergent sequences of solutions, with no perturbations and with vanishing perturbations, respectively, are still solutions. The class of well-posed hybrid systems includes the Krasovskii regularization of a general hybrid system and, more generally, it includes every hybrid system meeting some mild regularity assumptions (motivated in part by the developments of Chapter 4) on the data.

6.1 NOMINALLY WELL-POSED HYBRID SYSTEMS

From essentially every sequence of solutions to a hybrid system, a graphically convergent sequence can be extracted. More precisely, Theorem 5.7, restated for hybrid arcs, says the following:

Theorem 6.1. (Graphically convergent subsequences of hybrid arcs) *For every sequence $\{\phi_i\}_{i=1}^{\infty}$ of hybrid arcs $\phi_i : \operatorname{dom} \phi_i \to \mathbb{R}^n$ for which there exists a compact set $K \subset \mathbb{R}^n$ such that $\phi_i(0,0) \in K$ for all $i \in \mathbb{N}$ there exists a subsequence $\{\phi_{i_k}\}_{k=1}^{\infty}$ which converges graphically to some mapping $M : \mathbb{R}^2 \rightrightarrows \mathbb{R}^n$ with $(0,0) \in \operatorname{dom} M$.*

In general, the graphical limit of a graphically convergent sequence of solutions to a hybrid system may fail to be a solution to that system.

The general idea behind a hybrid system being nominally well-posed is that the limit of a graphically convergent sequence of solutions to $\mathcal{H}$ having a mild boundedness property should also be a solution to $\mathcal{H}$, while a graphically convergent sequence of solutions to $\mathcal{H}$ without that boundedness property should lead to a solution to $\mathcal{H}$ that blows up in finite time.

Definition 6.2. (Nominally well-posed hybrid system) *A hybrid system $\mathcal{H}$ is called* nominally well-posed *if the following property holds: for every graph-*

ically convergent sequence $\{\phi_i\}_{i=1}^{\infty}$ *of solutions to* $\mathcal{H}$ *with* $\lim_{i\to\infty}\phi_i(0,0) = \xi$ *for some* $\xi \in \mathbb{R}^n$,

(a) *if the sequence* $\{\phi_i\}_{i=1}^{\infty}$ *is locally eventually bounded then the sequence* $\{\text{length}(\phi_i)\}_{i=1}^{\infty}$ *is either convergent or properly divergent to* ∞ *and*

$$\phi = \underset{i\to\infty}{\text{gph-lim}}\,\phi_i$$

is a solution to $\mathcal{H}$ *with* $\phi(0,0) = \xi$ *and* $\text{length}(\phi) = \lim_{i\to\infty}\text{length}(\phi_i)$;

(b) *if the sequence* $\{\phi_i\}_{i=1}^{\infty}$ *is not locally eventually bounded then there exists a number* $m \in (0,\infty)$ *for which there exist* $(t_i, j_i) \in \text{dom}\,\phi_i$, $i = 1, 2, \ldots$, *such that* $\lim_{i\to\infty}|\phi_i(t_i, j_i)| = \infty$ *and*

$$\phi = \left(\underset{i\to\infty}{\text{gph-lim}}\,\phi_i\right)\Big|_{t+j<m}$$

is a maximal solution to $\mathcal{H}$ *with* $\text{length}(\phi) = m$ *and*

$$\lim_{t\to\sup_t \text{dom}\,\phi}|\phi(t, \sup_j \text{dom}\,\phi)| = \infty.$$

Illustrating the two cases in Definition 6.2 is possible in the setting of differential equations where the right-hand side is not globally Lipschitz continuous; this is done in Example 6.3. A general result, giving sufficient conditions for a hybrid system to be nominally well-posed and subsuming the case of differential equations with continuous right-hand sides, will be given in Section 6.2 as Theorem 6.8.

Example 6.3. (Nominally well-posed differential equation) On $\mathbb{R}$, consider the differential equation $\dot{z} = z^2$. For the initial condition $z(0) = 0$, the unique maximal solution is $z(t) = 0$ for $t \in [0,\infty)$. For an initial condition $z(0) > 0$, the unique maximal solution is $z(t) = -(t - 1/z(0))^{-1}$ for $t \in [0, 1/z(0))$; this solution blows up when $t \to 1/z(0)$: $\lim_{t\nearrow 1/z(0)} z(t) = \infty$.

Consider a sequence $\{z_i\}_{i=1}^{\infty}$ with $z_i(0) > 0$ and $z_i(0) \searrow 0$. Each element of the sequence is unbounded, but the sequence is locally eventually bounded (with respect to $\mathbb{R}$). Indeed, for every $T > 0$ there exists i_0 such that for every $i > i_0$, $T + 1 < 1/z_i(0)$. It follows that $z_i(t) < 1$ for every $i > i_0$, and every $t \in [0, T)$. A very similar argument shows that the sequence $\{z_i\}_{i=1}^{\infty}$ converges uniformly on compact subsets of $[0,\infty)$ (but not uniformly on $[0,\infty)$) to the function z identically equal to 0 on $[0,\infty)$. The sequence also converges graphically to z. This z is a solution to the differential equation on $[0,\infty)$. Note furthermore that the length of the domain of z, which is equal to ∞, is the limit of the lengths of domains of z_i, each of which is equal to $1/z_i(0)$. This illustrates case (a) of Definition 6.2.

Now consider a sequence $\{z_i\}_{i=1}^{\infty}$ with $z_i(0) = 1 + a_i$, $a_i > -1$, and $a_i \to 0$ as $i \to \infty$. This sequence is not locally eventually bounded. Graphically, $\{z_i\}_{i=1}^{\infty}$

converges to z given by $z(t) = -(t-1)^{-1}$ for $t \in [0,1)$, which is a solution to the differential equation. For every i, let $t_i = (1-a_i)/(1+a_i)$ to get $z_i(t_i) = 1 + 1/a_i \to \infty$ as $i \to \infty$. Furthermore, $(1-a_i)/(1+a_i) \to 1$ as $i \to \infty$, which agrees with the length of z being equal to 1. This illustrates case (b) of Definition 6.2. In this illustration it was not necessary to truncate the graphical limit of $\{z_i\}_{i=1}^{\infty}$ to obtain a solution.

Numerous examples of hybrid systems that fail to be nominally well-posed can be given by using data that does not meet the technical assumptions posed in the next section. A very simple example is provided by a system with F identically equal to 0, C being a set that is not closed, and $D = \emptyset$. Then, for every point $z^0 \in \partial C \setminus C$, every sequence $\{z_i\}_{i=1}^{\infty}$ of constant and complete solutions with $z_i(0) \to z^0$ as $i \to \infty$ is locally eventually bounded (in fact, globally uniformly bounded) and graphically (and uniformly) convergent, but the limit is not a solution. This violates (a) in Definition 6.2. A more interesting example, addressing (b), is given below.

Example 6.4. (Hybrid system not nominally well-posed) Consider a hybrid system $\mathcal{H}$ in $\mathbb{R}^3$ given by

$$C = \{x \in \mathbb{R}^3 : x_1 \geq 1,\ x_2 \in (0, 1/x_1],\ x_3 \in \{-1, 1\}\} \qquad f(x) = \begin{pmatrix} x_1^2 x_3 \\ 0 \\ 0 \end{pmatrix}$$

$$D = \{x \in \mathbb{R}^3 : x_1 \geq 1,\ x_2 = 1/x_1,\ x_3 = 1\} \qquad g(x) = \begin{pmatrix} x_1 \\ x_2 \\ -1 \end{pmatrix}.$$

For $\varepsilon \in (0,1)$, the unique maximal solution from the initial point $(1, \varepsilon, 1)$ is given by

$$x_\varepsilon(t, 0) = \begin{pmatrix} (1-t)^{-1} \\ \varepsilon \\ 1 \end{pmatrix} \quad \text{for } t \in [0, 1 - \varepsilon],$$

$$x_\varepsilon(t, 1) = \begin{pmatrix} (2\varepsilon - 1 + t)^{-1} \\ \varepsilon \\ -1 \end{pmatrix} \quad \text{for } t \in [1 - \varepsilon, \infty).$$

Every sequence of solutions $\{\varepsilon_i\}_{i=1}^{\infty}$ with $\varepsilon_i \in (0,1)$, $\varepsilon_i \searrow 0$ when $i \to \infty$, is convergent graphically. The limit is a function given by

$$x(t, 0) = \begin{pmatrix} (1-t)^{-1} \\ 0 \\ 1 \end{pmatrix} \quad \text{for } t \in [0, 1),$$

$$x(t,1) = \begin{pmatrix} (-1+t)^{-1} \\ 0 \\ -1 \end{pmatrix} \text{ for } t \in (1,\infty).$$

The domain of x, $([0,1) \times \{0\}) \cup ((1,\infty) \times \{1\})$, is not a hybrid time domain, x is not a hybrid arc, and – of course – x is not a solution to $\mathcal{H}$. This does not show yet that $\mathcal{H}$ is not nominally well-posed, because the sequence $\{x_{\varepsilon_i}\}_{i=1}^{\infty}$ is not locally eventually bounded and the conclusions of (a) in Definition 6.2 do not need to be met. However, for $\mathcal{H}$ to be nominally well-posed, (b) of Definition 6.2 requires that the truncation of x to $[0,1) \times \{0\}$ be a solution to $\mathcal{H}$. This fails because all values of x on that set are outside of C. Thus $\mathcal{H}$ is not nominally well-posed.

Note though that if C was replaced by its closure, the truncation of x just discussed would be a solution to the resulting hybrid system. In fact, the system with the closure of C would be nominally well-posed, as the general results of the next section will imply.

Figure 6.1 depicts sequences of solutions that converge to different types of solutions. Figure 6.1(b) shows a sequence of Zeno solutions converging to an eventually discrete solution, and Figure 6.1(c) depicts a nonlocally bounded sequence of solutions converging to an unbounded solution.

6.2 BASIC ASSUMPTIONS ON THE DATA

Verifying whether a hybrid system is nominally well-posed directly through the definition can be a cumbersome process. Conveniently, a set of simple conditions on the data of the system turns out to be sufficient for the system to be nominally well-posed and, in fact, well-posed.

Assumption 6.5. (Hybrid basic conditions)

(A1) C and D are closed subsets of $\mathbb{R}^n$;

(A2) $F : \mathbb{R}^n \rightrightarrows \mathbb{R}^n$ is outer semicontinuous and locally bounded relative to C, $C \subset \operatorname{dom} F$, and $F(x)$ is convex for every $x \in C$;

(A3) $G : \mathbb{R}^n \rightrightarrows \mathbb{R}^n$ is outer semicontinuous and locally bounded relative to D, and $D \subset \operatorname{dom} G$.

When $f : \mathbb{R}^n \to \mathbb{R}^n$ is a continuous function, the differential equation $\dot{z} = f(z)$ can be identified with a hybrid system satisfying hybrid basic conditions. A similar comment applies to a difference equation given by a continuous function. Constraining solutions to such a differential or difference equation to a closed set still leads to a system satisfying hybrid basic conditions. A broad class of (truly hybrid) systems that satisfy the hybrid basic conditions comes from Krasovskii regularization of a hybrid system which meets only a mild local boundedness condition.

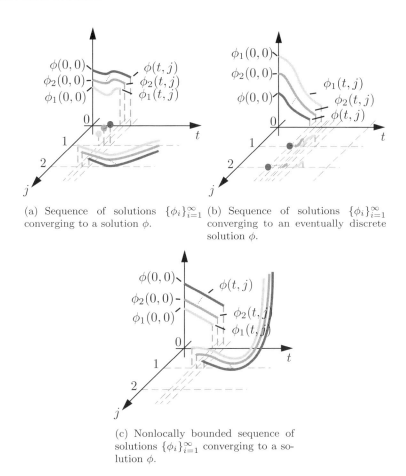

(a) Sequence of solutions $\{\phi_i\}_{i=1}^{\infty}$ converging to a solution ϕ.

(b) Sequence of solutions $\{\phi_i\}_{i=1}^{\infty}$ converging to an eventually discrete solution ϕ.

(c) Nonlocally bounded sequence of solutions $\{\phi_i\}_{i=1}^{\infty}$ converging to a solution ϕ.

Figure 6.1: Graphical convergence of sequence of solutions.

Example 6.6. (Regularization and hybrid basic conditions) Consider the hybrid system $\mathcal{H} = (C, F, D, G)$ with F and G locally bounded, but for which Assumption 6.5 may not be satisfied. In fact, it is sufficient that F and G be locally bounded relative to $\overline{C}$ and $\overline{D}$, respectively. Let the hybrid system $\widehat{\mathcal{H}}$ be the "Krasovskii regularization" of $\mathcal{H}$ as in Definition 4.13. Then $\widehat{\mathcal{H}}$ satisfies Assumption 6.5. The flow set $\widehat{C}$ and the jump set $\widehat{D}$ for $\widehat{\mathcal{H}}$ are relatively closed by their definitions; the flow map $\widehat{F}$ and the jump map $\widehat{G}$ for $\widehat{\mathcal{H}}$ have the needed properties thanks to Lemma 5.16. Indeed, $\widehat{F}$ is given by M_2 from Lemma 5.16 with M given by $M(x) = F(x)$ if $x \in C$, $M(x) = \emptyset$ if $x \notin C$. The same arguments apply to $\widehat{G}$.

Example 6.7. (Hybrid automata and hybrid basic conditions) Section 1.4 showed how a hybrid automaton can be modeled as a hybrid system of the form (1.1). The following assumptions, on the data of a hybrid automaton, ensure that the resulting hybrid system (1.1) satisfies Assumption 6.5. For every $q \in Q = \{1, 2, \ldots, q_{max}\}$,

- the set $\mathrm{Domain}(q)$ is closed;

- the flow map $z \mapsto f(q, z)$ is continuous on $\mathrm{Domain}(q)$;

while for every edge $(q, q') \in \mathrm{Edges}$,

- the guard set $\mathrm{Guard}(q, q')$ is closed;

- the reset map $z \mapsto \mathrm{Reset}(q, q', z)$ is continuous on $\mathrm{Guard}(q, q')$.

This can be justified directly from the constructions (1.10) and (1.12), by using the simple fact that the union of finitely many closed sets is closed and by using the closed graph property of outer semicontinous mappings (in particular, of continuous functions), as stated in Lemma 5.10.

The key consequence of the hybrid basic conditions is that they guarantee that the hybrid system satisfying them is nominally well-posed, as the following result states. The proof of the result is not included, as it is a special (but very important) case of Theorem 6.30.

Theorem 6.8. (Basic conditions and nominal well-posedness) *If the hybrid system $\mathcal{H} = (C, F, D, G)$ satisfies Assumption 6.5 then it is nominally well-posed.*

Nominally well-posed hybrid systems already appeared in previous chapters. These include the hybrid systems in Examples 4.14 and 4.15 resulting from a Krasovskii regularization, which, as argued in Example 6.6, have data satisfying Assumption 6.5. In particular, for the bouncing ball system in Example 4.14 with data $(\hat{C}, \hat{F}, \hat{D}, \hat{G})$, a sequence of solutions ϕ_i with initial conditions $\xi_i \searrow 0$, converges to the solution ϕ given by $\phi(t, j) = 0$ for all $(t, j) \in \mathrm{dom}\,\phi = \{0\} \times \mathbb{N}$; see Figure 6.2. The hybrid model for zero-cross detection in Example 4.18 is also nominally well-posed by construction.

Some, but not all, of the hybrid basic conditions are also necessary for a hybrid system to be nominally well-posed. It can be shown that a nominally well-posed hybrid system has a closed jump set and its jump map satisfies (A3).

Lemma 6.9. (A necessary condition for nominal well-posedness) *Let the hybrid system $\mathcal{H} = (C, F, D, G)$ be nominally well-posed. Then D is closed and G satisfies (A3) of the hybrid basic conditions.*

PROOF. Consider a convergent sequence $\{x_i\}_{i=1}^{\infty}$ of points $x_i \in D$ and any sequence $\{y_i\}_{i=1}^{\infty}$ with $y_i \in G(x_i)$. Suppose that $\lim_{i \to \infty} |y_i| = \infty$. Then the sequence $\{\phi_i\}_{i=1}^{\infty}$ of solutions to $\mathcal{H}$ with domains $(0, 0) \cup (0, 1)$ and given by $\phi_i(0, 0) = x_i$, $\phi_i(0, 1) = y_i$ is graphically convergent and not locally eventually bounded. By (b) of Definition 6.2, there exists a solution ϕ to $\mathcal{H}$ with length $\phi = 2$ which blows up, which is impossible. Hence G is locally bounded relative to

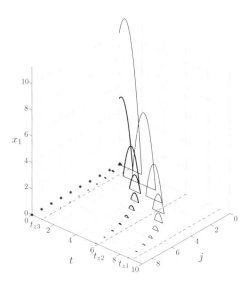

Figure 6.2: The height variable of a sequence of solutions to the bouncing ball system in Example 4.14 with data $(\hat{C}, \hat{F}, \hat{D}, \hat{G})$ converging to the always jumping solution at the origin.

not just D but in fact $\overline{D}$. Since $\lim_{i\to\infty} |y_i|$ is not ∞, and after passing to a subsequence if necessary, it can be assumed that $\lim_{i\to\infty} y_i$ exists. Then the sequence $\{\phi_i\}_{i=1}^{\infty}$, as defined before, is graphically convergent and bounded, in particular locally eventually bounded. By (a) of Definition 6.2, the hybrid arc ϕ given by $\phi(0,0) = \lim_{i\to\infty} x_i$, $\phi(0,1) = \lim_{i\to\infty} y_i$ is a solution to $\mathcal{H}$. This implies that $\lim_{i\to\infty} x_i \in D$ and so D is closed. $\square$

Similar conclusions cannot be made about C and F. For example, a system on $\mathbb{R}$ with $C = \mathbb{R}$, $F(x) = 1$ if $x < 0$, $F(x) = 2$ if $x \geq 0$, and empty D is nominally well-posed, as can be checked by inspection. Furthermore, a system on $\mathbb{R}$ with $C = (-\infty, 0)$, $F(x) = 1$ for all $x \in C$, $D = [0, \infty)$, and any G satisfying (A3) is nominally well-posed. To see that the two mentioned systems are nominally well-posed, even when F is not outer semicontinuous in the first system and C is not closed in the second system, it is enough to observe that solutions to these two systems are the same as the Krasovskii solutions to these two systems.

6.2.1 Definition and existence of solutions, revisited

Under Assumption 6.5, the definition of a solution to a hybrid system, Definition 2.6, simplifies somewhat. The requirement in Definition 2.6 that $\phi(t, j) \in C$ for all $t \in \operatorname{int} I$, where I is an interval with nonempty interior, is equivalent to $\phi(t, j) \in C$ for all $t \in I$ if C is closed. In turn, $\phi(t, j) \in C$ for all $t \in I$ is equivalent to $\phi(t, j) \in C$ for almost all $t \in I$, by continuity of $\phi(\cdot, j)$. Thus, Definition 2.6 can be restated as follows: a hybrid arc ϕ is a *solution to the hybrid system* $\mathcal{H}$ if $\phi(0, 0) \in C \cup D$ and

(S1*) for every $j \in \mathbb{N}$,

$$\begin{aligned} \phi(t, j) &\in C, \\ \dot{\phi}(t, j) &\in F(\phi(t, j)), \end{aligned} \qquad \text{for almost all } t \in I^j; \qquad (6.1)$$

(S2*) for every $(t, j) \in \operatorname{dom} \phi$ such that $(t, j + 1) \in \operatorname{dom} \phi$,

$$\begin{aligned} \phi(t, j) &\in D, \\ \phi(t, j + 1) &\in G(\phi(t, j)). \end{aligned} \qquad (6.2)$$

Of course, in (S1*), one can only place the condition on those $j \in \mathbb{N}$ for which I^j has nonempty interior.

Based on Lemma 5.26, the basic existence of solutions result, Proposition 2.10, can be stated with assumptions involving the data of $\mathcal{H}$.

Proposition 6.10. (Basic existence of solutions revisited) *Let* $\mathcal{H} = (C, F, D, G)$ *satisfy Assumption 6.5. Take an arbitrary* $\xi \in C \cup D$. *If* $\xi \in D$ *or*

(VC) *there exists a neighborhood* U *of* ξ *such that for every* $x \in U \cap C$,

$$F(x) \cap T_C(x) \neq \emptyset,$$

then there exists a nontrivial solution ϕ *to* $\mathcal{H}$ *with* $\phi(0, 0) = \xi$. *If (VC) holds for every* $\xi \in C \setminus D$, *then there exists a nontrivial solution to* $\mathcal{H}$ *from every initial point in* $C \cup D$, *and every* $\phi \in \mathcal{S}_{\mathcal{H}}$ *satisfies exactly one of the following conditions:*

(a) ϕ *is complete;*

(b) $\operatorname{dom} \phi$ *is bounded and the interval* I^J, *where* $J = \sup_j \operatorname{dom} \phi$, *has nonempty interior and* $t \mapsto \phi(t, J)$ *is a maximal solution to* $\dot{z} \in F(z)$, *in fact* $\lim_{t \to T} |\phi(t, J)| = \infty$, *where* $T = \sup_t \operatorname{dom} \phi$;

(c) $\phi(T, J) \notin C \cup D$, *where* $(T, J) = \sup \operatorname{dom} \phi$.

Furthermore, if $G(D) \subset C \cup D$, *then (c) above does not occur.*

Example 6.11. (Bouncing ball — existence and maximal solutions) Consider the Krasovskii regularization of the bouncing ball system, given in Example 4.14. The hybrid basic conditions are satisfied. To verify sufficient conditions for the existence of nontrivial solutions from an initial point in $\widehat{C} \cup \widehat{D}$, it is enough to show that $\widehat{F}(x) \in T_{\widehat{C}}(x)$ for every $x \in \widehat{C} \setminus \widehat{D}$. For $x \in \widehat{C}$ such that $x_1 > 0$, $T_{\widehat{C}}(x) = \mathbb{R}^2$. Consequently, for $x \in \widehat{C} \setminus \widehat{D}$ with $x_1 > 0$, $\widehat{F}(x) \in T_{\widehat{C}}(x)$ trivially holds. For $x \in \widehat{C}$ with $x_1 = 0$, $T_{\widehat{C}}(x) = \mathbb{R}_{\geq 0} \times \mathbb{R}$, that is, the tangent cone is the right-half plane. For $x \in \widehat{C} \setminus \widehat{D}$ with $x_1 = 0$ one also has $x_2 > 0$, and consequently $\widehat{F}(x) \in T_{\widehat{C}}(x)$ holds. In summary, (VC) holds at each point $x \in \widehat{C} \setminus \widehat{D}$, and nontrivial solutions to the hybrid system exist from each point $\xi \in \widehat{C} \cup \widehat{D}$. Note though that $\widehat{F}(\xi) \notin T_{\widehat{C}}(\xi)$ for $\xi \in \widehat{C} \cap \widehat{D}$.

Some additional arguments are needed to show that every maximal solution is complete. Because $\widehat{G}(\widehat{D}) \subset \widehat{C} \cup \widehat{D}$, case (c) above does not occur. Case (b) can also be excluded via additional arguments which show that every solution is bounded.

6.3 CONSEQUENCES OF NOMINAL WELL-POSEDNESS

Several properties of nominally well-posed hybrid systems are now established. Local boundedness of sets of solutions and the dependence of the sets of solutions on initial conditions is addressed first, reachable sets are discussed later, and the section concludes with some invariance properties of ω-limit sets. A detailed analysis of asymptotic stability for nominally well-posed hybrid systems is in Chapter 7. Application of the invariance properties established here to stability analysis is in Chapter 8.

6.3.1 Structure of sets of solutions

For a nominally well-posed hybrid system, the property of pre-forward completeness, which is defined below and essentially means that solutions do not blow up in finite time, ensures a boundedness property for sets of solutions.

Definition 6.12. (Pre-forward completeness) *Given a set $S \subset \mathbb{R}^n$, a hybrid system $\mathcal{H}$ on $\mathbb{R}^n$ is pre-forward complete from S if every $\phi \in \mathcal{S}_{\mathcal{H}}(S)$ is either bounded or complete.*

Proposition 6.13. (Local boundedness of sets of solutions) *Let $\mathcal{H}$ be nominally well-posed. Suppose that $\mathcal{H}$ is pre-forward complete from a compact set $K \subset \mathbb{R}^n$. Then, for every $m > 0$, there exists $\delta > 0$ and a compact set $K' \subset \mathbb{R}^n$ such that, for every $\phi \in \mathcal{S}_{\mathcal{H}}(K + \delta \mathbb{B})$ and every $(t, j) \in \operatorname{dom} \phi$ with $t + j < m$, $\phi(t, j) \in K'$.*

PROOF. Contradicting the conclusion leads to a sequence $\{\phi_i\}_{i=1}^{\infty}$ of solutions to $\mathcal{H}$ that is not locally eventually bounded while $\phi_i(0, 0) \to K$ when $i \to \infty$. Extracting a graphically convergent subsequence leads, thanks to (b) in

Definition 6.2, to a maximal solution to $\mathcal{H}$ that blows up in finite hybrid time. This contradicts that K is pre-forward complete. ☐

This proposition implies in particular that, for K as in the assumption, every sequence $\{\phi_i\}_{i=1}^{\infty}$ of solutions $\phi_i \in \mathcal{S}_{\mathcal{H}}$ with $\lim_{i \to \infty} \phi_i(0,0) \in K$ is locally eventually bounded. Furthermore, since $K \subset K + \delta \mathbb{B}$ for every $\delta > 0$, it implies that the set $\mathcal{S}_{\mathcal{H}}(K)$ is locally uniformly bounded in the following sense: for every $m > 0$ there exists a compact set $K' \subset \mathbb{R}^n$ such that $\phi(t,j) \in K'$ for every $\phi \in \mathcal{S}_{\mathcal{H}}(K)$ and every $(t,j) \in \operatorname{dom} \phi$ with $t + j < m$.

The next result describes the outer/upper semicontinuous dependence of solutions on initial points, in terms of (τ, ε)-closeness. That is, some uniformity in the distance between solutions from nearby initial points can be expected, under a pre-forward completeness condition.

Proposition 6.14. (Dependence on initial conditions) *Let $\mathcal{H}$ be nominally well-posed. Suppose that $\mathcal{H}$ is pre-forward complete from a compact set $K \subset \mathbb{R}^n$. Then for every $\varepsilon > 0$ and $\tau \geq 0$ there exists $\delta > 0$ with the following property: for every solution $\phi_\delta \in \mathcal{S}_{\mathcal{H}}(K + \delta \mathbb{B})$ there exists a solution ϕ to $\mathcal{H}$ with $\phi(0,0) \in K$ such that ϕ_δ and ϕ are (τ, ε)-close.*

PROOF. Suppose otherwise, that for some $\varepsilon > 0$ and $\tau \geq 0$ there exist, for each $i \in \mathbb{N}$, $\phi_i \in \mathcal{S}_{\mathcal{H}}(K + 1/i\mathbb{B})$ such that no solution to $\mathcal{H}$ from K is (τ, ε)-close to ϕ_i. The sequence $\{\phi_i\}_{i=1}^{\infty}$ is locally eventually bounded by Proposition 6.13. By Theorem 6.1, without loss of generality one can assume that the sequence is graphically convergent. Because $\mathcal{H}$ is nominally well-posed, the graphical limit of the sequence, say ϕ, is a solution to $\mathcal{H}$ with $\phi(0,0) \in K$. Now, by Theorem 5.25, ϕ_i are (τ, ε)-close to ϕ for all large enough i's, which is a contradiction. ☐

Figure 6.3 shows the dependence on initial conditions property stated in Proposition 6.14.

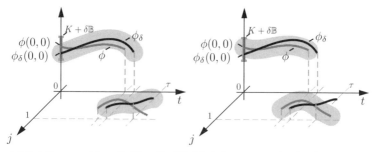

(a) Solution x_δ with $x_\delta(0,0) \in K + \delta \mathbb{B}$ and ε-neighborhood around it.

(b) Solution x with $x(0,0) \in K + \delta \mathbb{B}$ and ε-neighborhood around it showing (τ, ε)-closeness with x_δ.

Figure 6.3: Dependence on initial conditions.

6.3.2   Reachable sets

Definition 6.15. (Reachable sets) *Given an arbitrary $S \subset \mathbb{R}^n$ and $\tau \in \mathbb{R}_{\geq 0}$, the reachable set from S in hybrid time less or equal to τ is the set*

$$\mathcal{R}_{\leq \tau}(S) = \{\phi(t,j) \,:\, \phi \text{ is a solution to } \mathcal{H}, \ \phi(0,0) \in S, \ t + j \leq \tau\}.$$

Sets $\mathcal{R}_{<\tau}(S)$, $\mathcal{R}_{\geq \tau}(S)$, etc. should be understood analogously. The reachable set from S in infinite hybrid time is the set

$$\mathcal{R}(S) = \bigcup_{\tau \in \mathbb{R}_{\geq 0}} \mathcal{R}_{\tau}(S).$$

The properties shown below, along with other results in Chapter 7 which rely on additional asymptotic stability properties of some compact set, are useful when proving robust stability of compact sets for hybrid systems. Moreover, reachable sets are also useful in verification and validation of hybrid systems, where reachable sets are approximated using numerical algorithms.

Lemma 6.16. (Reachable set properties) *Let $\mathcal{H}$ be nominally well-posed.*

(a) *For any two compact sets $K, K' \subset \mathbb{R}^n$ with $K \subset \operatorname{int} K'$ there exists $\tau \in (0,1)$ such that*

$$\mathcal{R}_{\leq \tau}(K) \subset K'.$$

(b) *If $\mathcal{H}$ is pre-forward complete from a compact $K \subset \mathbb{R}^n$, then the reachable set $\mathcal{R}_{\leq \tau}(K)$ is compact for every $\tau \in \mathbb{R}_{\geq 0}$.*

PROOF. If the conclusion of (a) was to fail, there would exist a compact $K \subset \mathbb{R}^n$, $\varepsilon > 0$, a sequence $\{\tau'_i\}_{i=1}^{\infty}$ of numbers in $(0,1)$ converging to 0 and a sequence $\{\phi'_i\}_{i=1}^{\infty}$ of solutions $\phi'_i \in \mathcal{S}_{\mathcal{H}}(K)$ with $(\tau'_i, 0) \in \operatorname{dom} \phi_i$ and $|\phi'_i(\tau'_i, 0)|_K > \varepsilon$. For $i = 1, 2, \dots$, let $\tau_i \in (0, \tau'_i)$ be the minimum of all $t > 0$ such that $|\phi'_i(t, 0)|_K = \varepsilon$, and let ϕ_i be the truncation of ϕ'_i to $\operatorname{dom} \phi_i := [0, \tau_i] \times \{0\}$. Then $\{\phi_i\}_{i=1}^{\infty}$ is a locally eventually bounded sequence of solutions to $\mathcal{H}$; without loss of generality it can be assumed that it is graphically convergent and that, furthermore, the sequences of points $\phi_i(0,0)$ and $\phi_i(\tau_i, 0)$ both converge. But then the graphical limit, say ϕ, is such that $\lim_{i \to \infty} \phi_i(0,0) \in \phi(0,0)$, $\lim_{i \to \infty} \phi_i(\tau_i, 0) \in \phi(0,0)$ where $\lim_{i \to \infty} \phi_i(0,0) \in K$ while $|\lim_{i \to \infty} \phi_i(\tau_i, 0)|_K = \varepsilon$. This is only possible if ϕ is set-valued, and this contradicts that ϕ be a solution to $\mathcal{H}$, as required by $\mathcal{H}$ being nominally well-posed.

The statement (b) is immediate from $\mathcal{H}$ being nominally well-posed and Proposition 6.13.   □

6.3.3 ω-limit sets

Definition 6.17. (ω-limit set of a hybrid arc) *The ω-limit set of a hybrid arc $\phi : \operatorname{dom} \phi \to \mathbb{R}^n$, denoted $\Omega(\phi)$, is the set of all points $x \in \mathbb{R}^n$ for which there exists a sequence $\{(t,j)_i\}_{i=1}^{\infty}$ of points $(t_i, j_i) \in \operatorname{dom} \phi$ with $\lim_{i \to \infty} t_i + j_i = \infty$ and $\lim_{i \to \infty} \phi(t_i, j_i) = x$. Every such point x is an ω-limit point of ϕ.*

In other words, the ω-limit set of a hybrid arc ϕ consists of all accumulation points of ϕ. The following example illustrates this concept.

Example 6.18. (ω-limit sets) The ω-limit set for each of the following hybrid arcs is computed.

1. Given the hybrid arc ϕ

$$\phi(t, j) = \begin{cases} t & \text{if } (t, j) \in [0, 1] \times \{0\}, \\ \exp(-t) & \text{if } (t, j) \in [1, \infty) \times \{1\} , \end{cases}$$

for every unbounded and increasing sequence $(t_i, j_i) \in \operatorname{dom} \phi$, $\lim_{i \to \infty} \phi(t_i, j_i) = 0$. Then, $\Omega(\phi) := \{0\}$.

2. For the hybrid arc ϕ

$$\phi(t, j) = t \qquad \forall (t, j) \in [0, 1] \times \{0\} ,$$

$\Omega(\phi) := \emptyset$.

3. Given the hybrid arc ϕ

$$\phi(t, j) = t - j \qquad (t, j) \in [j, j + 1] \times \{j\} , \ j \in \mathbb{N},$$

for each $x \in [0, 1]$ there exists an unbounded and increasing sequence $(t_i, j_i) \in \operatorname{dom} \phi$ such that $\lim_{i \to \infty} \phi(t_i, j_i) = x$. Thus $\Omega(\phi) := [0, 1]$.

Directly from the definition, it can be argued that for every hybrid arc ϕ, its ω-limit set $\Omega(\phi)$ is closed, that if ϕ is complete then $\Omega(\phi)$ is nonempty, unless $|\phi(t, j)| \to \infty$ when $t + j \to \infty$, and finally, if ϕ is precompact, that is, is complete and bounded, then $\Omega(\phi)$ is nonempty and compact. These conclusions also come from general properties of limits of sequences of sets, thanks to the following:

$$\Omega(\phi) = \lim_{i \to \infty} \{\phi(t, j) : t + j \geq \tau_i\} \tag{6.3}$$

for every sequence $\{\tau_i\}_{i=1}^{\infty}$ such that $\lim_{i \to \infty} \tau_i = \infty$.

Definition 6.19. (Weak invariance) *Given a hybrid system $\mathcal{H}$, a set $S \subset \mathbb{R}^n$ is said to be*

- *weakly forward invariant if for every $\xi \in S$ there exists at least one complete $\phi \in \mathcal{S}_{\mathcal{H}}(\xi)$ with $\operatorname{rge} \phi \subset S$;*

- *weakly backward invariant if for every $\xi \in S$, every $\tau > 0$, there exists at least one $\phi \in \mathcal{S}_{\mathcal{H}}(S)$ such that for some $(t^*, j^*) \in \operatorname{dom} \phi$, $t^* + j^* \geq \tau$, it is the case that $\phi(t^*, j^*) = \xi$ and $\phi(t, j) \in S$ for all $(t, j) \in \operatorname{dom} \phi$ with $t + j \leq t^* + j^*$;*

- *weakly invariant if it is both weakly forward invariant and weakly backward invariant.*

Note that weak invariance of S amounts to the following: for every $\xi \in S$, every $\tau > 0$, there exists at least one complete $\phi \in \mathcal{S}_{\mathcal{H}}(S)$ with $\mathrm{rge}\,\phi \subset S$ such that for some $(t^*, j^*) \in \mathrm{dom}\,\phi$ with $t^* + j^* \geq \tau$, it is the case that $\phi(t^*, j^*) = \xi$. Further discussion of weak invariance properties is in Section 8.2.

Example 6.20. (Invariance of a set) Consider a hybrid system $\mathcal{H}$ in $\mathbb{R}^2$ given by

$$C = \left\{x \in \mathbb{R}^2 \,:\, |x| = 1, x_2 \geq 0\right\} \qquad f(x) = \begin{pmatrix} x_2 \\ -x_1 \end{pmatrix}$$

$$D = \left\{x \in \mathbb{R}^2 \,:\, |x| = 1, x_2 \leq 0\right\} \qquad g(x) = \begin{pmatrix} x_2 \\ -x_1 \end{pmatrix}.$$

Clearly, the set

$$S_1 = \left\{x \in \mathbb{R}^2 \,:\, |x| = 1\right\}$$

is weakly forward and weakly backward invariant, and hence, weakly invariant (the invariance properties are in fact "strong" as they hold for every solution to $\mathcal{H}$). The set

$$S_2 = \left(\left\{x \in \mathbb{R}^2 \,:\, |x| = 1\right\} \cap C\right) \cup g(D)$$

is weakly forward invariant, but not weakly backward invariant since solutions cannot reach points in the interior of the third quadrant while staying in S_2. The set

$$S_3 = \left(\left\{x \in \mathbb{R}^2 \,:\, |x| = 1\right\} \cap C\right) \cup \{(0, -1)\}$$

is both weakly forward and backward invariant. Figure 6.4 depicts these cases.

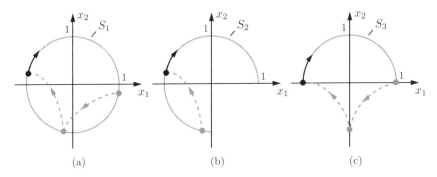

Figure 6.4: Sets in Example 6.20.

Proposition 6.21. (Weak invariance of $\Omega(\phi)$) *Let $\mathcal{H}$ be nominally well-posed. Suppose that $\phi \in \mathcal{S}_{\mathcal{H}}$ is precompact. Then $\Omega(\phi)$ is weakly invariant, both forward and backward weak invariance can be verified by complete solutions, and $|\phi(t, j)|_{\Omega(\phi)} \to 0$ as $t + j \to \infty$, $(t, j) \in \mathrm{dom}\,\phi$.*

PROOF. Pick an arbitrary $\xi \in \Omega(\phi)$, $\tau > 0$, and let $(t_i, j_i) \in \operatorname{dom} \phi$, $i \in \mathbb{N}$, be such that $\phi(t_i, j_i) \to \xi$. For all large enough i, there exist $(t'_i, j'_i) \in \operatorname{dom} \phi$ such that $t_i + j_i - \tau - 1 \le t'_i + j'_i \le t_i + j_i - \tau$. Consider complete $\phi_i \in \mathcal{S}_{\mathcal{H}}$ given by $\phi_i(t, j) = \phi(t + t'_i, j + j'_i)$, $i \in \mathbb{N}$, and note that $\phi_i(t_i - t'_i, j_i - j'_i) \to \xi$. As $\operatorname{rge} \phi_i \subset \operatorname{rge} \phi$, the sequence $\{\phi_i\}_{i=1}^{\infty}$ is locally eventually bounded. Thanks to Theorem 6.1, without loss of generality it can be assumed that it is graphically convergent. Furthermore, it can be assumed that $(t_i - t'_i, j_i - j'_i)$ converge to some (t^*, j^*) with $\tau - 1 \le t^* + j^* \le \tau$. Let $\phi^* = \operatorname{gph-lim}_{i \to \infty} \phi_i$. Because $\mathcal{H}$ is nominally well-posed, $\phi^* \in \mathcal{S}_{\mathcal{H}}$, and in fact ϕ^* is complete. As it was noted in Example 5.19, $\operatorname{rge} \phi \subset \lim_{i \to \infty} \operatorname{rge} \phi_i$, while by (6.3), $\lim_{i \to \infty} \operatorname{rge} \phi_i$ is exactly $\Omega(\phi)$. By the very definition of graphical convergence, $\phi^*(t^*, j^*) = \xi$. Thus ϕ^* verifies weak backward invariance, while the weak forward invariance is verified by a complete $\phi^{\#} \in \mathcal{S}_{\mathcal{H}}(\xi)$ given by $\phi^{\#}(t, j) = \phi^*(t + t^*, j + j^*)$. The conclusion about $|\phi(t, j)|_{\Omega(\phi)} \to 0$ follows from representation (6.3) and Theorem 5.4. $\qquad \square$

The properties of $\Omega(\phi)$ established by Proposition 6.21 permit the generation of invariance principles for hybrid systems $\mathcal{H}$. These will be introduced in Chapter 8.

Corollary 6.22. (Special cases of weak invariance of $\Omega(\phi)$) *Let $\mathcal{H}$ be nominally well-posed. Suppose that $\phi \in \mathcal{S}_{\mathcal{H}}$ is precompact.*

(a) *If $\sup_j \operatorname{dom} \phi = \infty$ and $\sup \{|t - t'| : (t, j), (t', j) \in \operatorname{dom} \phi\} \to 0$ when $j \to \infty$, then both forward and backward weak invariance can be verified by discrete solutions.*

(b) *If ϕ is such that, for some $\tau > 0$ and all but a finite number of $j \in \mathbb{N}$, $\sup\{|t - t'| \,|\, (t, j), (t', j) \in \operatorname{dom} \phi\} \ge \tau$, then, for every $\xi \in \Omega(\phi)$, either weak forward invariance of $\Omega(\phi)$ at ξ can be verified by a solution that flows from ξ or backward weak invariance of $\Omega(\phi)$ can be verified by a solution that flows to ξ.*

The following definition introduces the concept of an ω-limit not just for one solution to a hybrid system, but for all solutions from a given set of initial conditions. The concept is useful when describing the asymptotic behavior of a system given a set of initial conditions.

Definition 6.23. (ω-limit set of a set) *The ω-limit set of a set $S \subset \mathbb{R}^n$, denoted $\Omega(S)$, is the set of all $x \in \mathbb{R}^n$ for which there exists a sequence $\{\phi_i\}_{i=1}^{\infty}$ of solutions $\phi_i \in \mathcal{S}_{\mathcal{H}}(S)$ and a sequence $\{(t, j)_i\}_{i=1}^{\infty}$ of points $(t_i, j_i) \in \operatorname{dom} \phi_i$ such that $\lim_{i \to \infty} t_i + j_i = \infty$ and $\lim_{i \to \infty} \phi_i(t_i, j_i) = x$.*

As it was the case for ω-limit sets of hybrid arcs, it is convenient to think of ω-limit sets of sets in terms of limits of sequences of sets. More precisely,

$$\Omega(S) = \lim_{i \to \infty} \mathcal{R}_{\ge \tau_i}(S) \qquad (6.4)$$

for every sequence $\{\tau_i\}_{i=1}^{\infty}$ such that $\lim_{i \to \infty} \tau_i = \infty$. This immediately shows that $\Omega(S)$ is always closed.

Definition 6.24. (Uniform pre-attractivity) *A compact set $\mathcal{A} \subset \mathbb{R}^n$ is said to be uniformly pre-attractive from a set $S \subset \mathbb{R}^n$ if every $\phi \in \mathcal{S}_\mathcal{H}(S)$ is bounded and for every $\varepsilon > 0$ there exists $\tau > 0$ such that $|\phi(t, j)|_\mathcal{A} \le \varepsilon$ for every $\phi \in \mathcal{S}_\mathcal{H}(S)$ and $(t, j) \in \operatorname{dom}\phi$ with $t + j \ge \tau$.*

Definition 6.25. (Strong forward pre-invariance) *A set $S \subset \mathbb{R}^n$ is said to be strongly forward pre-invariant if for every $\phi \in \mathcal{S}_\mathcal{H}(S)$, $\operatorname{rge}\phi \subset S$.*

Proposition 6.26. (Invariance and attractivity of ω-limit sets of sets) *Let $\mathcal{H}$ be nominally well-posed. Consider an arbitrary $S \subset \mathbb{R}^n$. Then $\Omega(S)$ is closed. If there exists $\tau_0 > 0$ such that $\mathcal{R}_{\ge\tau_0}(S)$ is bounded, then $\Omega(S)$ is compact. If additionally $\Omega(S) \ne \emptyset$, then $\Omega(S)$ is weakly backward invariant and uniformly pre-attractive from S. If additionally $\Omega(S) \subset S$, then $\Omega(S)$ is strongly forward pre-invariant.*

PROOF. It was already said that (6.4) implies that $\Omega(S)$ is closed. Suppose that $\mathcal{R}_{\ge\tau_0}(S) \subset K$, where $K \subset \mathbb{R}^n$ is compact. Then $\Omega(S) \subset K$ and so $\Omega(S)$ is compact.

Suppose $\Omega(S) \ne \emptyset$. To see weakly forward invariance, pick an arbitrary $\xi \in \Omega(S)$ and $m > 0$. There exist $\phi_i \in \mathcal{S}_\mathcal{H}(S)$ and $(t_i, j_i) \in \operatorname{dom}\phi_i$ with $t_i + j_i \to \infty$ such that $\phi_i(t_i, j_i) \to \xi$. Without loss of generality it can be assumed that $t_i + j_i \ge m + 1$ and so there exist $t_i' + j_i' \in \operatorname{dom}\phi_i$ with $t_i + j_i - m - 1 \le t_i' + j_i' \le t_i + j_i - m$. It can be also assumed that the sequence of points $(t_i - t_i', j_i - j_i')$ is convergent, say to (t^*, j^*); if not, one passes to a subsequence. Similarly, without loss of generality it can be assumed that $\tau_i := t_i' + j_i' \ge \tau_0$. For $i = 1, 2, \ldots$, define hybrid arcs y_i by $y_i(t, j) = \phi_i(t + t_i', j + j_i')$ for all those (t, j) for which $(t + t_i', j + j_i') \in \operatorname{dom}\phi_i$. Then $\operatorname{rge} y_i \subset \mathcal{R}_{\ge\tau_i}(S) \subset \mathcal{R}_{\ge\tau_0}(S) \subset K$, and the sequence $\{y_i\}_{i=1}^\infty$ is locally eventually bounded. As $\mathcal{H}$ is nominally well-posed, and thanks to Theorem 6.1, without loss of generality it can be assumed that the sequence $\{y_i\}_{i=1}^\infty$ converges graphically to y, where y is a solution to $\mathcal{H}$. Furthermore, $y(t^*, j^*) = \xi$ and $t^* + j^* \ge m$. This verifies weak backward invariance.

As $\mathcal{R}_{\ge\tau_i}(S) \subset K$, (6.4) and Theorem 5.4 (b) imply that for every $\varepsilon > 0$ there exists i_0 such that for all $i > i_0$,

$$\mathcal{R}_{\ge\tau_i}(S) \subset \Omega(S) + \varepsilon\mathbb{B}.$$

This is exactly uniform pre-attractivity of $\Omega(S)$ from S.

Finally, suppose $\Omega(S) \subset S$. If $\Omega(S)$ is not strongly forward pre-invariant, then there exists $\phi \in \mathcal{S}_\mathcal{H}(\Omega(S)) \subset \mathcal{S}_\mathcal{H}(S)$ such that $\phi(0, 0) \in \Omega(S)$ but $\xi := \phi(t, j) \notin \Omega(S)$ for some $(t, j) \in \operatorname{dom}\phi$. By weak backward invariance of $\Omega(S)$ (at the point $\phi(0, 0)$), there exist compact solutions ϕ_i to $\mathcal{H}$ with $\phi_i(0, 0) \in \Omega(S) \subset S$, $\operatorname{length}\phi_i > i$, and $\phi_i(\max \operatorname{dom}\phi_i) = \phi(0, 0)$. Concatenating ϕ_i and ϕ yields a solution y_i to $\mathcal{H}$ with $y_i(0, 0) \in S$, $y_i(t_i, j_i) = \xi \notin \Omega(S)$, while $t_i + j_i > i$. This is a contradiction with the definition of $\Omega(S)$. Thus $\Omega(S)$ is strongly forward pre-invariant. $\qquad\square$

6.4   WELL-POSED HYBRID SYSTEMS

The definition of a nominally well-posed hybrid system placed conditions on sequences of solutions to a system. This section defines a narrower class of hybrid systems: systems that are well-posed. The distinction comes in considering sequences of solutions not to the nominal system, but solutions generated with vanishing state perturbations. It will turn out, in the subsequent chapters, that well-posed hybrid systems have good robustness properties.

Definition 6.27. (Perturbed hybrid system) *Given a hybrid system $\mathcal{H}$ and a function $\rho : \mathbb{R}^n \to \mathbb{R}_{\geq 0}$, the ρ-perturbation of $\mathcal{H}$, denoted $\mathcal{H}_\rho$, is the hybrid system*

$$\begin{cases} x \in C_\rho & \dot{x} \in F_\rho(x) \\ x \in D_\rho & x^+ \in G_\rho(x) \end{cases} \tag{6.5}$$

where

$$C_\rho = \{x \in \mathbb{R}^n : (x + \rho(x)\mathbb{B}) \cap C \neq \emptyset\} \ ,$$

$$F_\rho(x) = \overline{\mathrm{con}}F((x + \rho(x)\mathbb{B}) \cap C) + \rho(x)\mathbb{B} \quad \forall x \in \mathbb{R}^n \ ,$$

$$D_\rho = \{x \in \mathbb{R}^n : (x + \rho(x)\mathbb{B}) \cap D \neq \emptyset\} \ ,$$

$$G_\rho(x) = \{v \in \mathbb{R}^n : v \in g + \rho(g)\mathbb{B}, \ g \in G((x + \rho(x)\mathbb{B}) \cap D)\} \quad \forall x \in \mathbb{R}^n \ .$$

It is straightforward from the definitions that, given an arbitrary $\rho : \mathbb{R}^n \to \mathbb{R}_{>0}$, Krasovskii solutions to $\mathcal{H}$ are solutions to $\mathcal{H}_\rho$. In fact, $\widehat{C} \subset C_\rho$, $\widehat{D} \subset D_\rho$, $\widehat{F}(x) \subset F_\rho(x)$ for all $x \in \widehat{C}$, and $\widehat{G}(x) \subset G_\rho(x)$ for all $x \in \widehat{D}$, where $\widehat{C}$, $\widehat{D}$, $\widehat{F}$, $\widehat{G}$ are the data of the Krasovskii regularization $\widehat{\mathcal{H}}$ of $\mathcal{H}$, as given in Definition 4.13.

Proposition 6.28. (Hybrid basic conditions under perturbation) *Let $\mathcal{H}$ satisfy Assumption 6.5. If $\rho : \mathbb{R}^n \to \mathbb{R}_{\geq 0}$ is continuous, then the hybrid system $\mathcal{H}_\rho$ satisfies Assumption 6.5.*

PROOF. Showing that C_ρ and D_ρ are closed is straightforward. That $F_\rho(x)$ and $G_\rho(x)$ are nonempty on C_ρ and D_ρ, respectively, follows directly from the definitions and the fact that $F(x)$ and $G(x)$ are nonempty on C and D, respectively. Convexity of the values of F_ρ is obvious.

Values of the mapping F_ρ can be viewed as $M_2(x) + \rho(x)\mathbb{B}$, where M_2 is obtained as in Lemma 5.17 from the mapping $M : \mathbb{R}^n \rightrightarrows \mathbb{R}^n$ given by $M(x) = F(x)$ if $x \in C$, $M(x) = \emptyset$ if $x \notin C$. This M is osc and locally bounded, in part thanks to C being closed. Hence, by Lemma 5.17, M_2 is osc and locally bounded. It is now easy to show that F_ρ is osc and locally bounded, and thus osc and locally bounded relative to C_ρ.

The mapping G_ρ can be obtained by first considering $M : \mathbb{R}^n \rightrightarrows \mathbb{R}^n$ given by $M(x) = G(x)$ if $x \in D$, $M(x) =$ if $x \notin D$; constructing M_1 as in Lemma 5.17; and then constructing M_3 as in Lemma 5.17, not from M but rather from M_1. Lemma 5.17 concludes that G_ρ is osc and locally bounded. $\qquad\square$

Definition 6.29. (Well-posed hybrid system) *A hybrid system $\mathcal{H}$ is called well-posed if the following property holds: given an arbitrary continuous function $\rho : \mathbb{R}^n \to \mathbb{R}_{\geq 0}$, a decreasing sequence $\{\delta_i\}_{i=1}^{\infty}$ of numbers in $(0,1)$ with $\lim_{i\to\infty} \delta_i = 0$, and a graphically convergent sequence $\{\phi_i\}_{i=1}^{\infty}$ of solutions to $\mathcal{H}_{\delta_i\rho}$ with $\lim_{i\to\infty} \phi_i(0,0) = \xi \in \mathbb{R}^n$,*

(a) *if the sequence $\{\phi_i\}_{i=1}^{\infty}$ is locally eventually bounded, then the sequence $\{\mathrm{length}(\phi_i)\}_{i=1}^{\infty}$ either converges or properly diverges to ∞ and*

$$\phi = \underset{i\to\infty}{\mathrm{gph\text{-}lim}}\, \phi_i$$

is a solution to $\mathcal{H}$ with $\phi(0,0) = \xi$ and $\mathrm{length}(\phi) = \lim_{i\to\infty} \mathrm{length}(\phi_i)$;

(b) *if the sequence $\{\phi_i\}_{i=1}^{\infty}$ is not locally eventually bounded, then*

- *there exist the smallest $j^* \in \mathbb{N}$, $t^* \in \mathbb{R}_{>0}$ for which there exist $(t_i, j^*) \in \mathrm{dom}\, \phi_i$ for all large enough i such that $\lim_{i\to\infty} t_i = t^*$ and $\lim_{i\to\infty} |\phi_i(t_i, j^*)| = \infty$, and*
- *the mapping*

$$\phi = \left(\underset{i\to\infty}{\mathrm{gph\text{-}lim}}\, \phi_i \right) \Big|_{t+j < t^* + j^*}$$

is a maximal solution to $\mathcal{H}$ and

$$\lim_{t\to t^*} |\phi(t, j^*)| = \infty.$$

Obviously, a well-posed hybrid system is nominally well-posed. The converse statement is not true. A simple example showing this is a hybrid system (that reduces to a differential equation) on $\mathbb{R}$ given by $f(x) = -1$ if $x < 0$ and $f(x) = 1$ if $x \geq 0$. By inspection it can be verified that it is nominally well-posed. However, it is not well-posed. The function $z(t) = 0$ for all $t \in [0, \infty)$ is not a solution. However, it is a limit of solutions generated with vanishing state perturbations. This can be shown via arguments similar to those in Example 4.1. More generally, that such a system is not well-posed can be deduced from Theorem 6.31.

Theorem 6.30. (Basic conditions and well-posedness) *If a hybrid system $\mathcal{H} = (C, F, D, G)$ satisfies Assumption 6.5 then it is well-posed.*

PROOF. Suppose that $\{\phi_i\}_{i=1}^{\infty}$ is locally eventually bounded. Example 5.19 shows that $\lim_{i\to\infty} \mathrm{dom}\, \phi_i = \mathrm{dom}\, \phi$. Then Example 5.3 implies that $\mathrm{dom}\, \phi$ is a hybrid time domain and $\mathrm{length}(\mathrm{dom}\, \phi) = \lim_{i\to\infty} \mathrm{length}(\mathrm{dom}\, \phi_i)$.

Now, for each $j \in \mathbb{N}$ let $I^j = \{t \in \mathbb{R}_{\geq 0} : (t, j) \in \mathrm{dom}\, \phi\}$ and let I_i^j be defined similarly from ϕ_i. The very definition of set convergence implies that $\phi(\cdot, j)$ is the graphical limit of $\phi_i(\cdot, j)$'s. If I^j is a singleton $\{(t, j)\}$, then either $I_i^j = \{(t_i, j)\}$ for infinitely many i's, in which case $\phi(t, j) = \lim_{i\to\infty} \phi_i(t_i, j)$ is a singleton, or $\mathrm{int}\, I_i^j \neq \emptyset$ for infinitely many i's, in which case Lemma 5.29 implies $\phi(t, j)$ is a singleton. If int $I^j \neq \emptyset$, then for all large enough i, int $I_i^j \neq \emptyset$, and Lemma 5.29 (a) shows that $\phi(\cdot, j)$ is locally absolutely continuous and satisfies the differential inclusion in (6.1) of restated Definition 2.6 in Section 6.2.1. Also then, for each $t \in I^j$ there exist $t_i \in I_i^j$ such that $\phi_i(t_i, j) \to \phi(t, j)$, $\phi_i(t_i, j) \in C_{\delta_i \rho}$. Since $\lim_{i \to \infty} C_{\delta_i \rho} = C$, it follows that $\phi(t, j) \in C$. Thus ϕ is a hybrid arc that satisfies condition (S1*) in restated Definition 2.6.

To check condition (S2*), pick any $(t, j) \in \operatorname{dom} \phi$ so that $(t, j + 1) \in \operatorname{dom} \phi$. Then $(t, j, \phi(t, j)) = \lim_{i \to \infty}(t_i', j, \phi_i(t_i', j))$ for some sequence $(t_i', j) \in \operatorname{dom} \phi_i$, while for another sequence $(t_i'', j+1) \in \operatorname{dom} \phi_i$, $(t, j+1, \phi(t, j+1)) = \lim_{i \to \infty}(t_i'', j+1, \phi_i(t_i'', j+1))$. The properties of hybrid time domains imply that for some t_i with $t_i' \leq t_i \leq t_i''$ both (t_i, j) and $(t_i, j + 1)$ are in $\operatorname{dom} \phi_i$, and thus $\phi_i(t_i, j) \in D_{\delta_i \rho}$ and $\phi_i(t_i, j+1) \in G_{\delta_i \rho}(\phi_i(t_i, j))$. Also, as ϕ_i's are locally eventually bounded and ϕ is single valued, it must be that $\lim_{i \to \infty} \phi_i(t_i, j) = \phi(t, j)$, $\lim_{i \to \infty} \phi_i(t_i, j+1)$, while $\lim_{i \to \infty} t_i = t$. This, combined with $\lim_{i \to \infty} D_{\delta_i \rho} = D$, gph-$\lim_{i \to \infty} G_{\delta_i \rho} = G$, leads immediately to (6.2). Thus, (S2*) is satisfied by ϕ.

Now suppose that $\{\phi_i\}_{i=1}^\infty$ is not locally eventually bounded. Let j^* be the least of $J \in \mathbb{N}$ for which the sequence $\{\phi_i\}_{i=1}^\infty$ truncated to $\mathbb{R}_{\geq 0} \times \{0, 1, \dots, J\}$ is not locally eventually bounded. Let $a_i \in \mathbb{R}_{\geq 0}$ be the least of all t such that $(t, j^*) \in \operatorname{dom} \phi_i$, define a similarly from ϕ. Then the sequence of truncations of $\{\phi_i\}_{i=1}^\infty$ to $\mathbb{R}_{\geq 0} \times \{0, 1, \dots, j^* - 1\}$ is locally eventually bounded, and in particular, $\phi_i(a_i, j^* - 1) \to \phi(a, j^* - 1)$. Now

$$\phi_i(a_i, j^*) \in G_{\delta_i \rho}(\phi_i(a_i, j^* - 1)) \subset G_\rho(\phi_i(a_i, j^* - 1))$$

and the last set is compact. So it must be that $\phi_i(a_i, j^*)$ converge, and the limit must be $\phi(a, j^*)$. Thus Lemma 5.29 (b) implies the existence of the desired t^*. The remaining conclusions follow from the arguments carried out for the case of $\{\phi_i\}_{i=1}^\infty$ being locally eventually bounded and from Lemma 5.29 (b). $\qquad\square$

6.5   CONSEQUENCES OF WELL-POSEDNESS

Numerous consequences of a hybrid system being well-posed for asymptotic stability and its robustness will be presented in subsequent chapters. Below, some consequences not directly related to stability analysis are presented. The first theorem effectively justifies focusing on hybrid systems meeting the hybrid basic conditions, if one wants to study systems that are well-posed. Indeed, a well-posed system has the same solutions as its "Krasovskii regularization," and the latter does meet the hybrid basic conditions.

Theorem 6.31. (Generalized solutions and well-posed hybrid systems) *Let a hybrid system $\mathcal{H}$ be well-posed. Then a hybrid arc ϕ is a solution to $\mathcal{H}$ if and only if ϕ is a Hermes solution to $\mathcal{H}$. If F and G are locally bounded, then furthermore a hybrid arc ϕ is a solution to $\mathcal{H}$ if and only if ϕ is a Krasovskii solution to $\mathcal{H}$.*

PROOF. Obviously, every solution to $\mathcal{H}$ is a Hermes and a Krasovskii solution to $\mathcal{H}$. To show the first conclusion of the theorem, it is enough to consider a compact Hermes solution to $\mathcal{H}$ and show that it is a solution to $\mathcal{H}$. Let $\rho : \mathbb{R}^n \to \mathbb{R}_{>0}$ be continuous. Let $\{\phi_i\}_{i=1}^{\infty}$, $\{e_i\}_{i=1}^{\infty}$ be sequences of hybrid arcs and admissible state perturbations, respectively, coming from the definition of Hermes solutions, Definition 4.12. The conditions placed on $\{\phi_i\}_{i=1}^{\infty}$ by Definition 4.12 entail uniform boundedness of this sequence and imply, via Theorem 5.25, that the sequence converges graphically to ϕ. The uniform boundedness guarantees that there exists a decreasing sequence $\{\delta_i\}_{i=1}^{\infty}$ of numbers in $(0, 1)$ with $\lim_{i \to \infty} \delta_i = 0$ and such that ϕ_i is a solution to $\mathcal{H}_{\delta_i \rho}$. Indeed, one just needs to pick δ_i so that $e_i(t, j) \leq \delta_i \rho(\phi_i(t, j))$ for all $(t, j) \in \operatorname{dom} \phi_i$. Then, for each $i = 1, 2, \ldots$, ϕ_i is also a solution to $\mathcal{H}_{\delta_i \rho}$, where $\mathcal{H}_{\delta_i \rho}$ is the $\delta_i \rho$-perturbation of $\mathcal{H}$. Since $\mathcal{H}$ is well-posed, ϕ is a solution to $\mathcal{H}$.

To see the second conclusion, recall that Corollary 4.23 showed that under the boundedness assumptions, every Krasovskii solution to $\mathcal{H}$ is a Hermes solution. The arguments of the previous paragraph showed already that Hermes solutions to $\mathcal{H}$ are solutions to $\mathcal{H}$. □

Corollary 6.32 provides the promised earlier proof of one implication in Theorem 4.17: that Hermes solutions are Krasovskii solutions. The reverse implication was already shown in Corollary 4.23.

Corollary 6.32. (Hermes solutions are Krasovskii solutions) *Let $\mathcal{H}$ be a hybrid system with F and G locally bounded. If a hybrid arc ϕ is a Hermes solution to $\mathcal{H}$ then it is a Krasovskii solution to $\mathcal{H}$.*

PROOF. Let ϕ be a Hermes solution to $\mathcal{H}$. Under the boundedness assumption, the Krasovskii regularization $\widehat{\mathcal{H}}$ of $\mathcal{H}$ is well-posed, thanks to Example 6.6 and Theorem 6.30. Since the data of $\widehat{\mathcal{H}}$ contains the data of $\mathcal{H}$, ϕ is a Hermes solution to $\widehat{\mathcal{H}}$. Then, Theorem 6.31 implies that ϕ is a solution to $\widehat{\mathcal{H}}$. But this means that ϕ is a Krasovskii solution to $\mathcal{H}$. □

The next two results are slight generalizations of Propositions 6.13 and 6.14. The proofs are essentially identical, but directly rely on the systems being well-posed rather than nominally well-posed.

Proposition 6.33. (Local uniform boundedness of $\mathcal{S}_{\mathcal{H}_\rho}(K)$) *Let $\mathcal{H}$ be well-posed. Suppose that $\mathcal{H}$ is forward pre-complete from a compact set $K \subset \mathbb{R}^n$, $\rho : \mathbb{R}^n \to \mathbb{R}_{\geq 0}$ is continuous, and $\{\delta_i\}_{i=1}^{\infty}$ is a sequence of numbers in $(0, 1)$ with $\lim_{i \to \infty} \delta_i = 0$. Then, for each $m > 0$, there exists $\delta > 0$, $i_0 > 0$, and a compact set $K' \subset \mathbb{R}^n$ such that, for each solution ϕ to $\mathcal{H}_{\delta_i \rho}$ with $\phi(0, 0) \in K + \delta \mathbb{B}$, $i > i_0$, and all $(t, j) \in \operatorname{dom} \phi$ with $t + j < m$, $\phi(t, j) \in K'$.*

Proposition 6.34. (Dependence on initial conditions and perturbations) *Let $\mathcal{H}$ be well-posed. Suppose that $\mathcal{H}$ is pre-forward complete from a compact set $K \subset \mathbb{R}^n$ and $\rho : \mathbb{R}^n \to \mathbb{R}_{\geq 0}$. Then for every $\varepsilon > 0$ and $\tau \geq 0$ there exists $\delta > 0$ with the following property: for every solution ϕ_δ to $\mathcal{H}_{\delta \rho}$ with $\phi_\delta(0, 0) \in K + \delta \mathbb{B}$ there exists a solution ϕ to $\mathcal{H}$ with $\phi(0, 0) \in K$ such that ϕ_δ and ϕ are (τ, ε)-close.*

Proposition 6.35. (Robust uniform non-Zenoness and non-discreteness) Let $\mathcal{H}$ be well-posed and $\rho : \mathbb{R}^n \to \mathbb{R}_{\geq 0}$ be continuous. For every compact set $K \subset \mathbb{R}^n$ that is strongly forward pre-invariant for $\mathcal{H}_\rho$, the following are true:

(a) if there are no complete discrete (i.e., instantaneous Zeno) solutions $\phi \in \mathcal{S}_\mathcal{H}(K)$ then there exists $\delta > 0$ such that the set of all solutions ϕ to $\mathcal{H}_{\delta\rho}$ with $\phi(0,0) \in K$ is uniformly non-discrete (or uniformly non-Zeno), in the sense that there exist $T, J \geq 0$ such that, for every solution ϕ in that set of solutions, every $(t_1, j_1), (t_2, j_2) \in \operatorname{dom}\phi$, if $t_1 - t_2 \leq T$ then $j_1 - j_2 \leq J$;

(b) if there are no complete continuous solutions $\phi \in \mathcal{S}_\mathcal{H}(K)$ then there exists $\delta > 0$ such that the set of all solutions ϕ to $\mathcal{H}_{\delta\rho}$ with $\phi(0,0) \in K$ is uniformly non-continuous, in the sense that there exist $T, J \geq 0$ such that, for every solution ϕ in that set of solutions, every $(t', j'), (t'', j'') \in \operatorname{dom}\phi$, if $j' - j'' \leq J$ then $t' - t'' \leq T$.

PROOF. Only (a) is shown; the proof of (b) is analogous. If the conclusion of (a) were to fail, there would exist, for $i = 1, 2, \ldots,$ solutions ϕ_i to $\mathcal{H}_{\rho/i}$ with $\phi_i(0,0) \in K$ and $(t'_i, j'_i), (t''_i, j''_i) \in \operatorname{dom}\phi_i$ such that $t'_i - t''_i \leq 1/i$ and $j'_i - j''_i > i$. Because $\rho/i \leq \rho$ and K is strongly forward pre-invariant for $\mathcal{H}_\rho$, one can consider tails of solutions ϕ_i from (t''_i, j''_i), to obtain a new sequence of solutions ϕ_i to $\mathcal{H}_{\rho/i}$ with $\phi_i(0,0) \in K$ and $(t_i, j_i) \in \operatorname{dom}\phi_i$, where $(t_i, j_i) = (t'_i, j'_i) - (t''_i, j''_i)$, such that $t_i \leq 1/i$ and $j_i > i$. Compactness of K implies that the sequence $\{\phi_i\}_{i=1}^\infty$ is bounded, and thus locally eventually bounded; furthermore, that without loss of generality one can assume that points $\phi_i(0,0)$ converge. Theorem 6.1 guarantees that there exists a graphically convergent subsequence of $\{\phi_i\}_{i=1}^\infty$, the limit ϕ of which, with $\phi(0,0) \in K$, is a complete solution to $\mathcal{H}$, because $\mathcal{H}$ is well-posed and $\operatorname{length}\phi_i \geq i$. It is easy to show that in fact ϕ is discrete (following Examples 5.3 and 5.19, or directly). This contradicts the assumption. □

A corresponding result for a nominally well-posed system $\mathcal{H}$ would assume that K is strongly forward pre-invariant for $\mathcal{H}$, and, in (a), it would conclude that $\mathcal{S}_\mathcal{H}(K)$ is uniformly non-discrete. Proving this would involve considering $\rho \equiv 0$ in the proof above.

Note also that without further assumptions Proposition 6.35 can not conclude that the set of all solutions from a neighborhood of K is uniformly non-discrete from knowing that there are no complete discrete solutions in K, even when no perturbations are considered. Further assumptions may involve, for example, pre-asymptotic stability; see Proposition 7.13. Indeed, consider $\mathcal{H}$ in $\mathbb{R}$ with $C = [0, 1]$, $f(x) = x$, $D = \{1\}$, $g(1) = 1$. The compact set $K = \{0\}$ is strongly forward pre-invariant; the unique maximal solution from K is constant, complete, and continuous; but from every neighborhood of K there exist complete solutions which are eventually discrete.

6.6 NOTES

The regularity properties of the flow map and jump map in the data of a hybrid system imposed by Assumption 6.5 have already been recognized in the study of continuous-time and discrete-time systems; see, for example, Filippov [34], Aubin and Cellina [6], Aubin and Frankowska [7], and Kellett and Teel [61]. The importance of Assumption 6.5 in modeling, stability, and robustness of hybrid systems was investigated by Goebel and Teel [40] and Sanfelice et al. [105].

The notion of weak forward invariance in Definition 6.19 essentially agrees with the concept of viability used by Aubin et al. [9], and if one insists on uniqueness of trajectories, with invariance as used by Lygeros et al. [79]. A similar property to backward forward invariance in Definition 6.19 but for continuous-time system was given by Melnik and Valero [91] under the name "negative semi-invariance." Proposition 6.21 extends the results on ω-limit sets by Birkhoff [16, Chapter VII], LaSalle [71, Chapter 1 §5, Chapter 2 §5], and Filippov [35, Chapter 3 §12.4] to the hybrid setting. It can also be seen as a generalization of [79, Lemma IV.1].

The definition and results of ω-limit set of a set for hybrid trajectories were given by Cai et al. [23]. They follow the ideas proposed for classical dynamical systems, for example, by Hale et al. [48].

The outer perturbation of a hybrid system in Section 6.4 was proposed by Goebel and Teel [40].

Chapter Seven

Asymptotic stability, an in-depth treatment

Chapter 3 defined global uniform pre-asymptotic stability for a closed set in a hybrid system and gave numerous sufficient conditions for it. Those sufficient conditions did not require the system to be well-posed, as in Definition 6.29, or even nominally well-posed, as in Definition 6.2. This chapter defines local pre-asymptotic stability for a compact (closed and bounded) set and studies its properties for systems that are nominally well-posed or well-posed. For nominally well-posed hybrid systems, pre-asymptotic stability turns out to be equivalent to uniform pre-asymptotic stability. For well-posed systems, pre-asymptotic stability turns out to be equivalent to uniform, robust pre-asymptotic stability and implies the existence of a smooth Lyapunov function.

While Chapter 3 dealt with global (and uniform) pre-asymptotic stability, the more general local pre-asymptotic stability is studied in this chapter, although for the more restrictive case of compact sets.

Definition 7.1. (Local pre-asymptotic stability (LpAS)) *Let $\mathcal{H}$ be a hybrid system in $\mathbb{R}^n$. A compact set $\mathcal{A} \subset \mathbb{R}^n$ is said to be*

- *stable for $\mathcal{H}$ if for every $\varepsilon > 0$ there exists $\delta > 0$ such that every solution ϕ to $\mathcal{H}$ with $|\phi(0,0)|_{\mathcal{A}} \leq \delta$ satisfies $|\phi(t,j)|_{\mathcal{A}} \leq \varepsilon$ for all $(t,j) \in \operatorname{dom} \phi$;*

- *locally pre-attractive for $\mathcal{H}$ if there exists $\mu > 0$ such that every solution ϕ to $\mathcal{H}$ with $|\phi(0,0)|_{\mathcal{A}} \leq \mu$ is bounded and, if ϕ is complete, then also $\lim_{t+j\to\infty} |\phi(t,j)|_{\mathcal{A}} = 0$;*

- *locally pre-asymptotically stable for $\mathcal{H}$ if it is both stable and locally pre-attractive for $\mathcal{H}$.*

An example of a hybrid system with a locally, but not globally, pre-asymptotically stable origin is now given.

Example 7.2. (Local pre-AS) On $\mathbb{R}^2$ consider a hybrid system given by

$$C = \left\{ x \in \mathbb{R}^2 : x_2 \geq 0,\ x_1 \leq x_2^2 \right\} \qquad f(x) = \begin{pmatrix} 1 \\ 1 \end{pmatrix}$$

$$D = \left\{ x \in \mathbb{R}^2 : x_2 \geq 0,\ x_1 = x_2^2 \right\} \qquad g(x) = \begin{pmatrix} -x_2/2 \\ x_2^2 \end{pmatrix}.$$

See Figure 7.1. The unique maximal solution from the origin is constant and discrete. Consider the other initial points. Maximal solutions from near the origin, starting in $C \setminus D$, flow until they reach D, jump, flow again until they reach D, etc. Maximal solutions from near the origin, starting in D, jump first to $C \setminus D$ and then experience the behavior just described. Maximal solutions starting far from the origin flow, except possibly the initial jump, never reach D and, in fact, they diverge. One of the maximal solutions from $(-1/4, 0)$ is continuous, divergent, while grazing D at $t = 1/2$. The other maximal solution from $(-1/4, 0)$ exhibits behavior as shown in Figure 7.1(b) and is divergent. A careful analysis reveals a periodic maximal solution, from $(-1/8, 1/16)$, which flows to D along the line $x_2 = x_1 + 3/16$, jumps back from D to the starting point, and repeats this behavior infinitely many times. A rough estimate shows that for every $\delta \in (0, 3/(16\sqrt{2}))$, if $|\phi(0, 0)| \le \delta$ then $|\phi(t, j)| \le 3\delta$ for all $(t, j) \in \operatorname{dom} \phi$, for all solutions ϕ. Local stability follows. Furthermore, $|\phi(0, 0)| < 3/(16\sqrt{2})$ then $\phi(t, j) \to 0$ as $t + j \to \infty$. (In fact, for initial points below the line $x_2 = x_1 + 3/16$, maximal solutions converge to 0.) This shows local attractivity. Chapter 9 will suggest a quicker way to establish local pre-attractivity here; see Example 9.12.

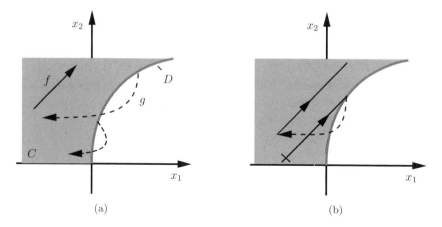

(a)   (b)

Figure 7.1: The hybrid system from Example 7.2. Solutions may flow in the set $C = \left\{x \in \mathbb{R}^2 : x_2 \ge 0, \ x_1 \le x_2^2\right\}$ and may jump from the set $D = \left\{x \in \mathbb{R}^2 : x_2 \ge 0, \ x_1 = x_2^2\right\}$. The solid arrow indicates the direction of flow, which is determined by $f(x) = (1, 1)$, and the dotted arrows indicate jumps, which are determined by $g(x) = (-x_2/2, x_2^2)$. A sample diverging solution starting from $\times$ is shown in (b).

7.1 PRE-ASYMPTOTIC STABILITY FOR NOMINALLY WELL-POSED SYSTEMS

This section considers nominally well-posed hybrid systems with locally pre-asymptotically stable compact sets. Properties of the basin of attraction and uniformity of convergence are analyzed, and some general examples of locally pre-asymptotically stable sets are given.

Definition 7.3. (Basin of pre-attraction) *Let $\mathcal{H}$ be a hybrid system on $\mathbb{R}^n$ and $\mathcal{A} \subset \mathbb{R}^n$ be locally pre-asymptotically stable for $\mathcal{H}$. The basin of pre-attraction of $\mathcal{A}$, denoted $\mathcal{B}^p_{\mathcal{A}}$, is the set of points $\xi \in \mathbb{R}^n$ such that every solution ϕ to $\mathcal{H}$ with $\phi(0,0) = \xi$ is bounded and, if it is complete, then also $\lim_{t+j\to\infty} |\phi(t,j)|_{\mathcal{A}} = 0$.*

Note that the basin of pre-attraction automatically includes all points outside of $C \cup D$. Indeed, the definition of the solution to a hybrid system implies that there are no solutions from $\xi \notin C \cup D$, and thus the condition that every solution from ξ be bounded, etc., is vacuously satisfied.

In Example 7.2, $\mathcal{B}^p_{\mathcal{A}} \cap (C \cup D) = \{x \in C \cup D : x_2 < x_1 + 3/16\}$, and thus

$$\mathcal{B}^p_{\mathcal{A}} = \left\{ x \in \mathbb{R}^2 : x_2 < 0 \text{ or } x_2 < x_1 + 3/16 \text{ or } x_1 > 0, x_2 < \sqrt{x_1} \right\}.$$

Note that $\mathcal{B}^p_{\mathcal{A}}$ is open. This feature turns out to be a general property of nominally well-posed hybrid systems.

Proposition 7.4. (Basin of pre-attraction is open) *Let $\mathcal{H}$ be a nominally well-posed hybrid system. If a compact set $\mathcal{A} \subset \mathbb{R}^n$ is locally pre-asymptotically stable for $\mathcal{H}$, then the basin of pre-attraction of $\mathcal{A}$ is an open set containing $\mathcal{A}$.*

PROOF. Suppose otherwise, that for some ξ in the basin of pre-attraction, $\mathcal{B}^p_{\mathcal{A}}$, there exists a sequence of points $\xi_i \in \mathbb{R}^n \setminus \mathcal{B}^p_{\mathcal{A}}$ and $\xi_i \to \xi$. For each $i \in \mathbb{N}$, there exists $\phi_i \in \mathcal{S}_{\mathcal{H}}(\xi_i)$ that either blows up in finite time or is complete and does not converge to $\mathcal{A}$. Taking $K = \{\xi\}$ in Proposition 6.13 implies that $\{\phi_i\}_{i=1}^\infty$ is locally eventually bounded. This in particular ensures that length(dom ϕ_i) $\to \infty$; indeed, if there are infinitely many ϕ_i's that blow up in finite time, local eventual boundedness ensures that the blow up times diverge to ∞. Without loss of generality, it can be assumed that $\{\phi_i\}_{i=1}^\infty$ converges graphically, thanks to Theorem 6.1. Since $\mathcal{H}$ is nominally well-posed, the graphical limit ϕ is a complete solution to $\mathcal{H}$, with $\phi(0,0) = \xi \in \mathcal{B}^p_{\mathcal{A}}$.

Now pick $\varepsilon > 0$ and let $\delta_1 > 0$ come from the definition of pre-stability of $\mathcal{A}$. Then every ϕ_i that blows up in finite time must satisfy $|\phi_i(t,j)|_{\mathcal{A}} \geq \delta_1$ for all $(t,j) \in \text{dom}\,\phi_i$, otherwise it could not blow up. Let δ_2 come from the definition of pre-attractivity of $\mathcal{A}$, and note that every ϕ_i that is complete must be such that $|\phi_i(t,j)|_{\mathcal{A}} \geq \delta_2$ for all $(t,j) \in \text{dom}\,\phi_i$, otherwise it would need to converge to $\mathcal{A}$. Then the graphical limit ϕ is such that $|\phi(t,j)|_{\mathcal{A}} \geq \min\{\delta_1, \delta_2\}$ for all $(t,j) \in \text{dom}\,\phi$. Since ϕ is complete, this contradicts $\phi(0,0) \in \mathcal{B}^p_{\mathcal{A}}$. $\qquad\square$

Before further analysis of the properties of pre-asymptotic stability in nominally well-posed systems, two general results on pre-asymptotically stable compact sets are given. One result pertains to a set that is both strongly forward pre-invariant and locally uniformly pre-attractive. The other result is a special case of the first, and involves the ω-limit of a set.

Proposition 7.5. (Stability from invariance plus uniform convergence) *Let $\mathcal{H}$ be nominally well-posed. Suppose that a compact set $\mathcal{A} \subset \mathbb{R}^n$ has the following properties:*

1. *it is strongly forward pre-invariant, and*

2. *it is uniformly pre-attractive from a neighborhood of itself, i.e., there exists $\mu > 0$ such that $\mathcal{A}$ is uniformly pre-attractive from $\mathcal{A} + \mu \mathbb{B}$.*

Then the compact set $\mathcal{A}$ is locally pre-asymptotically stable.

PROOF. Only the stability needs to be shown. Assume that $\mathcal{A}$ is not stable. Then, for some $\varepsilon > 0$ and $i = 1, 2, \ldots$, there exist $\phi_i \in \mathcal{S}_{\mathcal{H}}$ with $|\phi_i(0,0)|_{\mathcal{A}} < \mu$, $\lim_{i \to \infty} |\phi_i(0,0)|_{\mathcal{A}} = 0$, and $(t_i, j_i) \in \operatorname{dom} \phi_i$ such that $|\phi_i(t_i, j_i)|_{\mathcal{A}} > \varepsilon$. Considering $\varepsilon/2$ in the definition of uniform pre-attractivity of $\mathcal{A}$ from $\mathcal{A} + \mu \mathbb{B}$ yields $\tau > 0$ such that $t_i + j_i < \tau$, for $i = 1, 2, \ldots$. Because $\mathcal{A}$ is strongly forward pre-invariant and compact, the sequence $\{\phi_i\}_{i=1}^{\infty}$ is locally bounded; recall Proposition 6.13. Extracting a graphically convergent subsequence from $\{\phi_i\}_{i=1}^{\infty}$, via Theorem 6.1, and relying on $\mathcal{H}$ being nominally well-posed yields $\phi \in \mathcal{S}_{\mathcal{H}}$ with $\phi(0,0) \in \mathcal{A}$ and $|\phi(t,j)|_{\mathcal{A}} \geq \varepsilon$, hence $\phi(t,j) \notin \mathcal{A}$ for some $(t,j) \in \operatorname{dom} \phi$ with $t + j \leq \tau$. This contradicts strong forward pre-invariance of $\mathcal{A}$. $\qquad \square$

The preceding proposition is illustrated through the following example, which describes a finite-time observer for a linear system. An observer is useful in control systems when a feedback law is prescribed in terms of states that are not measured but are "observable." In this case, the observable but unmeasured states can be reconstructed using an observer.

Example 7.6. (Finite-time observers for linear systems) Consider a linear, continuous-time system $\dot{\xi} = F\xi + v$, where ξ belongs to a compact set $K_1 \subset \mathbb{R}^n$ and v belongs to a compact, convex set $K_2 \subset \mathbb{R}^n$. Let $H \in \mathbb{R}^{r \times n}$ and suppose that the measurements of the output vector $H\xi$ and the input vector v are available. Suppose also that the pair (H, F) is *observable*, that is, $\dot{\xi} = F\xi$ and $H\xi(t) \equiv 0$ imply $\xi(t) \equiv 0$. This property enables assigning the spectra of the matrix $F - LH$ arbitrarily through the matrix L. In particular, a classical dynamical system with state $\widehat{\xi}$ can be constructed so that $\xi(t) - \widehat{\xi}(t)$ approaches zero as $t \to \infty$. Such a dynamical system is called an observer. A classical observer has the form $\dot{\widehat{\xi}} = (F - LH)\widehat{\xi} + LH\xi + v$, where L is chosen so that $F - LH$ is Hurwitz, meaning that each eigenvalue of $F - LH$ has negative real part. This choice gives the observation error equation $\dot{e} = (F - LH)e$, where $e := \xi - \widehat{\xi}$. Since $F - LH$ is Hurwitz, the error e converges to zero exponentially.

Now consider a hybrid observer that reconstructs the state ξ in finite time. The first thing to note is that the observability of the pair (H, F) permits finding matrices L_1 and L_2 such that, for almost all $\delta > 0$, the matrix $I - \exp((F - L_2 H)\delta) \exp(-(F - L_1 H)\delta)$ is invertible. Define $F_i := F - L_i H$ and henceforth assume that $\delta > 0$, L_1 and L_2 are such that $I - \exp(F_2 \delta) \exp(-F_1 \delta)$ is invertible.

Consider a hybrid system with state $x = (\xi, \widehat{\xi}_1, \widehat{\xi}_2, \tau)$, flow set $C := K_1 \times \mathbb{R}^n \times \mathbb{R}^n \times [0, \delta]$, jump set $D := K_1 \times \mathbb{R}^n \times \mathbb{R}^n \times \{\delta\}$, flow map

$$F(x) = \left\{ \begin{bmatrix} F\xi + v \\ F_1 \widehat{\xi}_1 + (F - F_1)\xi + v \\ F_2 \widehat{\xi}_2 + (F - F_2)\xi + v \\ 1 \end{bmatrix} : v \in K_2 \right\},$$

and jump map

$$G(x) = \begin{bmatrix} \xi \\ G_1 \widehat{\xi}_1 + G_2 \widehat{\xi}_2 \\ G_1 \widehat{\xi}_1 + G_2 \widehat{\xi}_2 \\ 0 \end{bmatrix},$$

where

$$\begin{bmatrix} G_1 & G_2 \end{bmatrix} := (I - \exp(F_2 \delta) \exp(-F_1 \delta))^{-1} \begin{bmatrix} -\exp(F_2 \delta) \exp(-F_1 \delta) & I \end{bmatrix}.$$

This hybrid system contains two different continuous-time observers, of the form $\dot{\widehat{\xi}}_i = (F - L_i H)\widehat{\xi}_i + L_i H \xi + v$, the states of which make jumps every δ seconds according to the rule specified by the jump map.

The compact set $\mathcal{A} := \left\{ (\xi, \widehat{\xi}_1, \widehat{\xi}_2) \in K_1 \times \mathbb{R}^n \times \mathbb{R}^n : \xi = \widehat{\xi}_1 = \widehat{\xi}_2 \right\} \times [0, \delta]$ can be shown to be globally pre-asymptotically stable using Proposition 7.5. The set $\mathcal{A}$ is forward invariant since $G(\mathcal{A} \cap D) \subset \mathcal{A}$ and, during flows, the errors $e_i := \xi - \widehat{\xi}_i$ satisfy $\dot{e}_i = F_i e_i$. Moreover, the set $\mathcal{A}$ is globally uniformly pre-attractive. In particular, $(t, j) \in \mathrm{dom}\, x$ and $t \geq 2\delta$ imply $j \geq 2$ and $x(t, j) \in \mathcal{A}$. The implication $j \geq 2$ follows from the nature of the data of the hybrid system. The implication $x(t, j) \in \mathcal{A}$ follows from the fact that, when there exists w such that ξ, $\widehat{\xi}_1$, and $\widehat{\xi}_2$ satisfy

$$\widehat{\xi}_1 = \xi + \exp(F_1 \delta)w, \qquad \widehat{\xi}_2 = \xi + \exp(F_2 \delta)w,$$

then $G_1 \widehat{\xi}_1 + G_2 \widehat{\xi}_2 = \xi$ so that the jump map sends the state to $\mathcal{A}$. The given relations are satisfied after one jump followed by a flow interval of length δ. In this case, w is equal to the difference between ξ and $\widehat{\xi}_1$ (equivalently, between ξ and $\widehat{\xi}_2$) immediately after the jump. Thus, according to Proposition 7.5, the set $\mathcal{A}$ is globally pre-asymptotically stable. In fact, the analysis above shows that $(t, j) \in \mathrm{dom}\, x$ and either $t \geq 2\delta$ or $j \geq 2$ imply $x(t, j) \in \mathcal{A}$.

Proposition 7.5 together with results from the previous chapter yield the following corollary which is useful for asserting the existence of an asymptotically stable compact set even if the set is difficult to characterize explicitly.

Corollary 7.7. (Asymptotically stable Ω-limit sets) *Consider a nominally well-posed hybrid system $\mathcal{H}$ on $\mathbb{R}^n$. Let $K \subset \mathbb{R}^n$ be compact and suppose that*

1. *the reachable set from K is bounded,*

2. *the ω-limit set from K (recall Definition 6.23), denoted $\Omega(K)$, is nonempty and satisfies $\Omega(K) \subset \mathrm{int}(K)$.*

Then $\Omega(K)$ is locally pre-asymptotically stable with basin of pre-attraction containing K.

PROOF. According to Proposition 6.26, the set $\Omega(K)$ is compact, strongly forward pre-invariant and uniformly pre-attractive from K. Since K contains a neighborhood of $\Omega(K)$, the result follows from Proposition 7.5. $\qquad\square$

The following several results analyze the uniformity of pre-asymptotic stability. The main result here is Theorem 7.12, which states that pre-asymptotic stability, of a compact set in a nominally well-posed hybrid system, can be characterized by a $\mathcal{KL}$ bound. Some preliminary work is needed first. The first result below addresses uniform pre-attractivity (recall Definition 6.24) and the properties of the infinite-horizon reachable set (recall Definition 6.15).

Lemma 7.8. (Behavior from compact subsets of basin of pre-attraction) *Let $\mathcal{H}$ be a nominally well-posed hybrid system in $\mathbb{R}^n$ and $\mathcal{A} \subset \mathbb{R}^n$ be a compact set. If $\mathcal{A}$ is locally pre-asymptotically stable and K is a compact subset of $\mathcal{B}_{\mathcal{A}}^p$, the basin of pre-attraction of $\mathcal{A}$, then*

(a) *$\mathcal{A}$ is uniformly pre-attractive from K,*

(b) *$\mathcal{A} \cup \mathcal{R}(K)$ is a compact subset of $\mathcal{B}_{\mathcal{A}}^p$.*

PROOF. Let $\mathcal{B}_{\mathcal{A}}^p$ denote the basin of pre-attraction. For (a), suppose otherwise, that for some compact $K \subset \mathbb{R}^n$ and $\varepsilon > 0$, there exist $\phi_i \in \mathcal{S}_{\mathcal{H}}(K)$ with $(t_i, j_i) \in \mathrm{dom}\,\phi_i$, $t_i + j_i \geq i$, and $\phi_i(t_i, j_i) \notin \mathcal{A} + \varepsilon\mathbb{B}$. By pre-stability of $\mathcal{A}$ there exists $\delta > 0$ such that $\phi_i(t, j) \notin \mathcal{A} + \delta\mathbb{B}$ for all $(t, j) \in \mathrm{dom}\,\phi_i$, $t + j \leq i$. The sequence $\{\phi_i\}_{i=1}^{\infty}$ is locally uniformly bounded, by Proposition 6.13. Without loss of generality, it may be assumed that $\{\phi_i\}_{i=1}^{\infty}$ is graphically convergent. As $\mathcal{H}$ is nominally well-posed, the graphical limit ϕ is complete. But as $\phi_i(t, j) \notin \mathcal{A} + \delta\mathbb{B}$ for all $(t, j) \in \mathrm{dom}\,\phi_i$, $t + j \leq i$, it must be that $\mathrm{rge}\,\phi \cap \mathcal{A} + \delta\mathbb{B} = \emptyset$. Since $\phi(0, 0) \in K$, this contradicts $K \subset \mathcal{B}_{\mathcal{A}}^p$.

For (b), note that for any $\varepsilon > 0$ there exists $\tau > 0$, by part (a), such that

$$\mathcal{R}(K) \subset (\mathcal{A} + \varepsilon\mathbb{B}) \cup \mathcal{R}_\tau(K).$$

This implies that $\mathcal{R}(K)$ is bounded, by Lemma 6.16. Thus also $\mathcal{A} \cup \mathcal{R}(K)$ is bounded. The inclusion displayed above also implies that a convergent sequence of points $\xi_i \in \mathcal{A} \cup \mathcal{R}(K)$ either approaches $\mathcal{A}$, or remains in $\mathcal{R}_\tau(K)$, for some $\tau > 0$. Both $\mathcal{A}$ and $\mathcal{R}_\tau(K)$ are closed, hence the limit of ξ_i's is in $\mathcal{A} \cup \mathcal{R}(K)$, and so $\mathcal{A} \cup \mathcal{R}(K)$ is compact. As $K \subset \mathcal{B}_{\mathcal{A}}^p$, $\mathcal{R}(K) \subset \mathcal{B}_{\mathcal{A}}^p$ and so $\mathcal{A} \cup \mathcal{R}(K)$ is a compact subset of $\mathcal{B}_{\mathcal{A}}^p$. $\qquad\square$

When global pre-asymptotic stability of a compact set is discussed, as it was in Chapter 3, the distance from the set is adequate for measuring the uniformity of stability and pre-attractivity. However, the distance from the set is not adequate for local pre-asymptotic stability when characterizing behavior on the entire basin of pre-attraction. One reason for this inadequacy is that sets of the form $\{x \in \mathcal{B}_{\mathcal{A}}^p : |x|_{\mathcal{A}} \leq r\}$ need not be closed, unless r is sufficiently small or $\mathcal{B}_{\mathcal{A}}^p = \mathbb{R}^n$. This motivates the next definition.

Definition 7.9. (Proper indicator) *Let $\mathcal{U} \subset \mathbb{R}^n$ be an open set. A function $\omega : \mathcal{U} \to \mathbb{R}_{\geq 0}$ is a proper indicator on $\mathcal{U}$ if it is continuous and $\omega(x_i) \to \infty$ when $i \to \infty$ if either $|x_i| \to \infty$ or the sequence $\{x_i\}_{i=1}^\infty$ approaches the boundary of $\mathcal{U}$.*

Let $\mathcal{A} \subset \mathcal{U}$ be a compact set. A function $\omega : \mathcal{U} \to \mathbb{R}_{\geq 0}$ is a proper indicator of $\mathcal{A}$ on $\mathcal{U}$ if it is a proper indicator on $\mathcal{U}$ and $\omega(x) = 0$ if and only if $x \in \mathcal{A}$.

Simple examples of proper indicators on $\mathbb{R}^n$, and in fact of proper indicators of $\{0\}$ on $\mathbb{R}^n$, are provided by norms, for example the Euclidean norm $|\cdot|$. Given an arbitrary compact $\mathcal{A} \subset \mathbb{R}^n$, an example of a proper indicator of $\mathcal{A}$ on $\mathbb{R}^n$ is provided by the distance from $\mathcal{A}$, i.e., $\omega(x) = |x|_{\mathcal{A}}$. For a general open $\mathcal{U} \subset \mathbb{R}^n$, a proper indicator on $\mathcal{U}$ is provided by $\omega(x) = \left(|x|_{\mathbb{R}^n \setminus \mathcal{U}}\right)^{-1}$. Given a compact $\mathcal{A} \subset \mathcal{U}$, a proper indicator of $\mathcal{A}$ on $\mathcal{U}$ is provided by $\omega(x) = |x|_{\mathcal{A}} \left(|x|_{\mathbb{R}^n \setminus \mathcal{U}}\right)^{-1}$.

Definition 7.10. ($\mathcal{KL}$ pre-asymptotic stability) *Let $\mathcal{H}$ be a hybrid system in $\mathbb{R}^n$, $\mathcal{A} \subset \mathbb{R}^n$ be a compact set, and $\mathcal{U} \subset \mathbb{R}^n$ be an open set such that $\mathcal{A} \subset \mathcal{U}$. The set $\mathcal{A}$ is $\mathcal{KL}$ pre-asymptotically stable on $\mathcal{U}$ for $\mathcal{H}$ if for every proper indicator ω of $\mathcal{A}$ on $\mathcal{U}$ there exists a function $\beta \in \mathcal{KL}$ such that*

$$\omega(\phi(t,j)) \leq \beta\left(\omega(\phi(0,0)), t+j\right) \quad \text{for all} \ \ (t,j) \in \operatorname{dom} \phi \qquad (7.1)$$

for every $\phi \in \mathcal{S}_{\mathcal{H}}(\mathcal{U})$.

Global uniform pre-asymptotic stability, as studied in Chapter 3, allowed an equivalent characterization involving a $\mathcal{KL}$ function; recall Theorem 3.40. Relying on proper indicators of $\mathcal{B}_{\mathcal{A}}^p$ allows one to characterize the uniformity of local pre-asymptotic stability on the whole set $\mathcal{B}_{\mathcal{A}}^p$.

Lemma 7.11. ($\mathcal{KL}$ pAS from stability, uniform pre-attractivity and boundedness) *Let $\mathcal{H}$ be a hybrid system in $\mathbb{R}^n$, $\mathcal{A} \subset \mathbb{R}^n$ be a stable compact set, and $\mathcal{U} \subset \mathbb{R}^n$ be an open set such that $\mathcal{A} \subset \mathcal{U}$. Suppose that $\mathcal{A}$ is uniformly pre-attractive from compact subsets of $\mathcal{U}$ while for each compact $K \subset \mathcal{U}$, $\mathcal{R}(K)$ is bounded with respect to $\mathcal{U}$. Then $\mathcal{A}$ is $\mathcal{KL}$ pre-asymptotically stable on $\mathcal{U}$.*

PROOF. Given a proper indicator ω of $\mathcal{A}$ on $\mathcal{U}$, let

$$\beta(r,s) = \sup \left\{\omega(\phi(t,j)) \ : \ \phi \in \mathcal{S}_{\mathcal{H}}, \ \omega(\phi(0,0)) \leq r, \ t+j \geq s\right\}.$$

For every $r \geq 0$, consider the compact set $K = \{\xi \in \mathcal{U} : \omega(\xi) \leq r\}$. By assumption, $\mathcal{R}(K)$ is bounded with respect to $\mathcal{U}$, and so $\omega(\mathcal{R}(K))$ is bounded.

Consequently, $\beta(r,0)$ is finite. By the very definition, $s \to \beta(r,s)$ is nonincreasing, and so $\beta(r,s)$ is finite for all $s \geq 0$. By the uniform pre-attractivity from K, $\beta(r,s) \to 0$ as $s \to \infty$. The very definition of stability shows that for each $s \geq 0$, $\beta(r,s) \to 0$ as $r \to 0$. Thus, β has all of the properties required of a $\mathcal{KL}$ function. $\qquad\square$

The developments of this section so far are summarized in the theorem below.

Theorem 7.12. (From local pAS to uniform pAS) *Let $\mathcal{H}$ be nominally well-posed. Suppose that a compact set $\mathcal{A} \subset \mathbb{R}^n$ is locally pre-asymptotically stable. Then $\mathcal{B}_{\mathcal{A}}^p$ is open and $\mathcal{A}$ is $\mathcal{KL}$ pre-asymptotically stable on $\mathcal{B}_{\mathcal{A}}^p$.*

PROOF. Proposition 7.4 implies that $\mathcal{B}_{\mathcal{A}}^p$ is an open set. By Lemma 7.8 (a), $\mathcal{A}$ is uniformly pre-attractive from compact subsets of $\mathcal{B}_{\mathcal{A}}^p$. By Lemma 7.8 (b), $\mathcal{R}(K)$ is bounded with respect to $\mathcal{B}_{\mathcal{A}}^p$ for each compact $K \subset \mathcal{B}_{\mathcal{A}}^p$. Now, Lemma 7.11 applies, and the conclusion of the theorem follows. $\qquad\square$

Clearly, the bound (7.1) implies that the compact set $\mathcal{A}$ is locally pre-asymptotically stable. Thus, Theorem 7.12 provides an equivalent characterization of local pre-asymptotic stability in terms of a class-$\mathcal{KL}$ estimate, for hybrid systems that are nominally well-posed.

Uniformity of properties other than stability and pre-attractivity can be established for nominally well-posed systems having a pre-asymptotically stable compact set. An example is given in the next result, which states that the lack of continuous solutions implies uniform "non-continuity" of all solutions, while the lack of discrete solutions implies uniform "non-discreteness." It should be noted that Proposition 6.35 established a kind of robustness of uniform non-discreteness and uniform non-continuity. That is, these two properties hold for perturbations of well-posed systems, if the nominal systems lack continuous/discrete solutions.

Proposition 7.13. (Local pAS and uniform non-continuity, non-discreteness) *Let $\mathcal{H}$ be nominally well-posed. Suppose that a compact set $\mathcal{A} \subset \mathbb{R}^n$ is locally pre-asymptotically stable. For any compact set $K \subset \mathcal{B}_{\mathcal{A}}^p$, the following are true:*

(a) *if there are no complete discrete (i.e., instantaneous Zeno) solutions $\phi \in \mathcal{S}_{\mathcal{H}}(\mathcal{A})$ then the set $\mathcal{S}_{\mathcal{H}}(K)$ is uniformly non-discrete (or uniformly non-Zeno), in the sense that there exist $T, J \geq 0$ such that, for any $\phi \in \mathcal{S}_{\mathcal{H}}(K)$, any $(t_1, j_1), (t_2, j_2) \in \operatorname{dom}\phi$, if $t_1 - t_2 \leq T$ then $j_1 - j_2 \leq J$;*

(b) *if there are no complete continuous solutions $\phi \in \mathcal{S}_{\mathcal{H}}(\mathcal{A})$ then the set $\mathcal{S}_{\mathcal{H}}(K)$ is uniformly non-continuous, in the sense that there exist $T, J \geq 0$ such that, for any $\phi \in \mathcal{S}_{\mathcal{H}}(K)$, any $(t_1, j_1), (t_2, j_2) \in \operatorname{dom}\phi$, if $j_1 - j_2 \leq J$ then $t_1 - t_2 \leq T$.*

PROOF. Only (a) is shown; the proof of (b) is analogous. If the conclusion of (a) was to fail, there would exist, for $i = 1, 2, \ldots$, solutions $\phi_i \in \mathcal{S}_{\mathcal{H}}(K)$ and $t_i \in [0, 1/i]$ with $(t_i, 2i) \in \operatorname{dom}\phi_i$. By Theorem 7.12, there exists a proper indicator ω of $\mathcal{A}$ on $\mathcal{B}_\mathcal{A}^p$ and a $\mathcal{KL}$ function β such that the bound (7.1) holds for each solution to $\mathcal{H}$. In particular, the bound holds for ϕ_i' defined by $\phi_i'(t, j) = \phi_i(t - t_i^*, j - i)$, where $t_i^* \in [0, 1/i]$ is such that $(t_i^*, i) \in \mathrm{dom}\,\phi_i$. Thus the sequence $\{\phi_i'\}_{i=1}^\infty$ is uniformly bounded and $\phi_i'(0, 0)$ converge, as $i \to \infty$, to $\mathcal{A}$. Theorem 6.1 guarantees that there exists a graphically convergent subsequence of $\{\phi_i'\}_{i=1}^\infty$, the limit ϕ of which is a complete solution to $\mathcal{H}$, since $\mathcal{H}$ is nominally well-posed and length $\phi_i' \geq i$. It is easy to show that in fact ϕ is discrete (following Examples 5.3 and 5.19, or directly). Finally, $\phi(0, 0) \in \mathcal{A}$ and this contradicts the assumption. $\qquad\square$

To conclude the section, it is shown how a hybrid system with $C \cup D \neq \mathbb{R}^n$ can be augmented to give a system with the union of the flow and jump sets equal to $\mathbb{R}^n$, in fact with the jump set itself equal to $\mathbb{R}^n$, in a way that preserves pre-asymptotic stability of a compact set. Furthermore, pre-asymptotic stability turns to asymptotic stability in the augmented system, because the augmented system satisfies the existence conditions of Proposition 2.10 everywhere.

Lemma 7.14. (From pre-asymptotic to asymptotic stability) *Let* $\mathcal{H} = (C, F, D, G)$ *be a hybrid system on* $\mathbb{R}^n$ *and* $\mathcal{A} \subset \mathbb{R}^n$ *be compact. Consider an augmented hybrid system* $\mathcal{H}^*$ *on* $\mathbb{R}^n$ *with the flow map, the flow set, the jump map, and the jump set given, respectively, by*

$$F^* = F, \quad C^* = C, \quad G^* \text{ where } G^*(x) = \begin{cases} G(x) \cup \mathcal{A} & x \in D \\ \mathcal{A} & x \in \mathbb{R}^n \setminus D \end{cases}, \quad D^* = \mathbb{R}^n.$$

Then:

(a) *if* $\mathcal{H}$ *satisfies Assumption 6.5, then* $\mathcal{H}^*$ *satisfies Assumption 6.5;*

(b) *if* $\mathcal{A}$ *is* $\mathcal{KL}$ *pre-asymptotically stable, for* $\mathcal{H}$, *on an open set* $\mathcal{U}$, *then* $\mathcal{A}$ *is* $\mathcal{KL}$ *pre-asymptotically stable, for* $\mathcal{H}^*$, *on* $\mathcal{U}$.

PROOF. Obviously, D^* is closed. If $\mathcal{H}$ satisfies Assumption 6.5, then G^* is outer semicontinuous on D^* (thanks to $\mathcal{A}$ being closed), and G^* is locally bounded (thanks to $\mathcal{A}$ being bounded). Thus, $\mathcal{H}^*$ satisfies Assumption 6.5. Now suppose that ω is a proper indicator of $\mathcal{A}$ on $\mathcal{U}$, $\beta \in \mathcal{KL}$, and

$$\omega(\phi(t, j)) \leq \beta\left(\omega(\phi(0, 0)), t + j\right)$$

for all $(t, j) \in \mathrm{dom}\,\phi$, all solutions ϕ to $\mathcal{H}$ with $\phi(0, 0) \in \mathcal{U}$. If a solution ϕ to $\mathcal{H}^*$ is also a solution to $\mathcal{H}$, the bound displayed above holds for ϕ. If it is not a solution to $\mathcal{H}$, then it must be the case that for some $(t', j') \in \mathrm{dom}\,\phi$, $(t', j' + 1) \in \mathrm{dom}\,\phi$ and $\phi(t', j' + 1) \in \mathcal{A}$, while ϕ is a solution to $\mathcal{H}_\rho$ when restricted to $(t, j) \in \mathrm{dom}\,\phi$ with $t + j \leq t' + j'$. The latter fact implies that the $\mathcal{KL}$ bound is valid for all $t + j \leq t' + j'$. It is straightforward to establish that $\mathcal{A}$ is strongly forward invariant for $\mathcal{H}^*$. Then, since $\phi(t', j' + 1) \in \mathcal{A}$, $\omega(\phi(t, j)) = 0$ for all $(t, j) \in \mathrm{dom}\,\phi$ with $t + j \geq t' + j' + 1$. Thus, the $\mathcal{KL}$ bound holds for all $(t, j) \in \mathrm{dom}\,\phi$. The proof is finished. $\qquad\square$

7.2   ROBUSTNESS CONCEPTS

This section begins the study of robustness of pre-asymptotic stability in hybrid systems. Different concepts of robustness are considered, and some relationships between them are established. The results of this section do not depend on whether the hybrid system is nominally well-posed or well-posed. The first robustness concept considered follows.

Definition 7.15. (Robust pre-asymptotic stability) *Let $\mathcal{A} \subset \mathbb{R}^n$ be a compact set that is locally pre-asymptotically stable for $\mathcal{H}$, and let $\mathcal{U} \subset \mathcal{B}^p_{\mathcal{A}}$ be an open set such that $\mathcal{A} \subset \mathcal{U}$. The local pre-asymptotic stability of $\mathcal{A}$ is robust on $\mathcal{U}$ if there exists a continuous function $\rho : \mathbb{R}^n \to \mathbb{R}_{\geq 0}$ that is positive on $\mathcal{U} \setminus \mathcal{A}$ such that $\mathcal{A}$ is locally pre-asymptotically stable for $\mathcal{H}_\rho$, the ρ-perturbation of $\mathcal{H}$, and $\mathcal{U}$ is a subset of the basin of pre-attraction of $\mathcal{A}$ for $\mathcal{H}_\rho$.*

Example 7.16. (Robust, but not globally, global asymptotic stability) Let $\mathcal{H}$ be a hybrid system on $\mathbb{R}$ given by $C = (-\infty, 1]$, $F(x) = -x$ for all $x \in C$, $D = (1, \infty)$, $G(x) = 1$ for all $x \in D$. Clearly, the origin is locally asymptotically stable, because around the origin, the hybrid system reduces to the differential equation $\dot{x} = -x$. In fact, the basin of attraction of the origin is $\mathbb{R}$, and the origin is globally asymptotically stable. The local asymptotic stability of the origin is robust on the open set $\mathcal{U} = \mathbb{R} \setminus \{1\}$ but is not robust on $\mathbb{R}$. For the latter statement, note that for any function $\rho : \mathbb{R} \to \mathbb{R}_{\geq 0}$ with $\rho(1) > 1$, $1 \in D_\rho$ and $1 \in G_\rho(1)$, and then 1 is not in the basin of attraction of the origin. For the former statement, consider $\rho(x) = \min\{|x|, |x-1|\}/2$. This is a continuous function, positive on $\mathcal{U}$ except at the origin. Furthermore, $C_\rho = C$, $F_\rho(x) = [-x/2, -3x/2]$ if $x \leq 0$, $F_\rho(x) = [-3x/2, -x/2]$ if $0 < x < 1/2$, $F_\rho(x) \subset [-3x/2, -x/2]$ if $1/2 < x \leq 1$ $D_\rho = D$, $G_\rho = G$. It can be verified that the origin is asymptotically stable for $\mathcal{H}_\rho$. In fact, the behavior of solutions to $\mathcal{H}_\rho$ is similar to the behavior of solutions to $\mathcal{H}$, the difference being the rate of flow of solutions toward the origin.

It turns out that robustness of pre-asymptotic stability guarantees pre-asymptotic stability for the Krasovskii regularization of the hybrid system. This fact is established rigorously below. A Krasovskii regularization is well-posed, and in particular nominally well-posed, as argued in Example 6.6. Consequently, uniformity and $\mathcal{KL}$ characterizations of pre-asymptotic stability, as established in Section 7.1 for nominally well-posed systems, will become relevant in the analysis of robust pre-asymptotic stability.

Lemma 7.17. (Robustness and Krasovskii solutions) *Let $\mathcal{A} \subset \mathbb{R}^n$ be a compact set that is locally pre-asymptotically stable for $\mathcal{H}$, and let $\mathcal{U} \subset \mathcal{B}^p_{\mathcal{A}}$ be an open set such that $\mathcal{A} \subset \mathcal{B}^p_{\mathcal{A}}$. Suppose that the local pre-asymptotic stability of $\mathcal{A}$ is robust on $\mathcal{U}$. Then $\mathcal{A}$ is locally pre-asymptotically stable for $\widehat{\mathcal{H}}$, the Krasovskii regularization of $\mathcal{H}$, and $\mathcal{U}$ is a subset of the basin of pre-attraction of $\mathcal{A}$ for $\widehat{\mathcal{H}}$.*

PROOF. Let $\rho : \mathbb{R}^n \to \mathbb{R}_{\geq 0}$ come from Definition 7.15. It will be shown that Krasovskii solutions to $\mathcal{H}$ are solutions to $\mathcal{H}_\rho$ as long as they remain away from $\mathcal{A}$, and if they reach $\mathcal{A}$, they remain in $\mathcal{A}$ afterwards. This property immediately implies the conclusion of the lemma.

Because $\rho(x) > 0$ if $x \in \mathcal{U} \setminus \mathcal{A}$, $\widehat{C} \cap (\mathcal{U} \setminus \mathcal{A}) \subset C_\rho$, $\widehat{F}(x) \subset F_\rho(x)$ for all $x \in \widehat{C} \cap (\mathcal{U} \setminus \mathcal{A})$, $\widehat{D} \cap (\mathcal{U} \setminus \mathcal{A}) \subset D_\rho$, and $\widehat{G}(x) \subset G_\rho(x)$ for all $x \in \widehat{D} \cap (\mathcal{U} \setminus \mathcal{A})$. Let ϕ be a Krasovskii solution to $\mathcal{H}$ with $\phi(0,0) \in \mathcal{U}$. If $\phi(t,j) \notin \mathcal{A}$ for all $(t,j) \in \operatorname{dom} \phi$ then ϕ is also a solution to $\mathcal{H}_\rho$, because of the data inclusion just mentioned. In the opposite case, let $(T,J) \in \operatorname{dom} \phi$ be the first instance when $\phi(T,J) \in \mathcal{A}$. It will now be shown that $\phi(T,J) \in \mathcal{A}$ implies $\phi(t,j) \in \mathcal{A}$ for all $(t,j) \in \operatorname{dom} \phi$ with $t + j > T + J$. In other words, Krasovskii solutions do not flow out of $\mathcal{A}$ and do not jump out of $\mathcal{A}$.

First, consider a Krasovskii solution ψ with $[0,\tau] \times \{0\} = \operatorname{dom} \psi$, $\tau > 0$ and $\psi(0,0) \in \mathcal{A}$. Suppose that $\psi(t,0) \notin \mathcal{A}$ for $t \in (0,\tau]$. Then ψ is a solution to $\mathcal{H}_\rho$, with $\lim_{t \searrow 0} |\psi(t,0)|_{\mathcal{A}} = 0$ and $|\psi(\tau,0)|_{\mathcal{A}} > 0$. This contradicts local pre-asymptotic stability of $\mathcal{A}$ for $\mathcal{H}_\rho$. Thus, Krasovskii solutions do not flow out of $\mathcal{A}$. Second, note that for every $\delta > 0$, every $x \in \mathcal{A} \cap \widehat{D}$,

$$
\begin{aligned}
\widehat{G}(x) &\subset G\left((x + \delta \mathbb{B}) \cap D\right) + \delta \mathbb{B} \\
&= G\left[((x + \delta \mathbb{B}) \cap D) \cap \mathcal{A} \cup ((x + \delta \mathbb{B}) \cap D) \setminus \mathcal{A}\right] + \delta \mathbb{B} \\
&= G\left[((x + \delta \mathbb{B}) \cap D) \cap \mathcal{A}\right] \cup G\left[((x + \delta \mathbb{B}) \cap D) \setminus \mathcal{A}\right] + \delta \mathbb{B} \\
&\subset \mathcal{A} \cup G\left[((x + \delta \mathbb{B}) \cap D) \setminus \mathcal{A}\right] + \delta \mathbb{B},
\end{aligned}
$$

because for $x \in \mathcal{A} \cap D$, $G(x) \subset G_\rho(x) \subset \mathcal{A}$. Pick $\varepsilon > 0$. Local pre-asymptotic stability of $\mathcal{A}$ for $\mathcal{H}_\rho$, and hence for $\mathcal{H}$, implies that there exists $\delta > 0$ such that, for all $x \in \mathcal{A}$, $G\left((x + \delta \mathbb{B}) \cap D\right) \subset \mathcal{A} + \varepsilon \mathbb{B}$. Hence, for every $x \in \mathcal{A} \cap \widehat{D}$,

$$
\widehat{G}(x) \subset \mathcal{A} \cup (\mathcal{A} + \varepsilon \mathbb{B}) + \delta \mathbb{B} \subset \mathcal{A} + (\varepsilon + \delta) \mathbb{B}.
$$

Because δ can be chosen so that $\delta < \varepsilon$ and the inclusion above holds for all $\varepsilon > 0$, $\widehat{G}(x) \subset \mathcal{A}$ for every $x \in \mathcal{A} \cap \widehat{D}$. Thus, Krasovskii solutions do not jump out of $\mathcal{A}$, and the proof is finished. $\qquad \square$

Two concepts of robustness will be used in the study of robustness of $\mathcal{KL}$ pre-asymptotic stability. The definitions are below.

Definition 7.18. (Robust $\mathcal{KL}$ pre-asymptotic stability) *Let $\mathcal{A} \subset \mathbb{R}^n$ be a compact set and $\mathcal{U} \subset \mathbb{R}^n$ be an open set such that $\mathcal{A} \subset \mathcal{U}$.*

(a) *The set $\mathcal{A}$ is robustly $\mathcal{KL}$ pre-asymptotically stable on $\mathcal{U}$ for $\mathcal{H}$ if there exists a continuous function $\rho : \mathbb{R}^n \to \mathbb{R}_{\geq 0}$ that is positive on $\mathcal{U} \setminus \mathcal{A}$ such that $\mathcal{A}$ is $\mathcal{KL}$ pre-asymptotically stable on $\mathcal{U}$ for $\mathcal{H}_\rho$, the ρ-perturbation of $\mathcal{H}$.*

(b) *The set $\mathcal{A}$ is semiglobally practically robustly $\mathcal{KL}$ pre-asymptotically stable on $\mathcal{U}$ for $\mathcal{H}$ if for every ω that is a proper indicator of $\mathcal{A}$ on $\mathcal{U}$, every* continuous function $\rho : \mathbb{R}^n \to \mathbb{R}_{\geq 0}$ that is positive on $\mathcal{U} \setminus \mathcal{A}$, and every function $\beta \in \mathcal{KL}$ satisfying (7.1), the following holds: for every compact $K \subset \mathcal{U}$ and every $\varepsilon > 0$, there exists $\delta \in (0,1)$ such that every $\phi \in \mathcal{S}_{\mathcal{H}_{\delta\rho}}(K)$ satisfies

$$\omega(\phi(t,j)) \leq \beta(\omega(\phi(0,0)), t+j) + \varepsilon \quad \text{for all } (t,j) \in \text{dom } \phi.$$

It turns out that the semiglobal practical robustness of $\mathcal{KL}$ pre-asymptotic stability implies "global" robustness, in the sense of Definition 7.18 (a). This fact does not depend on the system being nominally well-posed or well-posed. Rather, what is involved is a construction of a continuous function ρ from the ε's and δ's coming from semiglobal practical robustness.

Lemma 7.19. (From semiglobal practical robustness to robustness) *Let $\mathcal{A} \subset \mathbb{R}^n$ be a compact set and $\mathcal{U} \subset \mathbb{R}^n$ be an open set such that $\mathcal{A} \subset \mathcal{U}$. If $\mathcal{A}$ is semiglobally practically robustly $\mathcal{KL}$ pre-asymptotically stable on $\mathcal{U}$, then $\mathcal{A}$ is robustly $\mathcal{KL}$ pre-asymptotically stable on $\mathcal{U}$.*

PROOF. Let ω be a proper indicator of $\mathcal{A}$ on $\mathcal{U}$, $\rho : \mathbb{R}^n \to \mathbb{R}_{\geq 0}$ be a continuous function, and $\beta \in \mathcal{KL}$ satisfy (7.1). Pick a sequence $\{r_n\}_{n\in\mathbb{Z}}$ such that $r_{n+1} \geq 4\beta(r_n, 0) \geq 4r_n > 0$ for each $n \in \mathbb{Z}$, $\lim_{n\to-\infty} r_n = 0$, and $\lim_{n\to\infty} r_n = \infty$. Consequently, for each $n \in \mathbb{Z}$, there exists $\delta_n \in (0,1)$ such that each solution ϕ to $\mathcal{H}_{\delta_n\rho}$ with $\omega(\phi(0,0)) \leq r_n$ satisfies $\omega(\phi(t,j)) \leq \beta(\omega(\phi(0,0)), t+j) + r_{n-1}/2$ (and hence $\omega(\phi(t,j)) \leq r_{n+1}/2$) for all $(t,j) \in \text{dom } \phi$, which implies the existence of $\tau_n > 0$ such that each solution ϕ to $\mathcal{H}_{\delta_n\rho}$ with $\omega(\phi(0,0)) \leq r_n$ satisfies $\omega(\phi(t',j')) \leq r_{n-1}$ for all $(t',j') \in \text{dom } \phi$ with $t' + j' \geq \tau_n$.

Now, pick any continuous function $\delta : \mathbb{R}^n \to \mathbb{R}_{\geq 0}$ that is positive on $\mathcal{U} \setminus \mathcal{A}$ and such that $\delta(x) \leq \min\{\delta_{n-1}, \delta_n, \delta_{n+1}\}\rho(x)$ for $r_{n-1} \leq \omega(x) \leq r_n$. Then, for every $n \in \mathbb{Z}$ and every solution ϕ to $\mathcal{H}_\delta$ with $\omega(\phi(0,0)) \leq r_n$, the following hold:

(i) $\omega(\phi(t,j)) \leq r_{n+1}/2$ for all $(t,j) \in \text{dom } \phi$, and thus $\mathcal{A}$ is stable and for each compact $K \subset \mathcal{U}$, the reachable set $\mathcal{R}(K)$ for the system $\mathcal{H}_\delta$ is bounded with respect to $\mathcal{U}$;

(ii) there exists $(t',j') \in \text{dom } \phi$ with $t' + j' \leq \tau_n$ such that $\omega(\phi(t',j')) \leq r_{n-1}$, and thus $\mathcal{A}$ is pre-attractive.

The stability in (i) and pre-attractivity in (ii) already imply that $\mathcal{A}$ is pre-asymptotically stable for $\mathcal{H}_\delta$. Furthermore, a combination of (i) and (ii) shows that, for each $n \in \mathbb{Z}$ and each ϕ to $\mathcal{H}_\delta$ with $\omega(\phi(0,0)) \leq r_n$, it is the case that $\omega(\phi(t,j)) \leq r_n/2$ for all $(t,j) \in \text{dom } \phi$ with $t + j \geq \tau_n$. This implies that $\mathcal{A}$ is uniformly pre-attractive from each compact subset of $\mathcal{U}$, for the hybrid system $\mathcal{H}_\delta$. Now, Lemma 7.11 implies that $\mathcal{A}$ is $\mathcal{KL}$ pre-asymptotically stable on $\mathcal{U}$, for the system $\mathcal{H}_\delta$. □

7.3   WELL-POSED SYSTEMS

In this section, well-posed hybrid systems with locally pre-asymptotically compact sets are considered. The uniformity and $\mathcal{KL}$ characterizations of local pre-asymptotic stability, shown in Section 7.1, do apply here, since well-posed systems are nominally well-posed. The goal of this section is to establish robustness properties of the $\mathcal{KL}$ pre-asymptotic stability. The main result of the section, Theorem 7.21, concludes that for well-posed hybrid systems local pre-asymptotic stability, and so $\mathcal{KL}$ pre-asymptotic stability, is robust.

The need for a hybrid system to be well-posed in order to establish robustness of pre-asymptotic stability is visible in the next lemma. A graphical limit of solutions generated with vanishing perturbations is considered, and property (a) in Definition 6.29 is relied upon in the proof below.

Lemma 7.20. (Semiglobal practical robustness of $\mathcal{KL}$ pre-asymptotic stability) *Let $\mathcal{H}$ be well-posed, $\mathcal{A} \subset \mathbb{R}^n$ be a compact set, and $\mathcal{U} \subset \mathbb{R}^n$ be an open set such that $\mathcal{A} \subset \mathcal{U}$. If $\mathcal{A}$ is $\mathcal{KL}$ pre-asymptotically stable on $\mathcal{U}$ then it is semiglobally practically robustly $\mathcal{KL}$ pre-asymptotically stable on $\mathcal{U}$.*

PROOF. Let ω be a proper indicator of $\mathcal{A}$ on $\mathcal{U}$, and let β be a $\mathcal{KL}$ function coming from the definition of $\mathcal{KL}$ pre-asymptotic stability of $\mathcal{A}$ on $\mathcal{U}$. Without loss of generality, it can be assumed that β is continuous. Fix $\varepsilon > 0$, a compact $K \subset \mathcal{U}$, and a continuous function $\rho : \mathbb{R}^n \to \mathbb{R}_{\geq 0}$ that is positive on $\mathcal{U} \setminus \mathcal{A}$. There exists $r > \varepsilon$ such that K is a subset of the compact set $\{x \in \mathcal{U} : \omega(x) \leq r\}$. Pick $\tau > 0$ large enough so that $\beta(m, t+j) \leq \varepsilon/2$ when $t + j \geq \tau$. It is first claimed that there exists $\delta > 0$ such that every solution ϕ to $\mathcal{H}_{\delta\rho}$ with $\omega(\phi(0,0)) \leq m$ satisfies

$$\omega(\phi(t,j)) \leq \beta(\omega(\phi(0,0)), t+j) + \epsilon/2 \tag{7.2}$$

for all $(t,j) \in \operatorname{dom} \phi$ with $t + j \leq 2\tau$. This implies $\omega(\phi(t,j)) \leq \epsilon$ for all $(t,j) \in \operatorname{dom} x$ with $\tau \leq t + j \leq 2\tau$. Using this fact recursively and relying on $m > \epsilon$ shows that $\omega(\phi(t,j)) \leq \epsilon$ when $t + j \geq \tau$. This, and (7.2), shows the inequality in the definition of semiglobal practical robustness. The radius of perturbation required by that definition can be taken to be $\delta\rho$.

To see that the claim is true, suppose otherwise: that there exists a sequence ϕ_i of solutions to $\mathcal{H}_{\delta_i\rho}$, where $\delta_i \searrow 0$, such that $\omega(\phi_i(0,0)) \leq m$ and points $(t_i, j_i) \in \operatorname{dom} \phi_i$ with $t_i + j_i \leq 2\tau$ so that (7.2) does not hold, i.e.,

$$\omega(\phi_i(t_i, j_i)) > \beta(\omega(\phi_i(0,0)), t_i + j_i) + \epsilon/2.$$

Since $\phi_i(0,0) \in \{x \in \mathcal{U} : \omega(x) \leq m\}$, which is a compact set, and $\phi_i(0,0) \in C_{\delta_i\rho} \cup D_{\delta_i\rho}$, one can assume that the points $\phi_i(0,0)$ converge to some point in $C \cup D$. At this point, $\mathcal{H}$ is pre-forward complete, and Proposition 6.33 implies that $\{\phi_i\}_{i=1}^{\infty}$ is locally eventually bounded. Since $\mathcal{H}$ is well-posed, the graphical limit of any graphically convergent subsequence of $\{\phi_i\}_{i=1}^{\infty}$, say ϕ, is a solution to $\mathcal{H}$. Without loss of generality, it can be assumed that (t_i, j_i)'s also converge, to some $(t, j) \in \operatorname{dom} \phi$. Continuity of β and ω shows that $\mathcal{KL}$ pre-asymptotic stability of $\mathcal{A}$ is violated by ϕ at $(t, j) \in \operatorname{dom} \phi$. This is a contradiction. $\qquad \square$

The developments of this section, as well as of Section 7.1 can be summarized in the following result.

Theorem 7.21. (Robustness of pre-asymptotic stability) *Let $\mathcal{H}$ be well-posed and $\mathcal{A} \subset \mathbb{R}^n$ be a compact set. If $\mathcal{A}$ locally pre-asymptotically stable then it is robustly $\mathcal{KL}$ pre-asymptotically stable on $\mathcal{B}_{\mathcal{A}}^p$.*

PROOF. Since $\mathcal{H}$ is well-posed, it is also nominally well-posed. Hence, Theorem 7.12 implies that $\mathcal{A}$ is $\mathcal{KL}$ pre-asymptotically stable on $\mathcal{B}_{\mathcal{A}}^p$. Lemma 7.20 implies that $\mathcal{A}$ is semiglobally practically robustly $\mathcal{KL}$ pre-asymptotically stable. Lemma 7.19 finishes the proof. $\qquad \square$

The next result describes how a hybrid system can be augmented, in a way that does not affect the basic assumptions on the system, and in a way that turns certain subsets of the basin of pre-attraction of a compact set to basins of pre-attraction. This trick is then employed in showing the existence of $\mathcal{KL}$ bounds for local pre-asymptotic stability, not necessarily on the whole basin of pre-attraction, but on its subsets.

Lemma 7.22. (Changing the basin of attraction) *Let $\mathcal{H} = (C, F, D, G)$ be a hybrid system on $\mathbb{R}^n$, $\mathcal{U} \subset \mathbb{R}^n$ be open, and $\mathcal{A} \subset \mathcal{U}$ be compact. Consider an augmented hybrid system $\mathcal{H}^*$ on $\mathbb{R}^n$ with the flow set, the flow map, the jump set, and the jump map given, respectively, by*

$$C^* = C, \quad F^* = F, \quad D^* = D \cup (\mathbb{R}^n \setminus \mathcal{U}), \quad G^*(x) = \begin{cases} G(x) & x \in D \\ G(x) \cup \{x\} & x \in \mathbb{R}^n \setminus \mathcal{U} \end{cases} .$$

Then

(a) *if $\mathcal{H}$ satisfies Assumption 6.5, then $\mathcal{H}^*$ satisfies Assumption 6.5;*

(b) *if $\mathcal{A}$ is locally pre-asymptotically stable for $\mathcal{H}$ and $\mathcal{U}$ is a strongly forward pre-invariant subset of the basin of pre-attraction of $\mathcal{A}$ for $\mathcal{H}$, then $\mathcal{A}$ is locally pre-asymptotically stable for $\mathcal{H}^*$ and the basin of pre-attraction of $\mathcal{A}$ for $\mathcal{H}^*$ equals $\mathcal{U}$.*

PROOF. Conclusion (a) is straightforward because $\mathcal{U}$ is open. Since $\mathcal{U}$ is a neighborhood of $\mathcal{A}$, and the construction of $\mathcal{H}^*$ does not alter the data in $\mathcal{U}$, local pre-asymptotic stability of $\mathcal{A}$ carries over from $\mathcal{H}$ to $\mathcal{H}^*$.

To see the second conclusion in (b), note that $\mathcal{A} \not\subset \mathbb{R}^n \setminus \mathcal{U}$, and because $x \in G^*(x)$ for $x \in \mathbb{R}^n \setminus \mathcal{U}$, points $x \in \mathbb{R}^n \setminus \mathcal{U}$ are not in the basin of pre-attraction of $\mathcal{A}$ for $\mathcal{H}^*$. Because $\mathcal{U}$ is strongly forward pre-invariant, points $x \in \mathcal{U}$ are in the basin of pre-attraction of $\mathcal{A}$ for $\mathcal{H}^*$. Hence, the basin equals $\mathcal{U}$. $\qquad \square$

Corollary 7.23. (From robust to $\mathcal{KL}$ pAS) *Let $\mathcal{A} \subset \mathbb{R}^n$ be a compact set that is locally pre-asymptotically stable for $\mathcal{H}$. Suppose that the local pre-asymptotic stability is robust on $\mathcal{U} \subset \mathcal{B}_{\mathcal{A}}^p$, an open and strongly forward pre-invariant set containing $\mathcal{A}$. Then $\mathcal{A}$ is robustly $\mathcal{KL}$ pre-asymptotically stable on $\mathcal{U}$.*

PROOF. Consider the system $\mathcal{H}^*$ from Lemma 7.22. Then $\mathcal{A}$ is locally pre-asymptotically stable for $\mathcal{H}^*$ and the basin of pre-attraction of $\mathcal{A}$ for $\mathcal{H}^*$ equals $\mathcal{U}$. Let ρ be the function coming from the robustness of pAS of $\mathcal{A}$ for $\mathcal{H}$, as in Definition 7.15. Without loss of generality it can be assumed that, for all $x \in \mathcal{U}$, $x + \rho(x)\mathbb{B} \subset \mathcal{U}$. Then the data of $\mathcal{H}_\rho$ agrees with the data of $(\mathcal{H}^*)_\rho$ on $\mathcal{U}$, and hence $\mathcal{A}$ is locally pre-asymptotically stable for $\mathcal{H}^*$, robustly on $\mathcal{U}$. Let $\widehat{\mathcal{H}^*}$ be the Krasovskii regularization of $\mathcal{H}^*$. Lemma 7.17 implies that $\mathcal{A}$ is locally pAS for $\widehat{\mathcal{H}^*}$ and the basin of pre-attraction of $\mathcal{A}$ for $\widehat{\mathcal{H}^*}$ equals $\mathcal{U}$. The conclusion of the corollary now comes from Theorem 7.21. □

In particular, considering $\mathcal{U} = \mathcal{B}_{\mathcal{A}}^p$ in Corollary 7.23 shows that robustness of pre-asymptotic stability on the basin of pre-attraction is equivalent to robust $\mathcal{KL}$ stability.

7.4 ROBUSTNESS COROLLARIES

The first corollary of robustness described here contains a reduction principle that can be used to simplify the analysis of a hybrid system.

Corollary 7.24. (Reduction principle) *Let $\mathcal{H} = (C, F, D, G)$ be well-posed. Suppose for $\mathcal{H}$ that the compact set $\mathcal{A}_1$ is locally pre-asymptotically stable with basin of pre-attraction $\mathcal{B}_{\mathcal{A}_1}^p$. In addition, suppose that, for the hybrid system $\mathcal{H}^r = (C \cap \mathcal{A}_1, F, D \cap \mathcal{A}_1, G)$, the compact set $\mathcal{A}_2 \subset \mathcal{A}_1$ is globally pre-asymptotically stable. Then, for the system $\mathcal{H}$, the compact set $\mathcal{A}_2$ is locally pre-asymptotically stable with basin of pre-attraction $\mathcal{B}_{\mathcal{A}_2}^p = \mathcal{B}_{\mathcal{A}_1}^p$.*

PROOF. Let $\varepsilon > 0$ be arbitrary. According to Lemma 7.20, for the system $\mathcal{H}^r$, the compact set $\mathcal{A}_2$ is semiglobally practically robustly $\mathcal{KL}$ pre-asymptotically stable on $\mathbb{R}^n$. In particular, there exists $\delta > 0$ such that each $\psi \in \mathcal{S}_{\mathcal{H}_\delta^r}(\mathcal{A}_2 + \delta\mathbb{B})$ satisfies $|\psi(t, j)|_{\mathcal{A}_2} \leq \varepsilon$ for all $(t, j) \in \operatorname{dom}\psi$. In addition, since the set $\mathcal{A}_1$ is stable for $\mathcal{H}$, there exists $\rho \in (0, \delta]$ such that each $\phi \in \mathcal{S}_{\mathcal{H}}(\mathcal{A}_1 + \rho\mathbb{B})$ satisfies $|\phi(t, j)|_{\mathcal{A}_1} \leq \delta$ for all $(t, j) \in \operatorname{dom}\phi$. Now consider $\phi \in \mathcal{S}_{\mathcal{H}}(\mathcal{A}_2 + \rho\mathbb{B})$. Since $\mathcal{A}_2 \subset \mathcal{A}_1$, it follows that $|\phi(0, 0)|_{\mathcal{A}_1} \leq \rho$ and, in turn, that $|\phi(t, j)|_{\mathcal{A}_1} \leq \delta$ for all $(t, j) \in \operatorname{dom}\phi$. In particular, from the definition of $\mathcal{H}_\delta^r$ and the fact that $\rho \leq \delta$, we have $\phi \in \mathcal{S}_{\mathcal{H}_\delta^r}(\mathcal{A}_2 + \delta\mathbb{B})$. Then, from the discussion above, it follows that $|\phi(t, j)|_{\mathcal{A}_2} \leq \varepsilon$ for all $(t, j) \in \operatorname{dom}\phi$. This bound establishes stability of $\mathcal{A}_2$ for the system $\mathcal{H}$.

Now let $\phi \in \mathcal{S}_{\mathcal{H}}(\mathcal{B}_{\mathcal{A}_1}^p)$. By definition, ϕ is bounded and if it is complete then it converges to $\mathcal{A}_1$. To prove that if it is complete it converges to $\mathcal{A}_2$ it is enough to prove that for each $\varepsilon > 0$ there exists $T > 0$ such that $|\phi(t, j)|_{\mathcal{A}_2} \leq \varepsilon$

for all $(t,j) \in \operatorname{dom}\phi$ satisfying $t + j \geq T$. To establish the latter property, the stability discussion above implies that it is enough to have that, for each $\rho > 0$, $|\phi(t,j)|_{\mathcal{A}_1} \leq \rho$ for some $(t,j) \in \operatorname{dom}\phi$. This property follows from convergence to $\mathcal{A}_1$. Therefore, if ϕ is complete then it converges to $\mathcal{A}_2$. Consequently, $\mathcal{B}^p_{\mathcal{A}_1} \subset \mathcal{B}^p_{\mathcal{A}_2}$. On the other hand, $\mathcal{B}^p_{\mathcal{A}_1} = \mathcal{B}^p_{\mathcal{A}_2}$ since $\mathcal{A}_2 \subset \mathcal{A}_1$. Thus, the result is established. $\qquad\square$

Example 7.25. Let $k > 0$, let $\sigma : \mathbb{R}^2 \to \mathbb{R}$ be continuous, and consider the hybrid system with state $x \in \mathbb{R}^2$ with data (C, F, D, G) given as

$$C = \{x \in \mathbb{R}^2 : x_1 x_2 \leq -0.25x_2^2, |x_1| \leq k\}$$

$$F(x) = \begin{bmatrix} x_2\sigma(x) - 0.2x_1 \\ x_1 \end{bmatrix}$$

$$D = \{x \in \mathbb{R}^2 : x_1 x_2 \geq -0.25x_2^2, |x_1| \leq k\}$$

and

$$G(x) = \begin{bmatrix} 0.9x_1 + x_2\sigma(x) \\ -0.8x_2 \end{bmatrix}.$$

The Lyapunov function $V(x) = \frac{1}{2}x_2^2$ can be used to verify that the set $\mathcal{A}_1 := [-k,k] \times \{0\}$ is globally pre-asymptotically stable. Then, restricting attention to flows in $C \cap \mathcal{A}_1$ and jumps in $D \cap \mathcal{A}_1$, the Lyapunov function $\frac{1}{2}x_1^2$ can be used to verify that the system with data $(C \cap \mathcal{A}_1, F, D \cap \mathcal{A}_1, G)$ has the origin globally pre-asymptotically stable. It follows from Corollary 7.24 that the origin of the original system is globally pre-asymptotically stable. From this property, it follows that if the constraint $|x_1| \leq k$ is removed in the original system then the resulting system has origin locally pre-asymptotically stable. In the case that the function σ is bounded, the resulting system has the origin globally pre-asymptotically stable. This fact follows from the observation that each solution of the resulting system is bounded, so that each solution is captured by the system with the constraint $|x_1| \leq k$ by picking k sufficiently large.

Example 7.26. (Linear feedback using a finite-time observer) For the control system $\dot{\xi} = F\xi + Ju$, where (F, J) is stabilizable, consider using the feedback $u = K\hat{\xi}_1$ where $F + JK$ is Hurwitz and $\hat{\xi}_1$ comes from the finite-time observer in Example 7.6. In Example 7.6, take the compact set K_1 to be the closed ball of radius $M > 0$, take the compact, convex set K_2 to be the closed ball of radius $\mu > 0$ where $\mu \geq \max_{\xi \in M\mathbb{B}} |JK\xi|$, and set

$$v = \mu JK\hat{\xi}_1 / \max\left\{\mu, |JK\hat{\xi}_1|\right\}$$

so that $v \in \mu\mathbb{B}$. With these choices, the closed-loop control system of this example, but with ξ restricted to the set K_1, matches the form of system in Example 7.6. In particular, each solution of the current system with ξ restricted to the set K_1 is a solution of the system in Example 7.6. Thus, it follows from the discussion in that example that the compact set

$$\mathcal{A}_1 := \left\{ (\xi, \hat{\xi}_1, \hat{\xi}_2) \in K_1 \times \mathbb{R}^n \times \mathbb{R}^n : \xi = \hat{\xi}_1 = \hat{\xi}_2 \right\} \times [0, \delta]$$

is globally pre-asymptotically stable for the present system with ξ restricted to K_1. Now, inspired by Corollary 7.24, further restrict the flow set and jump set by intersecting them with $\mathcal{A}_1$. This forces $\hat{\xi}_1 = \xi$ so that $v = \mu JK\xi / \max\{\mu, |JK\xi|\}$ and, since $\xi \in K_1$, from the condition on μ it follows that $v = JK\xi$. Then, since $F + JK$ is Hurwitz, it follows from Corollary 7.24 that the set

$$\mathcal{A}_2 := \{0\} \times \{0\} \times \{0\} \times [0, \delta]$$

is globally pre-asymptotically stable for the closed-loop system with ξ restricted to K_1. Since K_1 contains a neighborhood of the origin, this implies that the compact set $\mathcal{A}_2$ is locally asymptotically stable when the constraint $\xi \in K_1$ is removed and $v = JK\hat{\xi}_1$, in other words, the saturation level μ is removed. In fact, the closed-loop system without the constraint $\xi \in K_1$ has the set $\mathcal{A}_2$ globally asymptotically stable because this set is locally asymptotically stable and the system has appropriate homogeneity properties, which are discussed in more detail in Chapter 9.

The next corollary follows immediately from Theorem 7.12 and Lemma 7.20.

Corollary 7.27. (Mildly changing parameters) *Suppose for the well-posed hybrid system $\mathcal{H}$ with state (x, p) and data (C, F, D, G) that imposes $\dot{p} = 0$ and $p^+ = p$, the compact set $\mathcal{A}$ is globally pre-asymptotically stable. Then the hybrid system with data (C, F_ρ, D, G_ρ) that imposes $\dot{p} \in \rho\mathbb{B}$ and $p^+ \in p + \rho\mathbb{B}$ while leaving $\dot{x}$ and x^+ unchanged has the compact set $\mathcal{A}$ semiglobally practically asymptotically stable in $\rho > 0$.*

The next corollary considers switching between a finite family of well-posed hybrid systems $\mathcal{H}_q$, $q \in \mathcal{Q} \subset \mathbb{R}$, with state $z \in \mathbb{R}^m$ and data $(\tilde{C}_q, \tilde{F}_q, \tilde{D}_q, \tilde{G}_q)$ under a sufficiently small average dwell-time constraint on the switching, parametrized by $\delta > 0$ and $N \geq 1$, and with resets $z^+ \in R_q(z)$ allowed at switches. The overall model has state $x = (z, q, \tau)$ with data

$$C = \left\{ (z, q, \tau) \in \mathbb{R}^m \times \mathcal{Q} \times [0, N] : z \in \tilde{C}_q \right\} \tag{7.3}$$

$$F_\rho(x) = \begin{bmatrix} \tilde{F}_q(z) \\ 0 \\ [0, \rho] \end{bmatrix} \tag{7.4}$$

$$\begin{aligned}
D_a &= \left\{ (z, q, \tau) \in \mathbb{R}^m \times \mathcal{Q} \times [0, N] : z \in \tilde{D}_q \right\} \\
D_b &= \mathbb{R}^m \times \mathcal{Q} \times [1, N] \\
D &= D_a \cup D_b
\end{aligned} \tag{7.5}$$

and

$$G_a(x) = \begin{bmatrix} \tilde{G}_q(z) \\ q \\ \tau \end{bmatrix}$$

$$G_b(x) = \begin{bmatrix} R_q(z) \\ \mathcal{Q} \\ \tau - 1 \end{bmatrix} \tag{7.6}$$

and

$$G(x) = \begin{cases} G_a(x) & \forall x \in D_a \setminus D_b \\ G_b(x) & \forall x \in D_b \setminus D_a \\ G_a(x) \cap G_b(x) & \forall x \in D_a \cap D_b . \end{cases} \tag{7.7}$$

Suppose that the supplemental reset map R_q is outer semicontinuous, locally bounded, nonempty for each $z \in \mathbb{R}^m$, and such that $R_q(\mathcal{A}) \subset \mathcal{A}$. In this case, the hybrid system (C, F_0, D, G) has the compact set $\mathcal{A} \times \mathcal{Q} \times [0, N]$ globally pre-asymptotically stable since at most $\lfloor N \rfloor$ switches can occur in this case, and for each $\varepsilon > 0$ there exists $\delta > 0$ such that $R_q(\mathcal{A} + \delta \mathbb{B}) \subset \mathcal{A} + \varepsilon \mathbb{B}$. Consequently, the following corollary results from Theorem 7.12 and Lemma 7.20.

Corollary 7.28. (Slow average dwell-time switching) *Let $\mathcal{Q}$ be a finite subset of $\mathbb{R}$ and suppose that for each $q \in \mathcal{Q}$, the compact set $\mathcal{A}$ is globally pre-asymptotically stable for the well-posed hybrid system $\mathcal{H}_q$ with data $(\tilde{C}_q, \tilde{F}_q, \tilde{D}_q, \tilde{G}_q)$. Consider the hybrid system with data (C, F_ρ, D, G) constructed from the data of $\mathcal{H}_q$ and the reset map R_q as in (7.3)-(7.7). Moreover, suppose that the reset map R_q is outer semicontinuous, locally bounded, nonempty for each $z \in \mathbb{R}^m$ and $R_q(\mathcal{A}) \subset \mathcal{A}$. Under these conditions, the compact set $\mathcal{A} \times \mathcal{Q} \times [0, N]$ is semiglobally practically pre-asymptotically stable for the system (C, F_ρ, D, G).*

Additional robustness corollaries appear in Chapter 9.

7.5 SMOOTH LYAPUNOV FUNCTIONS

Lyapunov functions appeared first in Section 3.2, in sufficient conditions for uniform global pre-asymptotic stability. They appear again in Section 8.3 in relaxed sufficient conditions for uniform pre-asymptotic stability of a compact set for a well-posed hybrid system. In the current section, it is stated and illustrated that *smooth* Lyapunov functions are, in fact, guaranteed to exist for hybrid systems with a compact locally pre-asymptotically stable set, if the pre-asymptotic stability is uniform and robust. This is stated in Theorem 7.31. Since the developments in Chapter 6 established that local pre-asymptotic stability of a compact set, for a nominally well-posed hybrid system, is uniform, and for a well-posed system, it is furthermore robust, smooth Lyapunov functions are guaranteed to exist for pre-asymptotically stable compact sets in well-posed hybrid systems. This is stated in Corollary 7.32.

These converse Lyapunov theorems are useful for several reasons. First, they motivate looking for a Lyapunov function to certify asymptotic stability. Second, they provide a convenient way of summarizing robustness properties through bounds that do not explicitly involve solutions. Also, historically converse Lyapunov theorems have been used in the nonlinear control literature to illustrate control design principles like global asymptotic stabilization by backstepping or semiglobal practical stabilization by high-gain output feedback.

The initial discussion of Lyapunov functions was oriented towards global pre-asymptotic stability. Since this chapter is dealing with local pre-asymptotic stability, the concept of a Lyapunov function needs to be slightly modified. The distance from the pre-asymptotically stable set, as used in (3.2a), is replaced below, in (7.8), by a proper indicator.

Definition 7.29. (Lyapunov function) *Let $\mathcal{U} \subset \mathbb{R}^n$ be an open set such that $G(\mathcal{U}) \subset \mathcal{U}$ and let $\mathcal{A} \subset \mathcal{U}$ be compact. A function $V : \mathcal{U} \to \mathbb{R}_{\geq 0}$ is a Lyapunov function for $\mathcal{A}$ on $\mathcal{U}$ if it is continuously differentiable at every $x \in C \cap \mathcal{U}$ and there exists a proper indicator ω of $\mathcal{A}$ on $\mathcal{U}$ and $\underline{\alpha}, \overline{\alpha} \in \mathcal{K}_\infty$ such that:*

$$\underline{\alpha}(\omega(x)) \leq V(x) \leq \overline{\alpha}(\omega(x)) \qquad \text{for all} \ \ x \in (C \cup D \cup G(D)) \cap \mathcal{U}; \qquad (7.8)$$

$$\langle \nabla V(x), f \rangle \leq -V(x) \qquad \text{for all} \ \ x \in C \cap \mathcal{U}, \ f \in F(x); \qquad (7.9)$$

$$V(g) \leq V(x)/e \qquad \text{for all} \ \ x \in D \cap \mathcal{U}, \ g \in G(x). \qquad (7.10)$$

A smooth Lyapunov function for $\mathcal{A}$ on $\mathcal{U}$ is a Lyapunov function for $\mathcal{A}$ on $\mathcal{U}$ which is continuously differentiable on the whole set $\mathcal{U}$.

Smooth Lyapunov functions ensure not only pre-asymptotic stability, but also its robustness. Essentially, smoothness of V ensures that $\langle \nabla V(x), f \rangle$ is strictly negative, outside of $\mathcal{A}$, not just for $f \in F(x)$ but for $f \in F(y)$ for all y close enough to x, while continuity of V (ensured by smoothness) yields that $V(g) - V(x)$ is strictly negative, not just for $g \in G(x)$ but for $g \in G(y)$ for all y close enough to y. One can then easily conclude pre-asymptotic stability not just for $\mathcal{H}$ but for the Krasovskii regularization of $\mathcal{H}$. Because the Krasovskii regularization is well-posed, the stability turns out robust. This is precisely stated below, in Theorem 7.30.

What is perhaps more surprising and certainly harder to prove is that robustness of asymptotic stability implies the existence of smooth Lyapunov functions. In particular, smooth Lyapunov functions are guaranteed to exist for pre-asymptotically stable compact sets in well-posed hybrid systems This is stated in Theorem 7.31 and its corollaries.

Theorem 7.30. (Smooth Lyapunov functions imply robustness) *Let $\mathcal{H}$ be a hybrid system on $\mathbb{R}^n$, $\mathcal{U} \subset \mathbb{R}^n$ be an open set such that $G(\mathcal{U}) \subset \mathcal{U}$, and let $\mathcal{A} \subset \mathcal{U}$ be compact. Suppose that F is locally bounded. If there exists a smooth Lyapunov function for $\mathcal{A}$ on $\mathcal{U}$ then $\mathcal{A}$ is robustly $\mathcal{KL}$ pre-asymptotically stable on $\mathcal{U}$.*

PROOF. For every $x \in \mathcal{U} \setminus \mathcal{A}$ there exists $\delta > 0$ and $M > 0$ such that $V(y) \geq 2V(x)/3$, $|\nabla V(y) - \nabla V(x)| < V(x)/3M$, and $F(y) \subset M\mathbb{B}$ for all $y \in x + \delta\mathbb{B}$. Then, for any $x \in \left(\overline{C} \setminus \mathcal{A}\right) \cap \mathcal{U}$ and any $y \in (x + \delta\mathbb{B}) \cap C$, $f \in F(y)$ with δ chosen as above,

$$\langle \nabla V(x), f \rangle = \langle \nabla V(y), f \rangle + \langle (\nabla V(x) - \nabla V(y)), f \rangle \leq -V(y) + \frac{V(x)}{3M}M$$
$$\leq -\frac{2}{3}V(x) + \frac{1}{3}V(x) = -\frac{1}{3}V(x).$$

Consequently, $\langle \nabla V(x), f \rangle \leq -V(x)/3$ for all $f \in \overline{\mathrm{con}}F\left((x + \delta\mathbb{B}) \cap C\right)$, and in particular, for all $f \in \widehat{F}(x)$, where $\widehat{F}$ is the Krasovskii regularization of F. The inequality is in fact valid not only for $x \in \left(\overline{C} \setminus \mathcal{A}\right) \cap \mathcal{U}$ but for all $x \in \overline{C} \cap \mathcal{U}$, since $\nabla V(x) = 0$, $V(x) = 0$ when $x \in \mathcal{A}$.

Similarly, for every $x \in \mathcal{U} \setminus \mathcal{A}$ there exists $\delta > 0$ such that $V(y) \leq V(x)e^{2/3}$ for all $y \in x + \delta\mathbb{B}$. Then, for any $x \in \left(\overline{D} \setminus \mathcal{A}\right) \cap \mathcal{U}$ and any $y \in (x + \delta\mathbb{B}) \cap D$, $g \in G(y)$ with δ chosen as above,

$$V(g) \leq V(y)/e \leq V(x)/\sqrt[3]{e}.$$

Consequently, $V(g) \leq V(x)/\sqrt[3]{e}$ for all $x \in \left(\overline{D} \setminus \mathcal{A}\right) \cap \mathcal{U}$ and all $g \in \widehat{G}(x)$, where $\widehat{G}$ is the Krasovskii regularization of G. If $g \in \widehat{G}(x)$ for $x \in \mathcal{A}$, then, for every $\delta > 0$, $V(g) \leq \sup V\left(G((x + \delta\mathbb{B}) \cap D)\right) \leq \sup V\left((x + \delta\mathbb{B}) \cap D\right)/e$. Continuity of V then implies that $V(g) \leq 0$. Thus $V(g) \leq V(x)/\sqrt[3]{e}$ holds for all $x \in \overline{D} \cap \mathcal{U}$, all $g \in \widehat{G}(x)$.

Consequently, every Krasovskii solution ϕ to $\mathcal{H}$ with $\phi(0,0) \in \mathcal{U}$ satisfies

$$\underline{\alpha}(\omega(\phi(t,j))) \leq V(\phi(t,j)) \leq V(\phi(0,0))e^{-(t+j)/3} \leq \overline{\alpha}(\omega(\phi(0,0)))e^{-(t+j)/3},$$

which means that $\mathcal{A}$ is $\mathcal{KL}$ pre-asymptotically stable on $\mathcal{U}$ for $\widehat{\mathcal{H}}$, the Krasovskii regularization of $\mathcal{H}$. Example 6.6 and Theorem 6.30 imply that the Krasovskii regularization is well-posed. Lemma 7.19 and Lemma 7.20 imply that the $\mathcal{KL}$ pre-asymptotic stability of $\mathcal{A}$ on $\mathcal{U}$ for $\widehat{\mathcal{H}}$, and hence also for $\mathcal{H}$, is robust. $\qquad\square$

7.5.1 Existence

The following result and its corollaries are the main contribution of this chapter.

Theorem 7.31. (Robustness implies smooth Lyapunov functions) *Let $\mathcal{H}$ be a hybrid system on $\mathbb{R}^n$ with F locally bounded and let $\mathcal{A} \subset \mathbb{R}^n$ be a compact set. Suppose that $\mathcal{A}$ is locally pre-asymptotically stable for $\mathcal{H}$ and that $\mathcal{B}_{\mathcal{A}}^p$, the basin of attraction of $\mathcal{A}$ for $\mathcal{H}$, is open. If the local pre-asymptotic stability of $\mathcal{A}$ is robust on $\mathcal{B}_{\mathcal{A}}^p$ then there exists a smooth Lyapunov function for $\mathcal{A}$ on $\mathcal{B}_{\mathcal{A}}^p$.*

The proof of Theorem 7.31 is quite lengthy and is postponed until Section 7.6. It involves the following key steps:

- Noting that $\mathcal{A}$ is robustly $\mathcal{KL}$ pre-asymptotically stable for $\widehat{\mathcal{H}}$, the Krasovskii regularization of $\mathcal{H}$, and defining an upper semicontinuous preliminary Lyapunov function based on solutions to $\widehat{\mathcal{H}}$. The desired growth properties of this preliminary function come from the $\mathcal{KL}$ bounds and its upper semicontinuity comes from the fact that $\widehat{\mathcal{H}}$ is nominally well-posed.

- Smoothing the preliminary Lyapunov function. It is the robustness of pre-asymptotic stability for $\widehat{\mathcal{H}}$ that allows smoothing at points away from $\mathcal{A}$.

An important consequence of Theorem 7.31, Corollary 7.32, is stated below. It is obtained by combining Theorem 7.21 and Theorem 7.31 to yield the following corollary.

Corollary 7.32. (Smooth Lyapunov functions for well-posed systems) *Let $\mathcal{H}$ be a well-posed hybrid system on $\mathbb{R}^n$ and $\mathcal{A} \subset \mathbb{R}^n$ be a compact set that is locally pre-asymptotically stable for $\mathcal{H}$. Then, there exists a smooth Lyapunov function for $\mathcal{A}$ on $\mathcal{B}_{\mathcal{A}}^p$.*

Another consequence of the theorem above is the existence of smooth Lyapunov functions on certain subsets of the basin of pre-attraction. It is obtained by combining Corollary 7.23 with Theorem 7.31 to show the following corollary.

Corollary 7.33. (From robust to $\mathcal{KL}$ pAS) *Let $\mathcal{A} \subset \mathbb{R}^n$ be a compact set that is locally pre-asymptotically stable for $\mathcal{H}$. Suppose that the local pre-asymptotic stability is robust on $\mathcal{U} \subset \mathcal{B}_{\mathcal{A}}^p$, an open and strongly forward pre-invariant set containing $\mathcal{A}$. Then, there exists a smooth Lyapunov function for $\mathcal{A}$ on $\mathcal{U}$.*

7.5.2 Special Cases

The results on existence of smooth Lyapunov functions for well-posed hybrid systems with a pre-asymptotically stable compact set are now specialized to certain classes of systems. First, hybrid automata are considered. Recall that hybrid automata, including how to model them as a hybrid system, were considered in Section 1.4.2, and conditions on a hybrid automaton to ensure that the resulting hybrid system is well-posed were collected in Example 6.7.

Example 7.34. (Smooth Lyapunov functions for hybrid automata) Consider a hybrid automaton and let $\mathcal{H}$ be the hybrid system modeling it, as constructed in (1.10) and (1.12). Suppose that the hybrid automaton satisfies the assumptions listed in Example 6.7 — closed domains and guard sets, continuous flow maps and resets — and consequently, that $\mathcal{H}$ satisfies Assumption 6.5 and hence is well-posed. Suppose that

- for every $\varepsilon > 0$ there exists $\delta > 0$ such that $|z(0,0)| \leq \delta$ implies $|z(t,j)| \leq \varepsilon$ for all $(t,j) \in \mathrm{dom}\,\phi$, for every solution $\phi = (\sigma, z)$ to $\mathcal{H}$;

- there exists $\mu > 0$ such that z is bounded for every solution $\phi = (\sigma, z)$ to $\mathcal{H}$ with $|z(0,0)| \leq \mu$, and if, furthermore, ϕ is complete, then $\lim_{t+j \to \infty} z(t,j) = 0$.

This assumption is exactly the pre-asymptotic stability of the set $Q \times \{0\}$ for the hybrid system $\mathcal{H}$, and can be thought of as the pre-asymptotic stability of $\{0\}$ for the hybrid automaton, if one is not concerned about the behavior of the discrete state/logical mode. The conclusions of Proposition 7.4, regarding the basin of pre-attraction for $\mathcal{H}$ being open, and of Corollary 7.32, regarding the existence of a smooth Lyapunov function for $\mathcal{H}$, can be translated back to the hybrid automaton setting. They yield the following:

- For every $q \in Q$, let the set $\mathcal{B}_q$ consist of all $\xi \in \mathbb{R}^n$ such that, for every solution $\phi = (\sigma, z)$ to $\mathcal{H}$ with $\sigma(0,0) = q$, $z(0,0) = \xi$, z is bounded and if, furthermore, ϕ is complete, then $\lim_{t+j \to \infty} z(t,j) = 0$. Then $\mathcal{B}_q$ is open.

- For every $q \in Q$, there exists a continuously differentiable function $V_q : \mathcal{B}_q \to \mathbb{R}_{\geq 0}$, a proper indicator ω_q of $\{0\}$ on $\mathcal{B}_q$, and $\underline{\alpha}_q, \overline{\alpha}_q \in \mathcal{K}_\infty$ such that

$$\underline{\alpha}_q(|\xi|) \leq V_q(\xi) \leq \overline{\alpha}_q(|\xi|) \qquad \forall \xi \in S_q \cap \mathcal{B}_q$$

$$\langle \nabla V(\xi), f(q,\xi) \rangle \leq -V_q(\xi) \qquad \forall \xi \in \mathrm{Domain}(q) \cap \mathcal{B}_q$$

  where the set S_q consists of all $\xi \in \mathrm{Domain}(q)$, all $\xi \in \mathrm{Guard}(q,q')$ where q' is such that $(q,q') \in \mathrm{Edges}$, and all $\xi \in \mathrm{Reset}\,(q', q, \mathrm{Guard}(q', q))$ where q' is such that $(q', q) \in \mathrm{Edges}$.

- For every $(q, q') \in \mathrm{Edges}$,

$$V_{q'}\,(\mathrm{Reset}\,(q', q, \xi)) \leq V_q(\xi)/e \qquad \forall \xi \in \mathrm{Guard}(q, q').$$

In short, there exists a Lyapunov function V_q for every mode $q \in Q$ which strictly decreases during flow in mode q, and during a transition from mode q to mode q', the value of $V_q(\xi)$ strictly decreases to $V_{q'}(\xi')$, where ξ' represents the value of the continuous state after a reset.

Similarly, the existence of smooth Lyapunov functions can be deduced for switching systems, first modeled in the hybrid setting in Section 1.4.4. Consider a switching system with continuous dynamics f_q and subject to dwell-time, average dwell-time, or persistent dwell-time switching. Each of these cases can be modeled as a well-posed hybrid system, as shown in Section 2.4. Example 3.2 showed how to relate asymptotic stability of the origin for the switching system to asymptotic stability of an appropriate compact set for the hybrid system. Then, a smooth Lyapunov function for the hybrid system that models the switching system translates to a Lyapunov function for the switching system, one that depends not only on the state variable z but also on q and, furthermore, on the additional timer variable or variables. In contrast, natural sufficient conditions, for example, those described in Example 3.22 for the case of average dwell-time switching, usually involve Lyapunov functions depending on z and q.

A special case of quite different flavor is provided by constrained continuous-time systems which fail to have any complete solutions.

Example 7.35. (Perfectly incomplete systems) Consider a differential inclusion $\dot{z} \in F(z)$ constrained to C. Suppose that

- there does not exist a complete solution to $\dot{z} \in F(z)$ such that $z(t) \in C$ for all $t \geq 0$;

- C is compact and F satisfies Assumption 6.5.

Then, there exists a continuously differentiable $W : \mathbb{R}^n \to \mathbb{R}_{\geq 0}$ such that $W(z) > 0$ if $z \in C$ and

$$\langle W(z), f \rangle \leq -1 \quad \forall z \in C, \ f \in F(z).$$

To see this, note that any compact set $\mathcal{A}$ such that $\mathcal{A} \cap C = \emptyset$ is locally pre-asymptotically stable (for the hybrid system $\mathcal{H} = (C, F, D, G)$ with $D = \emptyset$). Indeed, local stability holds vacuously, since for some $\delta > 0$, there are no solutions starting in $\mathcal{A} + \delta \mathbb{B}$. Pre-attractivity holds because there are no complete solutions, and every solution is bounded due to C being bounded. Corollary 7.32 implies the existence of a smooth V, positive on C, and such that $\langle V(z), f \rangle \leq -V(z)$, for all $z \in C$, $f \in F(z)$. Then one can take $m = \min_{z \in C} V(z)$, which is positive, and consider $W = V/m$.

Note that W provides a smooth upper bound on the time that solutions to the differential inclusion remain in C, parameterized by initial points. That is, if $z : [0, T] \to \mathbb{R}$ is such that $\dot{z} \in F(z)$, $z \in C$ for almost all $t \in [0, T]$, then $T \leq W(z(0))$. The same conclusion can be reached in the case of C not compact but closed, and an additional condition ensuring that all solutions are bounded. A similar construction leads to a smooth upper bound on the number of jumps that a solution to $z^+ \in G(z)$, $z \in D$ may experience.

7.6 PROOF OF ROBUSTNESS IMPLIES SMOOTH LYAPUNOV FUNCTIONS

This section is devoted to proving Theorem 7.31.

7.6.0.1 Preliminary results

Two preliminary results are needed. The first one shows that given a state perturbation, smaller state perturbations can be constructed, with various useful properties. For example, Lemma 7.36 (a) shows that, given a hybrid system $\mathcal{H}$ and a state perturbation ρ, a smaller perturbation ρ' can be constructed so that every solution to $(\mathcal{H}_{\rho'})_{\rho'}$ is a solution of $\mathcal{H}_\rho$. Here $(\mathcal{H}_{\rho'})_{\rho'}$ is a ρ' perturbation of the ρ' perturbation $\mathcal{H}_{\rho'}$ of $\mathcal{H}$. The second preliminary result, Lemma 7.37, starts with a perturbation ρ and constructs a smaller one, ρ', so that for every compact solution ϕ to $\mathcal{H}_{\rho'}$, there exists a solution ψ to $\mathcal{H}_\rho$ between the endpoints of ϕ perturbed by small but otherwise arbitrary vectors.

Lemma 7.36. (Perturbations of perturbations) *Let* $\rho : \mathbb{R}^n \to \mathbb{R}_{\geq 0}$ *be a continuous function. There exists a continuous function* $\rho' : \mathbb{R}^n \to \mathbb{R}_{\geq 0}$ *such that* $\rho'(x) \leq \rho(x)$ *for all* $x \in \mathbb{R}^n$, $\rho'(x) > 0$ *if and only if* $\rho(x) > 0$, *and such that the following conditions are satisfied:*

(a)
$$\left(F_{\rho'}\right)_{\rho'}(x) \subset F_\rho(x) \quad \text{for all } x \in \left(C_{\rho'}\right)_{\rho'}, \qquad \left(C_{\rho'}\right)_{\rho'} \subset C_\rho,$$
$$\left(G_{\rho'}\right)_{\rho'}(x) \subset G_\rho(x) \quad \text{for all } x \in \left(D_{\rho'}\right)_{\rho'}, \qquad \left(D_{\rho'}\right)_{\rho'} \subset D_\rho,$$

where $\left(C_{\rho'}\right)_{\rho'}$, $\left(D_{\rho'}\right)_{\rho'}$, $\left(F_{\rho'}\right)_{\rho'}$, and $\left(G_{\rho'}\right)_{\rho'}$ are the data of $\left(\mathcal{H}_{\rho'}\right)_{\rho'}$, the ρ'-perturbation of the ρ'-perturbation $\mathcal{H}_{\rho'}$ of $\mathcal{H}$.

(b)
$$x + \rho'(x)\mathbb{B} \subset C_\rho \qquad \text{for all } x \in C_{\rho'},$$
$$x + \rho'(x)\mathbb{B} \subset D_\rho \qquad \text{for all } x \in D_{\rho'},$$
$$F_{\rho'}(x) + \rho'(x)\mathbb{B} \subset F_\rho(y) \qquad \text{for all } y \in x + \rho'(x)\mathbb{B},$$
$$g + \rho'(g)\mathbb{B} \subset G_\rho(y) \qquad \text{for all } g \in G_{\rho'}(x), \ y \in x + \rho'(x)\mathbb{B}.$$

PROOF. To begin, note that without loss of generality it can be assumed that $\rho(x) \leq 1$ and if $\rho(x) > 0$ then $\rho(y) > 0$ for all $y \in x + \rho(x)\mathbb{B}$. Indeed, in the opposite case, ρ can be replaced by $x \mapsto \min\{\rho(x), 1, |x|_S/2\}$ where $S = \{x \in \mathbb{R}^n : \rho(x) = 0\}$.

For (a), it is enough to pick ρ' that is positive whenever ρ is positive and so that for all $y \in x + \rho'(x)\mathbb{B}$, the inequality $\rho'(y) \leq \rho(x)/2$ holds, and consequently,

$$y + \rho'(y)\mathbb{B} \subset x + \rho(x)\mathbb{B} \qquad \text{for all } y \in x + \rho'(x)\mathbb{B}. \tag{7.11}$$

An example of such ρ' is $\rho'(x) = \min_{\{y \in \mathbb{R}^n : x \in y + \rho(y)\mathbb{B}\}} \rho(y)/2$. Then, if $x + \rho'(x)\mathbb{B}$ contains some y for which $y + \rho'(y)\mathbb{B} \cap C \neq \emptyset$, then (7.11) implies that $x + \rho(x)\mathbb{B} \cap C \neq \emptyset$. This shows that $\left(C_{\rho'}\right)_{\rho'} \subset C_\rho$, and the same reasoning shows that $\left(D_{\rho'}\right)_{\rho'} \subset D_\rho$. Regarding perturbations of F, for $x \in \left(C_{\rho'}\right)_{\rho'}$,

$$\begin{aligned}\left(F_{\rho'}\right)_{\rho'}(x) &= \overline{\mathrm{con}}F_{\rho'}\left((x + \rho'(x)\mathbb{B}) \cap C_{\rho'}\right) + \rho'(x)\mathbb{B} \\ &= \overline{\mathrm{con}} \bigcup_{y \in (x+\rho'(x)\mathbb{B}) \cap C_{\rho'}} \left[\overline{\mathrm{con}}F\left((y + \rho'(y)\mathbb{B}) \cap C\right) + \rho'(y)\mathbb{B}\right] + \rho'(x)\mathbb{B}\end{aligned}$$

which thanks to $\rho'(y) \leq \rho(x)/2$ for all $y \in x + \rho'(x)\mathbb{B}$, in particular $\rho'(x) \leq \rho(x)/2$, and to (7.11) is a subset of

$$\overline{\mathrm{con}} \bigcup_{y \in (x+\rho'(x)\mathbb{B}) \cap C_{\rho'}} \left[\overline{\mathrm{con}}F\left((x + \rho(x)\mathbb{B}) \cap C\right) + \frac{1}{2}\rho(x)\mathbb{B}\right] + \frac{1}{2}\rho(x)\mathbb{B}.$$

Since the sum of two convex sets is convex, this leads to $(F_{\rho'})_{\rho'}(x) \subset F_\rho(x)$. Regarding perturbations of G, for $x \in (D_{\rho'})_{\rho'} \subset D_\rho$, $(G_{\rho'})_{\rho'}(x) = \bigcup_{g \in S} g + \rho'(g)\mathbb{B}$ where $S = G_{\rho'}((x + \rho'(x)\mathbb{B}) \cap D_{\rho'})$. Now,

$$S = \bigcup_{y \in (x+\rho'(x)\mathbb{B}) \cap D_{\rho'}} G_{\rho'}(y) = \bigcup_{y \in (x+\rho'(x)\mathbb{B}) \cap D_{\rho'}} \bigcup_{z \in G((y+\rho'(y)\mathbb{B}) \cap D)} z + \rho'(z)\mathbb{B}$$

$$\subset \bigcup_{y \in (x+\rho'(x)\mathbb{B}) \cap D_{\rho'}} \bigcup_{z \in G((x+\rho(x)\mathbb{B}) \cap D)} z + \rho'(z)\mathbb{B} = \bigcup_{z \in G((x+\rho(x)\mathbb{B}) \cap D)} z + \rho'(z)\mathbb{B},$$

where the inclusion above comes from (7.11). Then, again thanks to (7.11), given any $z \in G((x + \rho(x)\mathbb{B}) \cap D)$, any $g \in z + \rho'(z)\mathbb{B}$, one has $g + \rho'(g)\mathbb{B} \subset z + \rho(z)\mathbb{B}$. This leads to $(G_{\rho'})_{\rho'}(x) \subset G_\rho(x)$.

For (b), it is additionally required that ρ' be such that for all $y \in x + \rho'(x)\mathbb{B}$, the inequality $\rho'(x) \le \rho(y)/2$ holds, and consequently,

$$x + \rho'(x)\mathbb{B} \subset y + \rho(y)\mathbb{B} \quad \text{for all} \quad y \in x + \rho'(x)\mathbb{B}. \tag{7.12}$$

To obtain such ρ', one can consider the pointwise minimum of ρ' having the properties used in the proof of (a) and the function $x \mapsto \min_{y \in x + \rho(x)\mathbb{B}} \rho(y)/2$. Given such a ρ', the proofs of inclusions in (b) are similar to those given for (a). $\qquad\square$

Lemma 7.37. (Perturbations of perturbed solutions) *Let $\rho : \mathbb{R}^n \to \mathbb{R}_{\ge 0}$ be a continuous function and set $S = \{x \in \mathbb{R}^n : \rho(x) > 0\}$. Suppose that F is locally bounded. There exists a continuous function $\rho' : \mathbb{R}^n \to \mathbb{R}_{\ge 0}$ such that $\rho'(x) \le \rho(x)$ for all $x \in \mathbb{R}^n$, $\rho'(x) > 0$ if and only if $x \in S$, and such that, for every solution ϕ to $\mathcal{H}_{\rho'}$ with $\operatorname{rge} \phi \subset S$, every $(t, j) \in \operatorname{dom} \phi$, and every $\xi \in \mathbb{B}$, there exists a solution ψ to $\mathcal{H}_\rho$ with $\operatorname{dom} \psi = \operatorname{dom} \phi$ and*

$$\psi(0,0) = \phi(0,0) + \rho'(\phi(0,0))\xi, \quad \psi(t,j) = \phi(t,j) + \rho'(\phi(t,j))\xi.$$

In fact, one can pick ρ' to be smooth on S and take $\psi(s,i) = \phi(s,i) + \rho'(\phi(s,i))\xi$ for all $(s,i) \in \operatorname{dom} \phi$, $s + i \le t + j$.

PROOF. Let $\rho_0 : \mathbb{R}^n \to \mathbb{R}_{\ge 0}$ be a continuous function such that $\rho_0(x) \le \rho(x)$ for all $x \in O$, $\rho_0(x) > 0$ if and only if $\rho(x) > 0$, and (b) of Lemma 7.36 holds (with ρ' replaced by ρ_0). It is straightforward to verify that any continuous function $\rho' : \mathbb{R}^n \to \mathbb{R}_{\ge 0}$ that is smooth on S, and such that $\rho'(x) \le \rho_0(x)$ for all $x \in O$, $\rho'(x) > 0$ if and only if $\rho_0(x) > 0$, and

$$|\nabla \rho'(x)| \le \frac{\rho_0(x)}{1 + \sup_{f \in F_{\rho_0}(x)} |f|} \quad \text{when } x \in C_{\rho_0},$$

meets the conclusions of the lemma, in fact the required solution ψ can be taken to be $\psi(s,i) = \phi(s,i) + \rho'(\phi(s,i))\xi$.

To find such a ρ', consider a locally finite open cover $\{\mathcal{U}_i\}_{i=1}^{\infty}$ of S, with $\overline{\mathcal{U}_i} \subset S$ compact, and a smooth partition of unity $\{\psi_i\}_{i=1}^{\infty}$ subordinate to this cover. For $x \in S$, let

$$\rho'(x) = \sum_{i=1}^{\infty} \frac{2^{-i} a_i}{\max_{z \in \mathcal{U}_i} \max\{\psi_i(z), \nabla\psi_i(z)\}} \psi_i(x),$$

where $a_i \in (0,1)$ is such that $a_i \leq \rho_0(x)$ for all $x \in \mathcal{U}_i$ and, if $\mathcal{U}_i \cap C_{\rho_0} \neq \emptyset$, such that $a_i \sup_{z \in \mathcal{U}_i \cap C_{\rho_0}} \max_{f \in F_{\rho_0}(z)} |f| \leq 1$. When $\rho(x) = 0$, set $\rho'(x) = 0$. □

7.6.0.2 Preliminary Lyapunov function

Let $\mathcal{H}$ be a hybrid system, $\mathcal{A}$ be a compact set that is locally pre-asymptotically stable for $\mathcal{H}$, robustly on the basin of pre-attraction, $\mathcal{B}_{\mathcal{A}}^p$, which is open. Lemma 7.17 implies that $\mathcal{A}$ is locally pre-asymptotically stable for $\widehat{\mathcal{H}}$, the Krasovskii regularization of $\mathcal{H}$, and furthermore, that the basin of pre-attraction of $\mathcal{A}$ for $\widehat{\mathcal{H}}$ equals $\mathcal{B}_{\mathcal{A}}^p$. Theorem 7.21 then implies that $\mathcal{A}$ is robustly $\mathcal{KL}$ pre-asymptotically stable on $\mathcal{B}_{\mathcal{A}}^p$, that is, that there exists a continuous function $\rho : \mathbb{R}^n \to \mathbb{R}_{\geq 0}$ that is positive on $\mathcal{B}_{\mathcal{A}}^p \setminus \mathcal{A}$ such that $\mathcal{A}$ is $\mathcal{KL}$ pre-asymptotically stable on $\mathcal{B}_{\mathcal{A}}^p$ for $\widehat{\mathcal{H}}_\rho$, the ρ perturbation of $\mathcal{H}$. By Lemma 3.41, there exists a proper indicator ω of $\mathcal{A}$ on $\mathcal{B}_{\mathcal{A}}^p$ and functions $a_1, a_2 \in \mathcal{K}_{\infty}$, $a_1 \leq a_2$, such that

$$a_1\left(\omega(\phi(t,j))\right) \leq a_2\left(\omega(\phi(0,0))\right) e^{-2(t+j)} \qquad \text{for all} \ \ (t,j) \in \operatorname{dom}\phi \qquad (7.13)$$

holds for every solution ϕ to $\widehat{\mathcal{H}}_\rho$ with $\phi(0,0) \in \mathcal{B}_{\mathcal{A}}^p$.

Consider the function $V_1 : \mathcal{B}_{\mathcal{A}}^p \to \mathbb{R}_{\geq 0}$ given, at each $\xi \in \mathcal{B}_{\mathcal{A}}^p$, by

$$V_0(\xi) = \begin{cases} \sup_{\phi \in \mathcal{S}_{\widehat{\mathcal{H}}}(\xi), \, (t,j) \in \operatorname{dom}\phi} a_1\left(\omega(\phi(t,j))\right) e^{t+j} & \xi \in \widehat{C} \cup \widehat{D} \\ a_1(\omega(\xi)) & \xi \notin \widehat{C} \cup \widehat{D}. \end{cases}$$

It is immediate from (7.13) that

$$a_1(\omega(\xi)) \leq V_0(\xi) \leq a_2(\omega(\xi)) \qquad \text{for all} \ \ \xi \in \mathcal{B}_{\mathcal{A}}^p. \qquad (7.14)$$

Furthermore, for every solution ϕ to $\widehat{\mathcal{H}}$ with $\phi(0,0) = \xi$,

$$V_0(\phi(t,j)) \leq V_0(\xi) e^{-t-j} \qquad \text{for all} \ \ (t,j) \in \operatorname{dom}\phi. \qquad (7.15)$$

Indeed, if $\phi(t,j) \notin \widehat{C} \cup \widehat{D}$, this comes directly from the definition of $V_0(\xi)$. If $\phi(t,j) \in \widehat{C} \cup \widehat{D}$, the inequality follows from the definitions of $V_0(\phi(t,j))$ and $V_0(\xi)$, by observing that any solution $\psi \in \mathcal{S}_{\widehat{\mathcal{H}}}(\phi(t,j))$ concatenated with ϕ yields an element of $\mathcal{S}_{\widehat{\mathcal{H}}}(\xi)$ given by $\phi(t',j')$ if $(t',j') \in \operatorname{dom}\phi$, $t' + j' < t + j$, $\psi(t'-t, j'-j)$ if $(t'-t, j'-j) \in \operatorname{dom}\psi$. Finally, V_0 is upper semicontinuous. Indeed, pick $\xi \in \mathcal{B}_{\mathcal{A}}^p \setminus \mathcal{A}$ and a sequence of points $\xi_i \in \mathcal{B}_{\mathcal{A}}^p$ converging to ξ

and, without loss of generality, such that $\lim_{i\to\infty} V_0(\xi_i)$ exists. If $\xi_i \notin \widehat{C} \cup \widehat{D}$ for infinitely many i's, then

$$\lim_{i\to\infty} V_0(\xi_i) = \lim_{i\to\infty} a_1(\omega(\xi_i)) = a_1(\omega(\xi)) \leq V_0(\xi).$$

In the opposite case, which can only occur if $\xi \in \widehat{C} \cup \widehat{D}$ since $\widehat{C}$, $\widehat{D}$ are closed, note that for all large enough i's, the suprema defining $V_0(\xi_i)$ can be taken over hybrid times (t, j) such that $e^{t+j} \leq a_2(\omega(\xi_i))/V_0(\xi_i)$. This follows from (7.13). Thus, (7.14) and continuity of a_1, a_2, ω imply that, for all all large enough i's, the suprema defining $V_0(\xi_i)$ can be taken over hybrid times (t, j) such that $e^{t+j} \leq 2a_2(\omega(\xi))/a_1(\omega(\xi))$. This bound and the fact that $\widehat{\mathcal{H}}$ is nominally well-posed (recall Example 6.6 and Theorem 6.8) imply both that the suprema are in fact maxima and that V_0 is upper semicontinuous. Essentially, the graphical limit of any graphically convergent subsequence of the sequence of arcs providing maxima in the definitions of $V_0(\xi_i)$ can be considered in the supremum defining $V_0(\xi)$.

7.6.0.3 Initial smoothing

Let $\rho' : \mathbb{R}^n \to \mathbb{R}_{\geq 0}$ be continuous and such that $\rho'(\xi) = 0$ if and only if $\xi \in \mathcal{A}$, $\xi + \rho'(\xi)\mathbb{B} \subset \mathcal{B}_{\mathcal{A}}^p \setminus \mathcal{A}$ for all $\xi \in \mathcal{B}_{\mathcal{A}}^p \setminus \mathcal{A}$,

$$a_1(\omega(\xi))/2 \leq V_0(\eta) \leq 2a_2(\omega(\xi)) \qquad \text{for all } \xi \in \mathcal{B}_{\mathcal{A}}^p, \ \eta \in \xi + \rho'(\xi)\mathbb{B}, \qquad (7.16)$$

and the conclusions of Lemma 7.37 hold. Pick any infinitely many times differentiable function $\psi : \mathbb{R}^n \to [0, 1]$ with $\psi(x) = 0$ if $x \notin \mathbb{B}$ and $\int \psi(x)\,dx = 1$. (Here and in what follows, integration is carried out over $\mathbb{R}^n$.) Consider the function $V_1 : \mathcal{B}_{\mathcal{A}}^p \to \mathbb{R}_{\geq 0}$ given, at each $\xi \in \mathcal{B}_{\mathcal{A}}^p$, by

$$V_1(\xi) = \int V_0(\xi + \rho'(\xi)\eta)\psi(\eta)\,d\eta. \qquad (7.17)$$

Directly from the definition of V_1 and from (7.16) it follows that

$$a_1(\omega(\xi))/2 \leq V_1(\xi) \leq 2a_2(\omega(\xi)) \qquad \text{for all } \xi \in \mathcal{B}_{\mathcal{A}}^p. \qquad (7.18)$$

The function V_1 is continuously differentiable on $\mathcal{B}_{\mathcal{A}}^p \setminus \mathcal{A}$. To see this, note that

$$V_1(\xi) = (\rho'(\xi))^{-n} \int k(\xi, \zeta)\,d\zeta$$

where $k(\xi, \zeta) = V_0(\zeta)\,\psi\left((\rho'(\xi))^{-1}(\zeta - \xi)\right)$. Now, continuous differentiability of the function $\xi \mapsto \int k(\xi, \zeta)\,d\zeta$ comes from the following lemma.

Lemma 7.38. (Differentiation under the integral sign) *Let $\mathcal{W} \subset \mathbb{R}^n$ be an open set and $k : \mathcal{W} \times \mathbb{R}^m \to \mathbb{R}$ be a function such that, for each $\xi \in \mathcal{W}$: (i) the function $\zeta \mapsto k(\xi, \zeta)$ is an L^1 function on $\mathbb{R}^m$; (ii) the function $\zeta \mapsto D_{\xi_i} k(\xi, \zeta)$,* where $D_{\xi_i} k$ denotes the partial derivative of k with respect to ξ_i, is an L^1 function on $\mathbb{R}^m$; and (iii) there exists an L^1 function b on $\mathbb{R}^m$ such that, for all ξ' sufficiently close to ξ, $|D_{\xi_i} k(\xi', \zeta)| \leq b(\zeta)$. Then the partial derivatives $D_{\xi_i} K$ of the function $K : \mathcal{W} \to \mathbb{R}$ given by

$$K(\xi) = \int k(\xi, \zeta)\, d\zeta$$

exist and $D_{\xi_i} K(\xi) = \int D_{\xi_i} k(\xi, \zeta)\, d\zeta$.

The assumptions of the lemma are easy to verify here. The function V_1 is upper semicontinuous, hence measurable; V_1 is locally bounded thanks to (7.18); the function

$$\zeta \mapsto \psi\left((\rho'(\xi))^{-1} (\zeta - \xi) \right)$$

and its partial derivatives with respect to ξ are nonzero only if $\zeta \in \xi + \rho'(\xi)\mathbb{B}$; and ρ' is such that $\xi + \rho'(\xi)\mathbb{B} \subset \mathcal{B}_{\mathcal{A}}^p \setminus \mathcal{A}$ for all $\xi \in \mathcal{B}_{\mathcal{A}}^p \setminus \mathcal{A}$.

The properties of ρ' stated in Lemma 7.37 imply, via (7.15), that $V_0(\phi(t, j) + \rho'(\phi(t, j))\eta) \leq V_0(\xi + \rho'(\xi)\eta)e^{-t-j}$ for every $\eta \in \mathbb{B}$ and every solution ϕ to $\widehat{\mathcal{H}}_{\rho'}$ with $\phi(0, 0) = \xi$, $\mathrm{rge}\, \phi \subset \mathcal{B}_{\mathcal{A}}^p \setminus \mathcal{A}$. This implies that

$$V_1(\phi(t, j)) \leq V_1(\xi)\, e^{-t-j} \qquad \text{for all} \quad (t, j) \in \mathrm{dom}\, \phi \tag{7.19}$$

for every such solution ϕ to $\widehat{\mathcal{H}}_{\rho'}$. This now implies that

$$\langle \nabla V_1(\xi), f \rangle \leq -V_1(\xi) \qquad \text{for all} \quad \xi \in C \setminus \mathcal{A},\ f \in F(\xi). \tag{7.20}$$

Indeed, as $\rho'(\xi) > 0$ for such a ξ, $\xi \in \mathrm{int}\, \widehat{C}_{\rho'}$ and each $f \in F(\xi)$ belongs to $F_{\rho'}(\eta)$ for all η close enough to ξ. This guarantees that $(t, 0) \mapsto \xi + tf$ is a solution to $\widehat{\mathcal{H}}_{\rho'}$ for small enough t, which combined with (7.19) implies (7.20). Similar, yet even simpler, arguments show that

$$V_1(g) \leq V_1(\xi)/e \qquad \text{for all} \quad \xi \in D \setminus \mathcal{A},\ g \in G(\xi) \tag{7.21}$$

as long as $g \notin \mathcal{A}$. But if $g \in \mathcal{A}$ then $V_1(g) = 0$ and so (7.21) holds.

7.6.0.4 Final smoothing

Pick $\varepsilon \in (0, 1]$ such that $\mathcal{A} + \varepsilon\mathbb{B} \subset \mathcal{B}_{\mathcal{A}}^p$. Let $b, c \in \mathcal{K}_\infty$ be such that

$$b(r) \leq \sup_{|\xi|_{\mathcal{A}} \leq r} \omega(\xi), \quad c(r) \leq \min_{\omega(\xi) = r} \frac{|\xi|_{\mathcal{A}}}{\max\{1, |\nabla V_1(\xi)|\}} \qquad \text{for all}\ \ r \in [0, \varepsilon].$$

Pick $s \in \mathcal{K}_\infty$ so that

$$s(r) \leq \left(b^{-1} \circ a_2^{-1}(r/2) \right)^2, \qquad s(r) \leq c \circ a_2^{-1}(r/2) \qquad \text{for all}\ \ r \in [0, \varepsilon]. \tag{7.22}$$

Let $S \in \mathcal{K}_\infty$ be the function given, for each $r \in \mathbb{R}_{\geq 0}$, by

$$S(r) = \int_0^r s(t)\,dt.$$

This function is continuously differentiable on $(0, \infty)$ with $S \in \mathcal{K}_\infty$, $\nabla S(r) = s(r)$, and as s is increasing,

$$S(r) \leq r s(r) \quad \text{and} \quad S(r/e) \leq S(r)/e \qquad \text{for all } r \geq 0. \tag{7.23}$$

Consider the function $V_2 : \mathcal{B}_\mathcal{A}^p \to \mathbb{R}_{\geq 0}$ given, for each $\xi \in \mathcal{B}_\mathcal{A}^p$, by

$$V_2(\xi) = S \circ V_1(\xi).$$

It is continuously differentiable on $\mathcal{B}_\mathcal{A}^p \setminus \mathcal{A}$. For all $\xi \in \mathcal{A} + \varepsilon \mathbb{B}$, it can be verified via (7.18), (7.22), (7.23) that

$$V_2(\xi) \leq s \circ V_1(\xi) \leq s \circ 2a_2 \circ \omega(\xi) \leq \left(b^{-1} \circ \omega(\xi) \right)^2 \leq |\xi|_\mathcal{A}^2,$$

and thus V_2 is differentiable at each $\xi \in \mathcal{A}$, with $\nabla V_2(\xi) = 0$, and also that

$$|\nabla V_2(\xi)| \leq s(V_1(\xi)) |\nabla V_1(\xi)| \leq c \circ \omega(\xi) |\nabla V_1(\xi)| \leq |\xi|_\mathcal{A},$$

and thus V_2 is continuously differentiable on $\mathcal{B}_\mathcal{A}^p$.

Thanks to (7.18), the function V_2 satisfies the bounds (7.8) required by the definition of a Lyapunov function with $\underline{\alpha}(r) = S\left(a_1(\omega(r))/2\right)$ and $\overline{\alpha}(r) = S\left(2a_2(\omega(r))\right)$. Condition (7.20) and the first property of S in (7.23) show that

$$\langle \nabla V_2(\xi), f \rangle \leq -s(V_1(\xi)) V_1(\xi) \leq -S(V_1(\xi)) = -V_2(\xi)$$

for all $\xi \in C \setminus \mathcal{A}$, all $f \in F(\xi)$. Consequently, the Lyapunov inequality (7.9) holds for V_2. Condition (7.21) and the second property of S in (7.23) show that

$$V_2(g) = S(V_1(g)) \leq S(V_1(\xi)/e) \leq S(V_1(\xi))/e = V_2(\xi)/e$$

for all $\xi \in D \setminus \mathcal{A}$, all $g \in G(\xi)$. Thus, (7.10) is satisfied by V_2.

7.7   NOTES

The analysis of uniformity in asymptotic stability for differential equations dates back to Massera [87]. For differential inclusions, uniformity and robustness was established by Clarke, Ledyaev, and Stern [27] for equilibria and by Teel and Praly [118] for sets; it was established for difference inclusions by Kellett and Teel [61]. Converse Lyapunov results for linear differential equations date back to Lyapunov [77]; for nonlinear differential equations, with different generality, to Massera [88], Kurzweil [65], Wilson [127], and more. For differential inclusions, converse results were given by Lin, Sontag, and Wang [74], and in a general setting, in [27] and [118]. Closely related results on converse theorems for switched systems were developed by Dayawansa and Martin [31] and by Mancilla-Aguiar and Garcia [84]. For difference inclusions, converse results were given by Jiang and Wang [58] and by Kellett and Teel [60], [61]. The link between robustness and smoothness of Lyapunov functions, used in [27], [118], and [61], appears to be first noted in [65]. Converse Lyapunov theorems for hybrid systems, without attention to smoothness issues, were developed by Ye, Michel, and Hou [130].

The results of Section 7.1 regarding the basin of attraction and the uniformity of pre-asymptotic stability were shown by Goebel and Teel [40]. Example 7.6 follows the presentation by Raff and Allgöwer in [99]. The trick in Lemma 7.14 was used in the original proof of the existence of smooth Lyapunov functions in Cai, Teel, and Goebel [25]. The semiglobal practical robustness of $\mathcal{KL}$ pre-asymptotic stability, as stated in Lemma 7.20, was established in [40] as well, and it was used by Cai et al. [25] to deduce robustness as stated here in Theorem 7.21. Corollary 7.24 appeared in Goebel, Sanfelice, and Teel [39] as Corollary 19 whereas Corollary 7.27 appeared in [25]. Generalizations of Corollary 7.28 were given in [39]. Additional consequences of robustness that were not presented here include robustness to temporal regularization described in [40, Example 5.6], robustness to simulation approximations discussed in Sanfelice and Teel [108], robustness to singular perturbations in Sanfelice and Teel [109], and robustness based on averaging, as described by Teel and Nešić [117]. The equivalence of robustness and the existence of a smooth Lyapunov function, included here in Theorems 7.30 and 7.31, was shown in a more general setting by Cai, Teel, and Goebel [24]. Lemma 7.38 comes from Lang [68, Chapter 13, Lemma 2.2].

Chapter Eight

Invariance principles

Invariance principles characterize the sets to which precompact solutions to a dynamical system must converge. They rely on invariance properties of ω-limit sets of solutions, as defined in Definition 6.17, and additionally on Lyapunov-like functions, which do not increase along solutions, or output functions. Invariance principles which rely on Lyapunov-like functions are presented in Section 8.2. Applications of these invariance principles to analysis of asymptotic stability are described in Section 8.3. Section 8.4 states an invariance principle involving not a Lyapunov-like function, but an output function having a certain property not along all solutions, but only along the solution whose behavior is being analyzed. Section 8.5 presents invariance principles for switching systems with dwell-time switching signals modeled as hybrid systems, as in Section 2.4.

Throughout this chapter, the following is assumed:

- the hybrid system $\mathcal{H}$ is nominally well-posed.

It is worth stressing that whether the system is well-posed is not relevant in this chapter.

8.1 INVARIANCE AND ω-LIMITS

The weak invariance property of a set was introduced in Definition 6.19. The adjective "weak" indicates that only one solution is required to remain in the set, in contrast to the "strong" notion in Definition 6.25. The definition of weak invariance of a set requires the set to be both weak forward and backward invariant. The following example illustrates some differences between the invariance concepts.

Example 8.1. (Weakly invariant sets) Consider a hybrid system in $\mathbb{R}$ given by

$$
\begin{aligned}
C &= [0, 2] & f(x) &= 1 \\
D &= [-2, -1] \cup [1, 2] & G(x) &= \begin{cases} x + 1 & \text{if } x \in [-2, -1] \cup (1, 2], \\ \{-1, 2\} & \text{if } x = 1. \end{cases}
\end{aligned}
$$

The largest weakly forward invariant subset of $\mathbb{R}$ for this system is

$$S_f = \{-2\} \cup \{-1\} \cup [0, 1].$$

Indeed, consider the solution ϕ given by $\phi(0,0) = -2$, $\phi(0,1) = -1$, $\phi(0,2) = 0$, $\phi(t,2) = t$ for $t \in [0,1]$, which then repeats the "jump to -1, jump to 0, flow to 1" behavior infinitely many times. It verifies forward invariance from -2 while its tails verify forward invariance from -1 and $[0,1]$. The largest weakly backward invariant subset of $\mathbb{R}$ is

$$S_b = \{-1\} \cup [0,3].$$

Indeed, the backward invariance can be verified by a solution that repeats the "jump from 1 to -1, jump to 0, flow from 0 to 1" behavior an arbitrary number of times and then, if needed, flows from 1 toward 2 and possibly jumps from 2 to 3. The largest weakly invariant, in other words both weakly forward and weakly backward invariant, subset turns out to be

$$S = \{-1\} \cup [0,1].$$

This set is not strongly invariant, in fact from every point in S there exists a solution that, eventually, flows out of S.

Proposition 6.21 concluded that for a precompact solution ϕ, its ω-limit set $\Omega(\phi)$ is weakly invariant and $|\phi(t,j)|_{\Omega(\phi)} \to 0$ when $t + j \to \infty$, $(t,j) \in \operatorname{dom}\phi$. Consequently, in Example 8.1, every complete solution, which is automatically precompact because $C \cup D$ is compact, converges to $S = \{-1\} \cup [0,1]$. In this example, though, all complete solutions are eventually the same, in the sense that they are all eventually periodic and repeat the "flow from 0 to 1, jump to -1, jump to 0" behavior infinitely many times. It is worth noting that the conclusion of convergence to S in the example is tight, in the sense that S is exactly the ω-limit set of every complete solution, so it is the smallest set to which the solutions can converge. Note also that relying only on weak forward or only on weak backward invariance would give a set larger than S.

8.2 INVARIANCE PRINCIPLES INVOLVING LYAPUNOV-LIKE FUNCTIONS

For a hybrid arc ϕ with domain $\operatorname{dom}\phi$, $t(j)$ denotes the least time t such that $(t,j) \in \operatorname{dom}\phi$, while $j(t)$ denotes the least index j such that $(t,j) \in \operatorname{dom}\phi$.

Given $V : \mathbb{R}^n \to \mathbb{R}$, any functions $u_c, u_d : \mathbb{R}^n \to [-\infty, \infty]$, and a set $U \subset \mathbb{R}^n$, it is said that the *growth of V along solutions to $\mathcal{H}$ is bounded by u_c, u_d on U* if for any $\phi \in \mathcal{S}_{\mathcal{H}}$ with $\operatorname{rge}\phi \subset U$,

$$V(\phi(\bar{t},\bar{j})) - V(\phi(\underline{t},\underline{j})) \le \int_{\underline{t}}^{\bar{t}} u_c(\phi(s,j(s)))\, ds + \sum_{j=\underline{j}+1}^{\bar{j}} u_d(\phi(t(j),j-1)) \quad (8.1)$$

for all $(\underline{t},\underline{j}), (\bar{t},\bar{j}) \in \operatorname{dom}\phi$ such that $(\underline{t},\underline{j}) \preceq (\bar{t},\bar{j})$. It is immediate that if the growth of V is bounded by u_c, u_d on U and u_c, u_d are nonpositive on U, then V

is *nonincreasing along* ϕ, for every solution $\phi \in \mathcal{S}_\mathcal{H}$, i.e., $V(\phi(\underline{t}, \underline{j})) \geq V(\phi(\overline{t}, \overline{j}))$. Along with the properties of the ω-limit set of a solution to a hybrid system $\mathcal{H}$, this fact leads to the following invariance principle.

Theorem 8.2. (Invariance principle involving a nonincreasing function) *Consider a continuous function* $V : \mathbb{R}^n \to \mathbb{R}$, *any functions* $u_c, u_d : \mathbb{R}^n \to [-\infty, \infty]$, *and a set* $U \subset \mathbb{R}^n$ *such that* $u_c(z) \leq 0$, $u_d(z) \leq 0$ *for every* $z \in U$ *and such that the growth of* V *along solutions to* $\mathcal{H}$ *is bounded by* u_c, u_d *on* U. *Let a precompact solution* $\phi^* \in \mathcal{S}_\mathcal{H}$ *be such that* $\overline{\mathrm{rge}\,\phi^*} \subset U$. *Then, for some* $r \in V(U)$, ϕ^* *approaches the nonempty set that is the largest weakly invariant subset of*

$$V^{-1}(r) \cap U \cap \left[\overline{u_c^{-1}(0)} \cup \left(u_d^{-1}(0) \cap G(u_d^{-1}(0)) \right) \right]. \tag{8.2}$$

PROOF. By Proposition 6.21, any precompact ϕ^* approaches its ω-limit, $\Omega(\phi^*)$, which is nonempty, compact, and weakly invariant. Precompactness of ϕ^*, continuity of V, and the fact that V is nonincreasing along ϕ^* implies that $V(\phi^*(t, j)) \to r$ as $t + j \to \infty$, $(t, j) \in \mathrm{dom}\,\phi^*$, for some $r \in \mathbb{R}$, which since $\overline{\mathrm{rge}\,\phi^*} \subset U$, is such that $r \in V(U)$.

Then, $V(\Omega(\phi^*)) = r$. Thus V is constant on solutions ϕ with $\mathrm{rge}\,\phi \subset \Omega(\phi^*)$, and so for any such solution and any $(\underline{t}, \underline{j}), (t, j), (\overline{t}, \overline{j}) \in \mathrm{dom}\,\phi$ with $(\underline{t}, \underline{j}) \preceq (t, j) \preceq (\overline{t}, \overline{j})$, one has

$$\int_{\underline{t}}^{\overline{t}} u_c(\phi(s, j(s)))\, ds + \sum_{i=\underline{j}+1}^{\overline{j}} u_d(\phi(t(i), i - 1)) = 0. \tag{8.3}$$

Pick $\xi \in \Omega(\phi^*)$. Since $\Omega(\phi^*)$ is weakly forward invariant, then there exists a complete solution $\phi \in \mathcal{S}_\mathcal{H}(\xi)$ with $\mathrm{rge}\,\phi \subset \Omega(\phi^*)$. If $(0, 1) \in \mathrm{dom}\,\phi$, applying (8.3) to $(\underline{t}, \underline{j}) = (0, 0)$, $(\overline{t}, \overline{j}) = (0, 1)$ yields $u_d(\phi(0, 0)) = 0$, and so $\xi \in u_d^{-1}(0)$. If $(T, 0) \in \mathrm{dom}\,\phi$ for some $T > 0$, then applying (8.3) to $(0, 0)$, $(T, 0)$ yields $\int_0^T u_c(\phi(s, 0))\, ds = 0$. Since u_c is nonpositive on S, $u_c(\phi(s, 0)) = 0$ for almost all $s \in [0, T]$. Hence, $\xi \in \overline{u_c^{-1}(0)}$. As $\Omega(\phi^*)$ is weakly backward invariant, then there exists $\phi \in \mathcal{S}_\mathcal{H}(\xi^*)$, $\xi^* \in \Omega(\phi^*)$, such that $\phi(t^*, j^*) = \xi$, $t^* + j^* \geq 1$, and $\phi(t, j) \in \Omega(\phi^*)$ for all $(t, j) \preceq (t^*, j^*)$. If $(t^*, j^* - 1) \in \mathrm{dom}\,\phi$, then (8.3) with $(\underline{t}, \underline{j}) = (t^*, j^* - 1)$, $(\overline{t}, \overline{j}) = (t^*, j^*)$ shows that $u_d(\phi(t^*, j^* - 1)) = 0$ and consequently $\xi = \phi(t^*, j^*) \in G(u_d^{-1}(0))$. If $(t^* - T, j^*) \in \mathrm{dom}\,\phi$ for some $T > 0$, then an argument similar to the one for forward invariance can be given. Thus

$$\Omega(\phi^*) \subset \left[\overline{u_c^{-1}(0)} \cup u_d^{-1}(0) \right] \cap \left[\overline{u_c^{-1}(0)} \cup G(u_d^{-1}(0)) \right]$$

$$= \overline{u_c^{-1}(0)} \cup \left(u_d^{-1}(0) \cap G(u_d^{-1}(0)) \right).$$

This finishes the proof. □

Example 8.3. (Planar oscillator with jumps) Consider the hybrid system (see also Example 6.20)

$$C = \{x \in \mathbb{R}^2 : x_2 \geq 0\} \qquad F(x) = \begin{pmatrix} x_2 \\ -x_1 \end{pmatrix}$$

$$D = \{x \in \mathbb{R}^2 : x_2 \leq 0\} \qquad G(x) = \begin{pmatrix} x_2 \\ -x_1 \end{pmatrix}.$$

The functions $V(x) = |x|$,

$$u_c(x) = \begin{cases} 0 & \text{if } x_2 \geq 0, \\ -\infty & \text{if } x_2 < 0, \end{cases} \qquad u_d(x) = \begin{cases} 0 & \text{if } x_2 \leq 0, \\ -\infty & \text{if } x_2 > 0, \end{cases}$$

are such that (8.1) holds for every solution to the hybrid system. For these functions, it follows that

$$\overline{u_c^{-1}(0)} = \mathbb{R} \times [0, +\infty), \ u_d^{-1}(0) = \mathbb{R} \times (-\infty, 0], \ G(u_d^{-1}(0)) = (-\infty, 0] \times \mathbb{R}.$$

Let $U = \mathbb{R}^2$. Theorem 8.2 states that every precompact solution to the hybrid system approaches the largest weakly invariant set contained in (8.2). Note that

$$\overline{u_c^{-1}(0)} \cup \left(u_d^{-1}(0) \cap G(u_d^{-1}(0))\right) = \overline{\mathbb{R}^2 \setminus ([0, \infty) \times (-\infty, 0])}, \qquad (8.4)$$

and that every solution ϕ to the hybrid system is precompact. Moreover, for all $(t, j) \in \operatorname{dom} \phi$, every such solution satisfies $V(\phi(t, j)) = V(\phi(0, 0)) = |\phi(0, 0)|$ and $\phi(t, j)$ belongs to (8.4). Then, the set (8.2) is weakly forward invariant for some $r \in [0, \infty)$ (cf. S_2 in Example 6.20); for this system, r is given by the norm of the initial condition of the solution under study. However, points in such set that are in the open third quadrant do not satisfy the conditions for weakly backward invariance. In fact, solutions can only reach that set at some (t, j) with $t + j > 0$ after a jump from a point in the open fourth quadrant, and such a point does not belong to (8.4). It is easy to check that the largest weakly invariant set in (8.2) is given by $\operatorname{rge} \phi$.

There are natural candidates for functions u_c, u_d that can be computed based on data of the hybrid system $\mathcal{H}$. First, note that if the function V is locally Lipschitz on a neighborhood of C, then

$$V(\phi(\overline{t}, \overline{j})) - V(\phi(\underline{t}, j)) =$$

$$\int_{\underline{t}}^{\overline{t}} \frac{d}{dt} V(\phi(t, j(t))) \, dt + \sum_{j=\underline{j}+1}^{\overline{j}} [V(\phi(t(j), j)) - V(\phi(t(j), j - 1))]. \ (8.5)$$

The integral above makes sense, and expresses the desired quantity, since $t \mapsto V(\phi(t, j(t)))$ is locally Lipschitz and absolutely continuous on every interval on which $t \mapsto j(t)$ is constant. Even without any regularity of V, the function $u_D : \mathbb{R}^n \to [-\infty, \infty)$

$$u_D(x) = \begin{cases} \max_{\xi \in G(x)} \{V(\xi) - V(x)\} & \text{if } x \in D, \\ -\infty & \text{otherwise,} \end{cases} \tag{8.6}$$

bounds the growth of V during jumps of solutions. Indeed, $V(\phi(t, j+1)) - V(\phi(t, j)) \le u_D(\phi(t, j))$ for any $\phi \in \mathcal{S}_{\mathcal{H}}$, any $(t, j), (t, j+1) \in \text{dom}\,\phi$. If V is continuously differentiable, the function $u_C : \mathbb{R}^n \to [-\infty, \infty)$

$$u_C(x) = \begin{cases} \max_{v \in F(x)} \langle \nabla V(x), v \rangle & \text{if } x \in C, \\ -\infty & \text{otherwise,} \end{cases} \tag{8.7}$$

bounds the growth of V during flows. Indeed, for any $\phi \in \mathcal{S}_{\mathcal{H}}$, $\frac{d}{dt} V(\phi(t, j(t)))$, where it exists, is bounded above by $u_C(\phi(t, j(t)))$.

Corollary 8.4. (Invariance principle with u_C and u_D functions) *Consider a continuous function $V : \mathbb{R}^n \to \mathbb{R}$, continuously differentiable on a neighborhood of C. Suppose that for a given set $U \subset \mathbb{R}^n$,*

$$u_C(z) \le 0, \quad u_D(z) \le 0 \text{ for all } z \in U . \tag{8.8}$$

Let a precompact $\phi^* \in \mathcal{S}_{\mathcal{H}}$ be such that $\overline{\text{rge}\,\phi^*} \subset U$. Then, for some $r \in V(U)$, ϕ^* approaches the nonempty set which is the largest weakly invariant subset of

$$V^{-1}(r) \cap U \cap \left[\overline{u_C^{-1}(0)} \cup \left(u_D^{-1}(0) \cap G(u_D^{-1}(0)) \right) \right] . \tag{8.9}$$

The invariance principle is now illustrated using the bouncing ball system. The principle can not be applied to the initial model in Example 1.1, since the initial model is not nominally well-posed. However, the principle can be applied to the Krasovskii regularization of the initial model.

Example 8.5. (Bouncing ball and invariance) Consider the Krasovskii regularization of the bouncing ball system from Example 1.1, as described in Example 4.14:

$$C = \{x \in \mathbb{R}^2 : x_1 \ge 0\} \qquad F(x) = \begin{cases} \begin{pmatrix} x_2 \\ -\gamma \end{pmatrix} & \text{if } x \in C, x \ne 0, \\ \begin{pmatrix} 0 \\ [0, -\gamma] \end{pmatrix} & \text{if } \quad x = 0, \end{cases}$$

$$D = \{x \in \mathbb{R}^2 : x_1 = 0, x_2 \le 0\} \qquad G(x) = \begin{pmatrix} 0 \\ -\lambda x_2 \end{pmatrix}$$

where $\gamma > 0$ and $\lambda \in [0, 1)$. A natural candidate for the function V is the energy function, that is, the continuously differentiable function $V : \mathbb{R}^n \to \mathbb{R}$ given by

$$V(x) = \frac{1}{2}x_2^2 + \gamma x_1.$$

Then $\langle \nabla V(x), F(x) \rangle = \langle (\gamma, x_2), (x_2, -\gamma) \rangle = 0$ for all $x \in C$, $x \neq 0$. A separate calculation verifies that $\langle \nabla V(x), f \rangle = 0$ for all $f \in F(0)$. Hence,

$$u_C(x) = 0$$

for all $x \in C$. In other words, the energy remains constant during flow. During jumps, the energy dissipates. That is, $V(G(x)) - V(x) = -\frac{1}{2}(1 - \lambda^2)x_2^2 - \gamma x_1 \leq 0$ for all $x \in D$, and

$$u_D(x) = -\frac{1}{2}(1 - \lambda^2)x_2^2 - \gamma x_1$$

for each $x \in D$. Since u_C and u_D are never positive, $U = \mathbb{R}^2$ satisfies the conditions in Corollary 8.4. Therefore, every precompact solution to the bouncing ball system converges to the nonempty set which is the largest weakly invariant set in (8.9) for some $r \in V(U)$. It can be easily argued that r must be 0. Indeed, the only weakly invariant subset of $V^{-1}(r)$ with $r > 0$ is the empty set, because jumps from nonzero points decrease the energy. With $r = 0$, and because

$$u_C^{-1}(0) = C, \quad G(u_D^{-1}(0)) = u_D^{-1}(0) = \{0\},$$

the set (8.9) turns out to be the origin. Hence, every precompact solution to the bouncing ball system converges to the origin. See Figure 8.1.

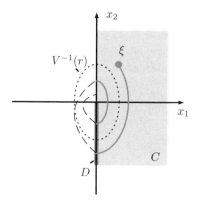

Figure 8.1: A solution to the bouncing ball system in the $x_1 - x_2$ plane and a level set of V (dashed). Solutions escape a level set with $r > 0$ in finite time.

Example 8.6. (Stabilization of attitude dynamics using quaternions) Consider the hybrid system resulting from controlling the attitude of a rigid body given by

$$C = \left\{ x \in \mathbb{S}^3 \times \mathbb{R}^3 \times \{-1, 1\} : q\eta \geq -\delta \right\}$$

$$f(x) = \begin{pmatrix} \frac{1}{2}\left(\begin{matrix} -\epsilon^\top \\ \eta I + S(\epsilon) \end{matrix} \right)\omega \\ J^{-1}\left(S(J\omega)\omega - c\,q\,\epsilon - K\omega \right) \\ 0 \end{pmatrix}$$

$$D = \left\{ x \in \mathbb{S}^3 \times \mathbb{R}^3 \times \{-1, 1\} : q\eta \leq -\delta \right\} \qquad G(x) = \begin{pmatrix} \eta \\ \epsilon \\ \omega \\ \overline{\mathrm{sign}(\eta)} \end{pmatrix},$$

where $x = \begin{pmatrix} \eta \\ \epsilon \\ \omega \\ q \end{pmatrix}$ is the state with components $\eta \in \mathbb{R}$ and $\epsilon \in \mathbb{R}^3$ defining the

quaternion vector $\begin{pmatrix} \eta \\ \epsilon \end{pmatrix} \in \mathbb{S}^3$, $\omega \in \mathbb{R}^3$ the angular velocity, and $q \in \{-1, 1\}$ defin-

ing a logic state; for each $u \in \mathbb{R}^3$, $S(u) = \begin{pmatrix} 0 & -u_3 & u_2 \\ u_3 & 0 & -u_1 \\ -u_2 & u_1 & 0 \end{pmatrix}$; J is the inertia

constant and $\delta \in (0, 1)$, $c > 0$, and $K = K^\top > 0$ are controller parameters. Consider the continuously differentiable function $V : \mathbb{R}^8 \to \mathbb{R}$ given by

$$V(x) = \frac{1}{2}\omega^\top J\omega + 2c(1 - q\eta)$$

and the set of desired attitudes

$$\mathcal{A} = \left\{ x \in \mathbb{S}^3 \times \mathbb{R}^3 \times \{-1, 1\} : \eta = q, \epsilon = 0, \omega = 0 \right\}.$$

Then, along flows, for each $x \in C$, we have

$$\langle \nabla V(x), f(x) \rangle = -\omega^\top K\omega,$$

and for such points, take $u_C(x) = -\omega^\top K\omega$. At jumps we have, for each $x \in D$ (at which points G is single valued),

$$V(G(x)) - V(x) = 2c(1 - (-q)\eta) - 2c(1 - q\eta) = 4\,c\,q\,\eta,$$

and for such points, take $u_D(x) = 4\,c\,q\,\eta$. Moreover, $u_D(x) < 0$ for all $x \in D \setminus \mathcal{A}$ since, for each point in D, we have $q\eta \leq -\delta$. It follows that

$$u_C^{-1}(0) = \left\{ x \in \mathbb{S}^3 \times \mathbb{R}^3 \times \{-1, 1\} : q\eta \geq -\delta, \omega = 0 \right\}, \quad u_D^{-1}(0) = \emptyset.$$

Then, every precompact solution to the hybrid system converges to the largest weakly invariant set W inside

$$\{x \in \mathbb{S}^3 \times \mathbb{R}^3 \times \{-1,1\} : \ q\eta \geq -\delta, \omega = 0\} \cap V^{-1}(r)$$

for some $r \in \mathbb{R}$. Note that for every point in $\mathbb{S}^3$, $\epsilon = 0$ implies $\eta = \pm 1$. Moreover, the closed-loop system is such that $\omega \equiv 0$ implies $\epsilon \equiv 0$. Then, since $\delta \in (0,1)$,

$$
\begin{aligned}
W \ &\subset \ \{x \in \mathbb{S}^3 \times \mathbb{R}^2 \times \{-1,1\} : \ q\eta \geq -\delta, \eta = \pm 1, \epsilon = 0, \omega = 0\} \cap V^{-1}(r) \\
&\subset \ (\{x : \ q = 1, \eta = 1, \epsilon = 0, \omega = 0\} \cup \{x : \ q = -1, \eta = -1, \epsilon = 0, \omega = 0\}) \\
&\quad \cap V^{-1}(r) \\
&\subset \ \{x \in \mathbb{S}^3 \times \mathbb{R}^2 \times \{-1,1\} : \ \eta = q, \epsilon = 0, \omega = 0\} \cap V^{-1}(0) \\
&\subset \ \mathcal{A}.
\end{aligned}
$$

Hence, since every solution to the hybrid system is complete and bounded, every solution converges to $\mathcal{A}$.

Theorem 8.2 and its proof, combined with Corollary 6.22, lead to the following result that is applicable when the hybrid time domain of a precompact solution falls into one of two special cases. The first case is where the lengths of intervals of flow decrease to 0. The second case is where the lengths of all but a finite number of intervals of flow are uniformly bounded below.

Corollary 8.7. (Special cases of the invariance principle involving a nonincreasing function) *Under the assumptions of Theorem 8.2,*

(a) *if* $\sup_j \operatorname{dom} \phi^* = \infty$ *and* $\sup \{|t - t'| : (t,j), (t',j) \in \operatorname{dom} \phi^*\} \to 0$ *when* $j \to \infty$, *then for some* $r \in V(U)$, ϕ^* *approaches the largest weakly invariant subset of*

$$V^{-1}(r) \cap U \cap u_d^{-1}(0) \cap G(u_d^{-1}(0));$$

(b) *if* ϕ^* *is such that, for some* $\tau > 0$ *and all but a finite number of* $j \in \mathbb{N}$, $\sup \{|t - t'| : (t,j), (t',j) \in \operatorname{dom} \phi^*\} \geq \tau$, *then, for some* $r \in V(U)$, ϕ^* *approaches the largest weakly invariant subset of*

$$V^{-1}(r) \cap U \cap \overline{u_c^{-1}(0)}.$$

8.3 STABILITY ANALYSIS USING INVARIANCE PRINCIPLES

The property of a compact set being pre-asymptotically stable can be established using Corollary 8.4 when there are no invariant sets nearby the set.

Theorem 8.8. (Barbasin-Krasovskii-LaSalle) *Let* $\mathcal{A} \subset \mathbb{R}^n$ *be compact. If*

(a) $G(D) \subset \overline{C} \cup D$ *and there exists a continuous function* $V : \mathbb{R}^n \to \mathbb{R}$ *which is positive definite on* $\overline{C} \cup D$ *with respect to* $\mathcal{A}$ *and continuously differentiable on a neighborhood of* C, *and a neighborhood* $U \subset \mathbb{R}^n$ *of* $\mathcal{A}$ *such that the bounds in* (8.8) *hold,* *then $\mathcal{A}$ is stable. If, additionally,*

(b) *there exists $r^* > 0$ such that, for every $r \in (0, r^*)$, the largest weakly invariant subset in (8.9) is empty,*

then $\mathcal{A}$ is locally pre-asymptotically stable.

PROOF. Without loss of generality, suppose that $G(x) = \emptyset$ if $x \notin D$. Assume (a) and let $\varepsilon > 0$ be small enough so that $\mathcal{A} + 2\varepsilon\mathbb{B} \subset U$. First, it is claimed that there exists an arbitrarily small r_ε such that the set

$$S = \{z \in \mathcal{A} + \varepsilon\mathbb{B} : V(z) \le r_\varepsilon\} \tag{8.10}$$

is strongly forward invariant for $\mathcal{H}$. Indeed, since V is positive definite on $\overline{C} \cup D$ with respect to $\mathcal{A}$, there exists $r_\varepsilon' > 0$ so that for $z \in (\mathcal{A} + 2\varepsilon\mathbb{B}) \cap (\overline{C} \cup D)$, $V(z) \le r_\varepsilon'$ implies $z \in (\mathcal{A} + \varepsilon\mathbb{B}) \cap (\overline{C} \cup D)$. Since $u_D(z) \le 0$ for all $z \in \mathcal{A}$, V is positive definite on $\overline{C} \cup D$ with respect to $\mathcal{A}$, and $G(D) \subset \overline{C} \cup D$, one has $G(\mathcal{A}) = G(\mathcal{A} \cap D) \subset \mathcal{A}$. By Lemma 6.9, D is closed and G satisfies (A3) of hybrid basic conditions. Then, by Lemma 5.15, there exists $\gamma > 0$ such that $G(\mathcal{A} + \gamma\mathbb{B}) \subset \mathcal{A} + \varepsilon\mathbb{B}$. By positive definiteness of V, again, there exists $r_\varepsilon'' > 0$ such that $z \in (\mathcal{A} + 2\varepsilon\mathbb{B}) \cap (\overline{C} \cup D)$ and $V(z) \le r_\varepsilon''$ imply $z \in (\mathcal{A} + \gamma\mathbb{B}) \cap (\overline{C} \cup D)$. Now, for $r_\varepsilon = \min\{r_\varepsilon', r_\varepsilon''\}$, the implication

$$V(z) \le r_\varepsilon, z \in (\mathcal{A} + 2\varepsilon\mathbb{B}) \cap D \Rightarrow \begin{cases} z \in (\mathcal{A} + \epsilon\mathbb{B}) \cap (\overline{C} \cup D), \\ G(z) \subset (\mathcal{A} + \epsilon\mathbb{B}) \cap (\overline{C} \cup D), \end{cases} \tag{8.11}$$

is true. This, and the fact that V is nonincreasing along flowing solutions to $\mathcal{H}$, imply that S is forward invariant. Now, by continuity of V, given any small enough $\varepsilon > 0$ and $r_\varepsilon > 0$ so that (8.11) holds, one can find $\delta \in (0, \varepsilon)$ so that $z \in (\mathcal{A} + \delta\mathbb{B}) \cap (\overline{C} \cup D)$ implies $V(z) \le r_\varepsilon$. Forward invariance of S implies that each maximal solution ϕ with $\phi(0,0) \in \mathcal{A} + \delta\mathbb{B}$ is such that $\mathrm{rge}\,\phi \subset \mathcal{A} + \varepsilon\mathbb{B}$. Thus, $\mathcal{A}$ is stable.

Now assume (a) and (b). In the construction above, one can take $r_\varepsilon \in (0, r^*)$. Then any solution ϕ to the system with $\phi(0,0) \in \mathcal{A} + \delta\mathbb{B}$ has $\mathrm{rge}\,\phi \subset S$ and, in particular, is bounded. If ϕ is complete, then Theorem 8.2 and assumption (b) imply that ϕ converges to largest invariant subset of (8.9) with $r = 0$. This subset is a subset of $\mathcal{A}$. Hence, $\mathcal{A}$ is pre-attractive. $\quad\square$

The first item of the following result states sufficient conditions for pre-asymptotic stability of compact sets that parallel the well-known Lyapunov conditions. The last two items correspond to special cases of solutions.

Corollary 8.9. (Lyapunov) *Let $\mathcal{A} \subset \mathbb{R}^n$ be compact. Suppose that assumption (a) of Theorem 8.8 holds. If any of the following three conditions holds,*

(i) $u_C(z) < 0$, $u_D(z) < 0$ *for all $z \in U \setminus \mathcal{A}$,*

(ii) $u_C(z) < 0$ *for each $z \in U \setminus \mathcal{A}$ and every complete discrete solution ϕ to $\mathcal{H}$ with $\mathrm{rge}\,\phi \subset U$ converges to $\mathcal{A}$,*

(iii) $u_D(z) < 0$ for each $z \in U \setminus \mathcal{A}$ and every complete continuous solution ϕ to $\mathcal{H}$ with $\mathrm{rge}\,\phi \subset U$ converges to $\mathcal{A}$,

then $\mathcal{A}$ is pre-asymptotically stable.

PROOF. It is enough to show that assumption (b) of Theorem 8.8 holds if either (i), (ii), or (iii) does. Pick $\varepsilon > 0$ and $r_\varepsilon > 0$ as in the first paragraph of the proof of Theorem 8.8, which is possible under assumption (a) of that theorem. Pick $r \in (0, r_\varepsilon)$ and suppose that the largest weakly invariant subset in (8.9) is nonempty. Hence, there exists a complete solution ϕ for which $V(\mathrm{rge}\,\phi) = r$. If (i) holds, then V is decreasing along ϕ and $V(\mathrm{rge}\,\phi) = r$ is impossible. If (ii) holds, then negativity of u_C implies that ϕ is discrete, and thus it converges to $\mathcal{A}$. This is impossible since $V(\mathrm{rge}\,\phi) = r$. If (iii) holds, then negativity of u_D implies that ϕ is continuous, and thus it converges to $\mathcal{A}$. This is also impossible. Hence the largest weakly invariant subset in (8.9) is empty and assumption (b) of Theorem 8.8 holds with r^* given by r_ε. $\qquad\square$

Corollary 8.9 helps in establishing pre-asymptotic stability for the bouncing ball system, with the argument shorter than what was provided in Examples 3.19 and 8.5.

Example 8.10. (Bouncing ball and invariance, revisited) Consider the Krasovskii regularization of the bouncing ball system, as recalled in Example 8.5 and let V be the energy function, as used in Example 8.5. Assumption (a) of Theorem 8.8 holds, with $U = \mathbb{R}^2$. Case (iii) of Corollary 8.9 holds, because the only complete continuous solution is identically equal to $\mathcal{A} = \{0\}$. Corollary 8.9 now establishes pre-asymptotic stability of $\mathcal{A}$.

8.4 MEAGRE-LIMSUP INVARIANCE PRINCIPLES

For general functions u_c and u_d, the growth condition in (8.1) for the invariance principle in Theorem 8.2 (and Corollary 8.7) needs to be satisfied by every solution x to $\mathcal{S}_\mathcal{H}$ with $\mathrm{rge}\,x \subset U$, for some set $U \subset \mathbb{R}^n$. In this section, invariance principles with conditions that involve only a single solution are introduced.

Below, a measurable function $f : \mathbb{R}_{\geq 0} \to \mathbb{R}$ is called *weakly meagre* if

$$\lim_{n \to \infty} \left(\inf_{t \in I_n} |f(t)| \right) = 0$$

for every family $\{I_n : n \in \mathbb{N}\}$ of nonempty and pairwise disjoint closed intervals I_n in $\mathbb{R}_{\geq 0}$ with lengths uniformly bounded below by a positive number. A sufficient condition for f to be weakly meagre is that $\lim_{M \to \infty} \int_M^{M+\tau} |f(t)|\,dt = 0$ for some $\tau > 0$. In particular, any L^1 function on $\mathbb{R}_{\geq 0}$ is weakly meagre.

Theorem 8.11. (Meagre-limsup conditions) *Let $\phi \in \mathcal{S}_\mathcal{H}$ be complete. Suppose that for some set $U \subset \mathbb{R}^n$ with $\overline{\mathrm{rge}\,\phi} \subset U$ there exist continuous functions $\ell_c, \ell_d : U \to [0, \infty]$ that satisfy the meagre-limsup conditions along ϕ given by*

(a) if $\sup_t \operatorname{dom} \phi = \infty$ then $t \mapsto \ell_c(\phi(t, j(t)))$ is weakly meagre;

(b) if $\sup_j \operatorname{dom} \phi = \infty$ then for all large enough j there exists $t_j^* \in [t(j), t(j+1)]$ such that $\limsup_{j \to \infty} \ell_d(\phi(t_j^*, j)) = 0$.

Then
$$\Omega(\phi) \subset E_{\phi, \ell_c} \cup E_{\phi, \ell_d},$$

where $E_{\phi, \ell_c} = \{z \in \overline{\operatorname{rge} \phi} : \ell_c(z) = 0\}$ and $E_{\phi, \ell_d} = \{z \in \overline{\operatorname{rge} \phi} : \ell_d(z) = 0\}$.

Furthermore, condition (b) can be replaced by the following: for all $\phi^* \in \mathcal{S}_{\mathcal{H}}$, if $(t, j-1), (t, j), (t, j+1) \in \operatorname{dom} \phi^*$, then $\ell_d(\phi^*(t, j)) = 0$. Finally, if for some $\tau > 0$ and all but a finite number of $j \in \mathbb{N}$, $\sup \{|t - t'| : (t, j), (t', j) \in \operatorname{dom} \phi\} \geq \tau$, then (b) is not needed to conclude that

$$\Omega(\phi) \subset E_{\phi, \ell_c}.$$

PROOF. Suppose that $\Omega(\phi) \subset E_{\phi, \ell_c} \cup E_{\phi, \ell_d}$ fails, i.e., that for some $\xi^* \in \Omega(\phi)$ and $\varepsilon, \delta > 0$,

$$\ell(\xi) := \min\{\ell_c(\xi), \ell_d(\xi)\} \geq \delta$$

for all $\xi \in \xi^* + \varepsilon \mathbb{B}$, $\xi \in \operatorname{rge} \phi$. By the definition of the ω-limit, there exists an increasing and unbounded sequence $(t_i, j_i) \in \operatorname{dom} \phi$ with $\phi(t_i, j_i) \to \xi^*$. It can be assumed that for all i, $t_i + j_i + 1 \leq t_{i+1} + j_{i+1}$. Ignoring initial terms if necessary, one obtains $\phi(t_i, j_i) \in \xi^* + \varepsilon/2\mathbb{B}$ for all $i \in \mathbb{N}$, and for some (arbitrarily small) $T \in (0, 1)$, $\phi(t, j_i) \in \xi^* + \varepsilon \mathbb{B}$ for all $t \in [t_i - T, t_i + T]$, $(t, j_i) \in \operatorname{dom} \phi$, all $i \in \mathbb{N}$.

For each i, either of the two conditions holds:

(α) either $t(j_i) \leq t_i - T$ or $t(j_i + 1) \geq t_i + T$ (ϕ flows for time T either before t_i or after t_i);

(β) $t(j_i) > t_i - T$ and $t(j_i + 1) < t_i + T$ (ϕ jumps within time T before and after t_i).

Either (α) or (β) has to occur infinitely many times. Suppose it is (α), and furthermore, that $t(j_i) \leq t_i - T$ for such i's (the other case is treated similarly). Note that then $\sup_t \operatorname{dom} \phi = \infty$. Then the fact that $\ell(\phi(t, j(t))) > \delta$ for any $t \in [t_i - T, t_i]$ for infinitely many i's contradicts weak meagreness of $t \mapsto \ell_c(\phi(t, j(t)))$ (note that intervals $[t_i - T, t_i]$ are disjoint). If (β) holds for infinitely many times, then $\sup_j \operatorname{dom} \phi = \infty$ and for infinitely many i's and all $t \in [t(j_i), t(j_i + 1)]$ one obtains $\ell_d(\phi(t, j_i)) > \delta$. This contradicts (b). Hence $\Omega(\phi) \subset E_{\phi, \ell_c} \cup E_{\phi, \ell_d}$.

Now suppose that for all $\phi^* \in \mathcal{S}_{\mathcal{H}}$, if $(t, j-1), (t, j), (t, j+1) \in \operatorname{dom} \phi^*$, then $\ell_d(\phi^*(t, j)) = 0$, Then (α) has to hold for all but a finite number of i's, possibly not for T but for some $\tau \in (0, T]$. Indeed, otherwise, for some sequence of $\tau_k \searrow 0$ and a subsequence t_{i_k} there is a jump at $t_-(k) \in [t_{i_k} - \tau_k, t_{i_k}]$ and at $t_+(k) \in [t_{i_k}, t_{i_k} + \tau_k]$, so that $(t_-(k), j_{i_k} - 1)$, (t_{i_k}, j_{i_k}), and $(t_+(k), j_{i_k} + 1)$ are all in $\operatorname{dom} x$. Now, consider a sequence of trajectories given by $x_k(t, j) = x(t + t_-(k), j + j_{i_k} - 1)$, and pick a graphically convergent subsequence. For the limit ϕ^* it must be that $(0,0), (0,1), (0,2) \in \operatorname{dom}\phi^*$, while $\phi^*(0,1) = \xi^*$. This contradicts $\ell_d(\phi^*(0,1)) = 0$.

Finally, suppose that for some $\tau > 0$ and all $j \in \mathbb{N}$,

$$\sup\{|t - t'| : (t,j),(t',j) \in \operatorname{dom}\phi\} \geq \tau$$

and that, similarly as above, $\ell_c(\xi) \geq \delta$. Choosing $T < \gamma/2$ above shows that (β) can not hold, and hence (α) has to hold many times. This leads to $\Omega(\phi) \subset E_{\phi,\ell_c}$. $\qquad\square$

The proof above applies, without change, to the situation where the functions ℓ_c and ℓ_d are not continuous. The required change, for this situation, is in the description of the sets E_{ϕ,ℓ_c} and E_{ϕ,ℓ_d}. One needs to take

$$E_{\phi,\ell_c} = \left\{ z \in \overline{\operatorname{rge}\phi} : \exists z_i \to z, z_i \in \operatorname{rge}\phi, \liminf_{i\to\infty} \ell_c(z_i) = 0 \right\},$$

$$E_{\phi,\ell_d} = \left\{ z \in \overline{\operatorname{rge}\phi} : \exists z_i \to z, z_i \in \operatorname{rge}\phi, \liminf_{i\to\infty} \ell_d(z_i) = 0 \right\}.$$

Corollary 8.12. (Continuous and discrete time meagre-limsup conditions) Let $\phi \in \mathcal{S}_\mathcal{H}$ be complete.

(a) If $\sup_j \operatorname{dom}\phi < \infty$ and there exists a function $\ell_c : \operatorname{rge}\phi \to [0,\infty]$ such that $t \mapsto \ell_c(\phi(t,j(t)))$ is weakly meagre, then

$$\Omega(\phi) \subset E_{\phi,\ell_c}.$$

(b) If $\sup_t \operatorname{dom}\phi < \infty$ and there exists a function $\ell_d : \operatorname{rge}\phi \to [0,\infty]$ such that, for all large enough j, there exists $t_j^* \in [t(j), t(j+1)]$ such that $\limsup_{j\to\infty} \ell_d(\phi(t_j^*,j)) = 0$, then

$$\Omega(\phi) \subset E_{\phi,\ell_d}.$$

PROOF. For (a) use $\ell_d(\xi) = r > 0$ for all $\xi \in \mathbb{R}^n$ in Theorem 8.11. For (b) use $\ell_c(\xi) = r > 0$. $\qquad\square$

Example 8.13. (Bouncing ball system and invariance, revisited) Let ϕ be a Zeno solution to the bouncing ball system in Example 8.5. Item (b) of Corollary 8.12 with $\ell_d(x) = |x_1|$ and $\{t_j^*\}_{j=0}^\infty$, $t_j^* = t(j+1)$, implies that the ω-limit set of ϕ is contained in $E_{\phi,\ell_d} = \{x \in \mathbb{R}^2 : x_1 = 0\}$. Then, $\lim_{t+j\to\infty} \phi_1(t,j) = 0$ from where convergence of ϕ to the origin follows.

Given a function $\ell : \mathbb{R}^n \to [-\infty,\infty]$, its *lower semicontinuous closure* $\underline{\ell} : \mathbb{R}^n \to [-\infty,\infty]$ is the greatest lower semicontinuous function on $\mathbb{R}^n$ bounded above by ℓ. Equivalently, for any $x \in \mathbb{R}^n$, $\underline{\ell}(x) = \liminf_{x_i \to x} \ell(x_i)$.

Corollary 8.14. (Invariance principle involving a meagre-limsup function) *Let $\phi \in S_{\mathcal{H}}$ be precompact. Suppose that there exist functions $\ell_c, \ell_d : \mathbb{R}^n \to [0, \infty]$ which satisfy the meagre-limsup conditions along ϕ. Then ϕ converges to the largest weakly invariant subset of*

$$\underline{\ell_c}^{-1}(0) \ \cup \ \underline{\ell_d}^{-1}(0),$$

where $\underline{\ell_c}$, respectively $\underline{\ell_d}$, is the lower semicontinuous closure of ℓ_c, respectively, of ℓ_d. If for some $\tau > 0$ and all $j \in \mathbb{N}$, $\sup \{|t - t'| : (t, j), (t', j) \in \mathrm{dom}\,\phi\} \geq \tau$, then ϕ converges to the largest weakly invariant subset of

$$\underline{\ell_c}^{-1}(0).$$

8.5   INVARIANCE PRINCIPLES FOR SWITCHING SYSTEMS

In Section 2.4, switching systems under certain classes of switching signals were modeled as hybrid systems. Based on these models, it is shown here how invariance principles for hybrid systems can be applied to switching systems.

Recall that the data of a switching system $\dot{z} = f_\sigma(z)$, originally stated in (2.5), is given by an index set $Q = \{1, 2, \ldots, q_{max}\}$, and for each $q \in Q$, a continuous function $f_q : \mathbb{R}^n \to \mathbb{R}^n$. The following assumptions will be used in what follows:

(S1) for each $q \in Q$, there exist a continuously differentiable function $V_q : \mathbb{R}^n \to \mathbb{R}$ and a continuous function $W_q : O \to \mathbb{R}_{\geq 0}$ such that

$$\langle \nabla V_q(z), f_q(z) \rangle \leq -W_q(z) \quad \text{for all } z \in \mathbb{R}^n;$$

(S2) the solution (z, q) to the switching system is such that, for each $q^* \in Q$, for any two consecutive intervals (t_j, t_{j+1}), (t_k, t_{k+1}) with $t_{j+1} \leq t_k$ and such that $q(t) = q^*$ for all $t \in (t_j, t_{j+1})$ and all $t \in (t_k, t_{k+1})$, one has

$$V_{q^*}(z(t_{j+1})) \geq V_{q^*}(z(t_k)).$$

The first assumption above implies that, given a solution (x, q) to the switching system, the function $t \mapsto V_{q^*(x(t))}$ is nonincreasing over every interval on which the switching signal has value q^*. The second assumption, combined with the first one, implies that the function $t \mapsto V_{q^*(x(t))}$ is, in fact, nondecreasing over the union of all intervals on which the switching signal has value q^*.

To obtain an invariance principle for dwell-time solutions, one can apply the hybrid invariance principles of Section 8.2 to the hybrid model of dwell-time switching systems in Example 2.13. A different route is chosen below, with the goal of presenting two techniques which may be of use in other settings. In the proof of Proposition 8.15, a solution to a switching system is restricted to intervals where the switching signal takes on a particular value q^*, and a solution to a nominally well-posed hybrid system is created from this restriction. Then, an invariance result from Section 8.2 is invoked. Proposition 8.16 utilizes a result from Section 8.4. In its proof, it is shown that the functions W_q from the bound in (S1) can be used to define a weakly meagre output.

Proposition 8.15. (Invariance principle for dwell-time solutions, take I) *Assume (S1) and that (S2) holds for every dwell-time solution to the switching system. Let (z, σ) be a complete and bounded dwell-time solution to the switching system, with dwell time $\tau_D > 0$. Then there exist $r_1, \ldots, r_{q_{max}} \in \mathbb{R}$ such that z approaches*

$$M = \bigcup_{q \in Q'} M_q(r_q, \tau_D), \tag{8.12}$$

where $M_q(r_q, \tau_D)$ is the largest subset of $V_q^{-1}(r_q) \cap W_q^{-1}(0)$ that is invariant in the following sense: for each $z_0 \in M_q(r_q, \tau_D)$ there exists a solution ξ to $\dot{z} = f_q(z)$ on $[0, \tau_D/2]$ such that $\xi(t) \in M_q(r_q, \tau_D)$ for all $t \in [0, \tau_D/2]$ and either $\xi(0) = z_0$ or $\xi(\tau_D/2) = z_0$.

PROOF. For each $q^* \in Q$ for which there are infinitely many disjoint time intervals $(t_j, t_j + \Delta t_j)$, $j = 0, 1, \ldots$, $\Delta t_j \geq \tau_D$, on which q equals q^*, consider the hybrid arc y with

$$\text{dom}\, y = \bigcup_{j=0}^{\infty} \left[\sum_{i=0}^{j-1} \Delta t_j, \sum_{i=0}^{j} \Delta t_j \right] \times \{j\}$$

(with the convention that $\sum_{i=0}^{-1} \Delta t_j = 0$) defined by

$$y(t, j) = z\left(t - \sum_{i=0}^{j-1} \Delta t_j + t_j \right)$$

for $t \in \left[\sum_{i=0}^{j-1} \Delta t_j, \sum_{i=0}^{j} \Delta t_j \right]$. Let $K \subset \mathbb{R}^n$ be a compact set with rge $y \subset K$, and

$$\kappa_{\tau_D}(\tau) = \begin{cases} 1 & \text{if } \tau < \tau_D, \\ [0, 1] & \text{if } \tau = \tau_D, \\ 0 & \text{if } \tau > \tau_D. \end{cases}$$

The hybrid arc y is a component of a precompact solution ϕ to the following hybrid system:

$$x = \begin{pmatrix} y \\ \tau \end{pmatrix} \in \mathbb{R}^{n+1} \qquad \begin{cases} \begin{aligned} \dot{y} &= f_q(y) \\ \dot{\tau} &\in \kappa_{\tau_D}(\tau) \end{aligned} \qquad & \tau \in [0, \tau_D] \\ \\ \begin{aligned} y^+ &\in K \\ \tau^+ &= 0 \end{aligned} \qquad & \tau = \tau_D. \end{cases} \tag{8.13}$$

The hybrid system in (8.13) is nominally well-posed. The assumptions of Corollary 8.4 are satisfied, with V replaced by V_{q^*}, thanks to (S1) and (S2): one takes

$u_C = W$ on the flow set, $u_D = 0$ on the jump set. Corollary 8.7, which involves special cases of Corollary 8.4, implies that, for some $r^* \in \mathbb{R}$, (y, τ) converges to M', the largest subset of

$$\left(V_{q^*}^{-1}(r^*) \cap K \cap W^{-1}(0)\right) \times \mathbb{R}$$

that is weakly backward invariant for (8.13) in the following sense: for each $(y_0, \tau_0) \in M'$, each $R \in \mathbb{R}_{\geq 0}$, there exists a complete solution (y^b, τ^b) to (8.13) such that $(y_0, \tau_0) = (y^b(t^*, j^*), \tau^b(t^*, j^*))$ for some $(t^*, j^*) \in \operatorname{dom}(y^b, \tau^b)$ with $t^* + j^* \geq R$ and $(y^b(t,j), \tau^b(t,j)) \in M'$ for all $(t,j) \in \operatorname{dom}(y^b, \tau^b)$.

Since (y, τ) approaches M', y approaches the projection of $M' \subset \mathbb{R}^n \times \mathbb{R}$ onto $\mathbb{R}^n$. It is thus sufficient to show that this projection, denoted by M'', is a subset of $M_{q^*}(r^*, \tau_D)$. Take any $y_0 \in M''$, and $\tau_0 \in \mathbb{R}$ such that $(y_0, \tau_0) \in M'$. Consider $R \geq \tau_D$ and let (y^b, τ^b) be a solution verifying that backward invariance at (y_0, τ_0). Then $(y^b(t^*, j^*), \tau^b(t^*, j^*)) = (y_0, \tau_0)$, where $t^* \in [t_{j^*}, t_{j^*+1}]$, $[t_{j^*}, t_{j^*+1}] \times \{j^*\} \subset \operatorname{dom}(x^b, \tau^b)$, and $t_{j^*+1} - t_{j^*} \geq \tau_D$. Either $t^* - t_{j^*} \geq \tau_D/2$ or $t_{j^*+1} - t^* \geq \tau_D/2$, and this leads to a solution (y', τ') to $\dot{y} = f(y)$, $\dot{\tau} = 1$ on $[0, \tau_D/2]$ such that $(y'(t), \tau'(t)) \in M'$ for all $t \in [0, \tau_D/2]$ and either $y(0) = y_0$ or $y(\tau_D/2) = y_0$. Then also $x'(t) \in M''$ for all $t \in [0, \tau_D/2]$, and so M'' is weakly backward invariant. Hence $M'' \subset M_{q^*}(r^*, \tau_D)$. □

Proposition 8.16. (Invariance principle for dwell-time solutions, take II) *Assume (S1) and let (z, σ) be a complete and bounded dwell-time solution to the switching system, with dwell time $\tau_D > 0$, that satisfies (S2). Then z approaches the largest subset N of*

$$\bigcup_{p \in Q} W_p^{-1}(0)$$

that is invariant in the following sense: for each $z_0 \in N$ there exist $q_1, q_2 \in Q$ such that $z_0 \in W_{p_1}^{-1}(0) \cup W_{p_2}^{-1}(0)$, $t_1, t_2 > 0$ with $t_1 + t_2 \geq \tau_D$, a solution $\xi_1 : [-t_1, 0] \to W_{p_1}^{-1}(0) \cap N$ to $\dot{\xi}_1 = f_{p_1}(\xi_1)$ such that $\xi_1(0) = z_0$, and a solution $\xi_2 : [0, t_2] \to W_{p_2}^{-1}(0) \cap N$ to $\dot{\xi}_2 = f_{p_2}(\xi_2)$ such that $\xi_2(0) = z_0$.

Proof. Let $\phi = (z, \sigma, \tau)$ be a solution to the hybrid system (2.6) corresponding to the z. (With some abuse of notation, z stands for both a component of the solution to a switching system and for a component of a solution to a hybrid system.) Let $\ell_c(\phi) = \ell_c(z, \sigma, \tau) = W_\sigma(z)$. Then $t \mapsto \ell_c(\phi(t, j(t)))$ is an L^1 function on $\mathbb{R}_{\geq 0}$, and thus it is weakly meagre. Indeed, for a $q^* \in Q$, let $[t_{j_i}, t_{j_i+1}] \times \{j_i\}$ be the sequence of all intervals in $\operatorname{dom} \phi$ on which $\sigma(t, j_i) = q^*$, with j_i increasing. (The sequence may be empty, finite, or infinite, and it is infinite for at least one $q^* \in Q$.) Let $\ell_{q^*} : \mathbb{R}_{\geq 0}$ be a function given by

$$\ell_{q^*}(t) = \begin{cases} W_{q^*}(x(t, j_i)) & \text{if } t \in (t_{j_i}, t_{j_i+1}), \\ 0 & \text{otherwise.} \end{cases}$$

Let $I_{q^*} = \bigcup(t_{j_i}, t_{j_i+1})$. By (S1) and (S2), for all $t \in I_{q^*}$,

$$V_{q^*}(z(t, j_i)) - V_{q^*}(z(t_{j_0}, t_{j_0+1})) \leq - \int_{s \in I_{q^*}, s \leq t} W_{q^*}(x(s, j(s))) \, ds$$

where $j(s)$ is such that $(s, j(s)) \in \operatorname{dom} \phi$. Since ϕ is precompact, $V_{q^*}(x(t, j(t)))$ is bounded over all $t \in I_{q^*}$. Then $w_{q^*} := \int_{s \in I_{q^*}} W_{q^*}(x(s, j(s))) \, ds$ exists. Since, for each $t \in \mathbb{R}$, $\int_0^t \ell(s) \, ds \leq \sum_{q^* \in Q} w_{q^*}$, the function $t \mapsto \ell_c(\phi(t, j(t)))$ is integrable on $\mathbb{R}$.

Corollary 8.14 implies that (z, σ, τ) approaches the largest subset N' of

$$\bigcup_{p \in Q} W_p^{-1}(0) \times \{p\} \times \mathbb{R}$$

that is invariant, with respect to (2.6), in the following sense: for each $(z^0, q^0, \tau^0) \in N'$, each $R \in \mathbb{R}$ there exists a complete solution (ξ, σ, τ) of (2.6) such that $(\xi, \sigma, \tau)(t, j) \in N'$ for all $(t, j) \in \operatorname{dom}(\xi, \sigma, \tau)$ and $(\xi, p, \sigma)(t^*, j^*) = (x^0, q^0, \tau^0)$ for some $(t, j) \in \operatorname{dom}(\xi, p, \sigma)$ with $t + j \geq R$. Thus x approaches the projection N'' of $N' \subset \mathbb{R}^n \times Q \times \mathbb{R}$ onto $\mathbb{R}^n$. It remains to show that N'' is invariant in the sense specified in the theorem. □

8.6 NOTES

The ideas behind invariance principles, involving a Lyapunov-like function that is nonincreasing along precompact solutions, date back to Barbashin and Krasovskii [14], Krasovskii [63], and LaSalle [69], [70]. Byrnes and Martin [22] provided a version of the invariance principle involving the zero-level set of an integrable output function, which was later extended to differential inclusions by Ryan [102] and to meagre output functions by Logemann and Ryan in [75]. Invariance principles for systems with discontinuous right-hand side were given by Shevitz and Paden [110] and Bacciotti and Ceragioli [10] for Filippov solutions, and by Bacciotti and Ceragioli [11] for Carathéodory solutions. A variety of invariance-like results for hybrid, impulsive, and switching systems can be found in the literature, for example, Lygeros et al. [79], Hespanha [54], Hespanha et al. [55], Bacciotti and Mazzi [12], and Mancilla-Aguilar and Garcia [85]. The main results of the chapter appeared, in more general forms, in Sanfelice et al. [104]. Further applications of these results appeared in [103, Chapter 4]. The hybrid controllers for robust, global stabilization of the attitude dynamics of a rigid body of Example 8.6 appeared in [90]. Applications to the theory of switching systems, as presented in Section 8.5, appeared in Goebel et al. [38]. The idea of considering functions ℓ_{q^*}, in the proof of Proposition 8.16, is similar to what is done in the proof of [55, Theorem 7].

Chapter Nine

Conical approximation and asymptotic stability

The goal of this chapter is to present a technique of approximating a hybrid system with a conical hybrid system: a system with conical flow and jump sets and with constant or linear flow and jump maps. The main result, Theorem 9.11, deduces pre-asymptotic stability for the original system from pre-asymptotic stability for the conical approximation. This result generalizes, to a hybrid system, the result that asymptotic stability for the linearization of a differential equation implies asymptotic stability for the differential equation. In many cases, the analysis of the conical approximation is simpler than of the original hybrid system; this is illustrated in several examples in Section 9.3.

Conical hybrid systems are a special case of homogeneous hybrid systems. Homogeneity property, defined and studied in Section 9.1, makes possible, for example, to deduce global properties, like global pre-asymptotic stability, from local properties. Homogeneity also has implications for the rate of convergence of solutions to pre-asymptotically stable sets. Furthermore, asymptotic stability in a homogeneous hybrid system is robust with respect to homogeneous perturbations. For the sake of generality, nonstandard dilations are considered in Section 9.1 and Section 9.2, while conical systems are homogeneous with respect to the standard dilation. It is noted here that more general approximations of hybrid systems, relying on nonstandard dilations, are possible.

9.1 HOMOGENEOUS HYBRID SYSTEMS

A *dilation* of $\mathbb{R}^n$, with coefficients $r_1, r_2, \ldots, r_n > 0$, is a family of linear mappings $M_\lambda : \mathbb{R}^n \to \mathbb{R}^n$, $\lambda > 0$, where each M_λ can be identified with the diagonal matrix

$$M_\lambda = \text{diag}\left\{\lambda^{r_1}, \lambda^{r_2}, \ldots, \lambda^{r_n}\right\}. \tag{9.1}$$

The term *standard dilation* will be used for a dilation with coefficients $r_i = 1$, $i = 1, 2, \ldots, n$, i.e., for the case of $M_\lambda x = \lambda x$ for each $x \in \mathbb{R}^n$.

Definition 9.1. (Homogeneous hybrid system) *A hybrid system $\mathcal{H}$ is homogeneous with respect to a dilation M_λ, with degree $d \in \mathbb{R}$, if, for all $\lambda > 0$,* the following conditions are satisfied.

$$
\begin{aligned}
F(M_\lambda x) &= \lambda^d M_\lambda F(x) \quad \forall x \in C, \\
C &= M_\lambda C, \\
G(M_\lambda x) &= M_\lambda G(x) \quad \forall x \in D, \\
D &= M_\lambda D.
\end{aligned}
$$

For example, a hybrid system for which the flow set C and the jump set D are cones and the flow map F and the jump map G are linear functions is homogeneous with respect to the standard dilation with degree $d = 0$. Similarly, a hybrid system for which C and D are cones, G is linear, and F is constant is homogeneous with respect to the standard dilation with degree $d = -1$. Systems with these homogeneity properties will be relied on in Section 9.3.

An example of a homogeneous hybrid system, with respect to a nonstandard dilation, is provided by the bouncing ball system in Example 1.1.

Example 9.2. (Bouncing ball) Recall the bouncing ball model from Example 1.1:

$$
C = \{x \in \mathbb{R}^2 : x_1 > 0 \text{ or } x_1 = 0, x_2 \geq 0\} \qquad F(x) = \begin{cases} \begin{pmatrix} x_2 \\ -g \end{pmatrix} & \text{if } x \in C, x \neq 0, \\ \begin{pmatrix} 0 \\ 0 \end{pmatrix} & \text{if } x = 0, \end{cases}
$$

$$
D = \{x \in \mathbb{R}^2 : x_1 = 0, x_2 < 0\} \qquad G(x) = \begin{pmatrix} 0 \\ -\gamma x_2 \end{pmatrix}.
$$

The hybrid system with such data is homogeneous with respect to the dilation

$$
M_\lambda = \begin{pmatrix} \lambda^2 & 0 \\ 0 & \lambda \end{pmatrix}
$$

and the degree $d = -1$. Indeed, $M_\lambda x = (\lambda^2 x_1, \lambda x_2)$, and if $x_1 > 0$ or $x_1 = 0$, $x_2 \geq 0$ then $\lambda^2 x_1 > 0$ or $\lambda^2 x_1 = 0$, $\lambda x_2 \geq 0$, thus $x \in C$ implies $M_\lambda x \in C$. Consequently, $M_\lambda C \subset C$, and a reverse inclusion follows from this one by considering λ^{-1}. Similar arguments can be made for D. For $x \in C$, $x \neq 0$, one has $M_\lambda x \in C$, $M_\lambda x \neq 0$, and

$$
\begin{aligned}
F(M_\lambda x) &= F\begin{pmatrix} \lambda^2 x_1 \\ \lambda x_2 \end{pmatrix} = \begin{pmatrix} \lambda x_2 \\ -g \end{pmatrix} = \begin{pmatrix} \lambda & 0 \\ 0 & 1 \end{pmatrix} \begin{pmatrix} x_2 \\ -g \end{pmatrix} \\
&= \lambda^{-1} \begin{pmatrix} \lambda^2 & 0 \\ 0 & \lambda \end{pmatrix} \begin{pmatrix} x_2 \\ -g \end{pmatrix} = \lambda^{-1} M_\lambda F(x).
\end{aligned}
$$

Furthermore, for $x = 0$, $M_\lambda x = 0$, and $F(x) = F(M_\lambda x)$. A similar calculation can be done for G.

Given $\lambda > 0$ and $d \in \mathbb{R}$, consider a hybrid system $\mathcal{H}_{\lambda,d}$ obtained from $\mathcal{H}$ by scaling the flow map F as follows:

$$\begin{cases} x \in C & \dot{x} \in \lambda^{-d}F(x) \\ x \in D & x^+ \in G(x). \end{cases} \tag{9.2}$$

In comparison to $\mathcal{H}$, solutions to (9.2) flow faster if $\lambda^{-d} > 1$ and slower if $\lambda^{-d} < 1$. A far closer relationship between solutions to $\mathcal{H}$ and $\mathcal{H}_{\lambda,d}$ is in fact true.

Lemma 9.3. (Dilation of solutions) *Let ϕ be a hybrid arc and M_λ be a dilation of $\mathbb{R}^n$. For every $\lambda > 0$, the function ψ defined at each $(t,j) \in \operatorname{dom} \phi$ by*

$$\psi(t,j) = M_\lambda \phi(t,j)$$

is a hybrid arc, with $\operatorname{dom} \psi = \operatorname{dom} \phi$. Moreover, if the hybrid system $\mathcal{H}$ is homogeneous with respect to the dilation M_λ, with degree $d \in \mathbb{R}$, then for each $\lambda > 0$, the following are equivalent:

- *ϕ is a solution to $\mathcal{H}$,*

- *ψ is a solution to $\mathcal{H}_{\lambda,d}$.*

PROOF. It is straightforward that ψ is a hybrid arc, with $\operatorname{dom} \psi = \operatorname{dom} \phi$. Now suppose that $\mathcal{H}$ is homogeneous with respect to M_λ, with degree d. If $\dot{\phi}(t,j) \in F(\phi(t,j))$, then

$$\dot{\psi}(t,j) = M_\lambda \dot{\phi}(t,j) \in M_\lambda F(\phi(t,j)) = \lambda^{-d} F(M_\lambda \phi(t,j)) = \lambda^{-d} F(\psi(t,j)),$$

and so $\dot{\psi}(t,j) \in \lambda^{-d} F(\psi(t,j))$. If $\phi(t,j) \in C$ then $M_\lambda \phi(t,j) \in M(\lambda(t,j))C = C$, so $\psi(t,j) \in C$ (similarly for D). If $\phi(t,j+1) \in G(\phi(t,j))$, then

$$M_\lambda \phi(t,j+1) \in M_\lambda G(\phi(t,j)) = G(M_\lambda \phi(t,j)) = G(\psi(t,j)),$$

and so $\psi(t,j+1) \in G(\psi(t,j))$. Hence, if ϕ is a solution to $\mathcal{H}$, then ψ is a solution to $\mathcal{H}_{\lambda,d}$. The reverse implication can be shown similarly. $\square$

Lemma 9.3 can be used to translate local properties of a homogeneous hybrid system to global properties. Below, Proposition 9.4 translates local information about solutions moving toward the origin to global pre-asymptotic stability. Later, related comments about the time it takes the solutions to reach the origin are given.

Given a dilation M_λ, a *homogeneous quasinorm* on $\mathbb{R}^n$ will denote a function $\omega : \mathbb{R}^n \to \mathbb{R}_{\geq 0}$ that is a proper indicator of 0 with respect to $\mathbb{R}^n$ and satisfies $\omega(M_\lambda x) = \lambda \omega(x)$ for all $\lambda > 0$, $x \in \mathbb{R}^n$. For the standard dilation, any norm is a homogeneous quasinorm. In general, an example is provided by

$$\omega(x) = \sqrt{|x_1|^{2/r_1} + \cdots + |x_n|^{2/r_n}}.$$

Given a homogeneous quasinorm ω, in what follows the set $\mathcal{B}_\omega$ is given by

$$\mathcal{B}_\omega = \{x \in \mathbb{R}^n \; : \; \omega(x) \leq 1\}.$$

Proposition 9.4. (From local information to global pAS) *Consider a hybrid system $\mathcal{H}$ such that*

(a) *$\mathcal{H}$ is homogeneous with respect to dilation M_λ with degree $d \in \mathbb{R}$;*

(b) *there exist $R > r > 0$, $m > 0$, and a homogeneous quasinorm ω such that for any solution ϕ to $\mathcal{H}$ with $\omega(\phi(0,0)) = r$ either*

    (i) *$\operatorname{length} \operatorname{dom} \phi \leq m$ and $\omega(\operatorname{rge} \phi) \leq R$, or*

    (ii) *there exists $(T, J) \in \operatorname{dom} \phi$ with $T + J \leq m$, $\omega(\phi(T, J)) \leq r/2$, and $\omega(\phi(t,j)) \leq R$ for all $(t,j) \in \operatorname{dom} \phi$, $t \leq T$, $j \leq J$.*

Then, 0 is pAS for $\mathcal{H}$.

PROOF. Let ψ be a solution to $\mathcal{H}$ with $2^{i-1}r \leq \omega(\psi(0,0)) \leq 2^i r$ for some $i \in \mathbb{Z}$. Pick $2^{-i} \leq \lambda \leq 2^{-i+1}$ such that $\lambda\omega(\psi(0,0)) = r$ and consider ϕ given by $\phi(t,j) = M_\lambda\psi(t,j)$. Lemma 9.3 implies that ϕ is a solution to the system (9.2), with $\omega(\phi(0,0)) = \omega(M_\lambda\psi(0,0)) = \lambda\omega(\psi(0,0)) = r$. Assumptions (i) and (ii) apply to ϕ, with m replaced by $m' = \lambda^d m$ if $\lambda^{-d} < 1$, $m' = m$ if $\lambda^{-d} \geq 1$. Translating this to ψ yields

(i') $\operatorname{dom} \psi$ is compact, with $t + j \leq m'$ for all $(t,j) \in \operatorname{dom} \psi$, and for all such (t,j), $\omega(\psi(t,j)) \leq R/\lambda \leq 2^i R$, or

(ii') there exists $(T, J) \in \operatorname{dom} \psi$ with $T + J \leq m'$, $\omega(\psi(T,J)) \leq r/(2\lambda) \leq 2^{i-1}r$, and such that $\omega(\psi(t,j)) \leq R/\lambda \leq 2^i R$ for all $(t,j) \in \operatorname{dom} \psi$, $t \leq T$, $j \leq J$.

These two statements are enough to conclude that 0 is pre-asymptotically stable for $\mathcal{H}$. □

For a homogeneous hybrid system with a pre-asymptotically stable origin, the degree of homogeneity has implications for the time it takes for solutions to converge to the origin. Recall first that for a differential equation $\dot{z} = f(z)$ with an asymptotically stable origin, the following can be said. If f is homogeneous with a negative degree, for example, if $f(z) = -z^{1/3}$ and the standard dilation is considered, then solutions converge to the origin in finite time. If f is homogeneous with degree 0, in particular when f is linear, then convergence is exponential: $|z(t)| \leq e^{-kt}|z(0)|$ for every solution z. If f is homogeneous with a positive degree, for example, if $f(z) = -z^3$, then convergence is slower than exponential and, in particular, solutions take an infinite amount of time to converge to the origin.

The degree d of homogeneity of a nominally well-posed hybrid system for which the origin is pre-asymptotically stable has similar consequences. First, define the following quantity which measures the time that the solutions remain away from the origin:

$$\mathcal{T}(\phi) = \sup \{t \in \mathbb{R}_{\geq 0} \,:\, \exists j \in \mathbb{N}(t, j) \in \operatorname{dom} \phi, \phi(t, j) \neq 0\}.$$

Lemma 9.3 and techniques similar to what is used in the proof of Proposition 9.4 below can be used to conclude the following facts.

If $d < 0$ then for every $\varepsilon > 0$ there exists $\delta > 0$ such that every solution with $|\phi(0,0| < \delta$ satisfies $\mathcal{T}(\phi) < \varepsilon$. In particular, every solution starting in the basin of attraction of the origin, except the origin itself, satisfies $\mathcal{T}(\phi) < \infty$. Under further conditions on the jumps that solutions must take, this fact can be used to show that solutions from nonzero initial conditions must be Zeno. Such arguments can be applied to the bouncing ball model — recall Example 9.2 — to deduce that solutions are Zeno.

If $d > 0$ and every solution ϕ from a nonzero initial condition satisfies $\mathcal{T}(\phi) > 0$, i.e., every such solution experiences some flow before reaching 0, then in fact every such solution satisfies $\mathcal{T}(\phi) = \infty$.

9.2 HOMOGENEITY AND PERTURBATIONS

A fundamental property of well-posed hybrid systems is that asymptotic stability is uniform and robust to perturbations, as stated in Theorem 7.21. For homogeneous hybrid systems, this result can be strengthened to allow for perturbations that are also homogeneous. Such perturbations are described in Definition 9.5 and the robustness result is given in Theorem 9.8.

Definition 9.5. (Homogeneous perturbation of a hybrid system) *Given a hybrid system $\mathcal{H}$, a dilation M_λ, $d \in \mathbb{R}$, and a homogeneous quasinorm ω, a homogeneous perturbation of $\mathcal{H}$ of size $\eta > 0$ is the hybrid system $\mathcal{H}_\eta^h$:*

$$\begin{cases} x \in C_\eta^h & \dot{x} \in F_\eta^h(x) \\ x \in D_\eta^h & x^+ \in G_\eta^h(x) \end{cases} \tag{9.3}$$

with the data given by

$$\begin{aligned}
C_\eta^h &= \{x \,:\, (x + \eta M(\omega(x))\mathcal{B}_\omega) \cap C \neq \emptyset\}, \\
F_\eta^h(x) &= \operatorname{con} F\left((x + \eta M(\omega(x))\mathcal{B}_\omega) \cap C\right) + \eta \omega^d(x) M(\omega(x))\mathcal{B}_\omega, \\
D_\eta^h &= \{x \,:\, (x + \eta M(\omega(x))\mathcal{B}_\omega) \cap D \neq \emptyset\}, \\
G_\eta^h(x) &= G\left((x + \eta M(\omega(x))\mathcal{B}_\omega) \cap D\right) + \eta M(\omega(x))\mathcal{B}_\omega.
\end{aligned}$$

Using the term "homogeneous" in the name of the perturbation in Definition 9.5 above is justified by the following result. Essentially, a homogeneous perturbation of a homogeneous hybrid system results in another homogeneous hybrid system.

Proposition 9.6. (Homogeneity in perturbation) *Let M_λ be a dilation and ω a homogeneous quasinorm. If the hybrid system $\mathcal{H}$ is homogeneous with respect to a dilation M_λ with degree $d \in \mathbb{R}$ then, for each $\eta > 0$, the hybrid system $\mathcal{H}_\eta^h$ is homogeneous with respect to the dilation M_λ with degree $d \in \mathbb{R}$.*

PROOF. Recall that $\omega(M_\lambda x) = \lambda\omega(x)$ for all $x \in \mathbb{R}^n$ and $\lambda > 0$ and note that

$$M(\omega(M_\lambda x)) = M_\lambda M(\omega(x)).$$

Then $F_\eta^h(M_\lambda x)$ turns to

$$\mathrm{con}F\left[M_\lambda x + \eta M_\lambda M(\omega(x))\mathcal{B}_\omega\right] + \eta\lambda^d\omega^d(x)M_\lambda M(\omega(x))\mathcal{B}_\omega$$

which is exactly $\lambda^d M_\lambda F_\eta^h(x)$. The case of G is similar; consider $d = 0$ above. Now, $M_\lambda C_\eta^h$ is

$$\{M_\lambda x \,:\, (x + \eta M(\omega(x))\mathcal{B}_\omega) \cap C \neq \emptyset\}$$

which, by taking $y = M_\lambda x$ and so $x = M(\lambda^{-1})y$, turns to

$$\left\{y \,:\, M(\lambda^{-1})(y + \eta M(\omega(y))\mathcal{B}_\omega) \cap C \neq \emptyset\right\}$$

and, since $M_\lambda C = C$, to

$$\left\{y \,:\, (y + \eta M(\omega(y))\mathcal{B}_\omega) \cap C \neq \emptyset\right\}.$$

This is exactly C_η^h. The case of D_η^h is parallel.   □

It needs to be noted that homogeneous perturbations of a hybrid system may lead to F_η that is not locally bounded in a neighborhood of 0. Indeed, when $d < 0$ it may happen that $\omega^d(x)M(\omega(x))$ is unbounded on a neighborhood of 0.

Example 9.7. (Homogeneous and unbounded perturbations) For $x = (x_1, x_2) \in \mathbb{R}^2$, consider

$$F(x) = \begin{pmatrix} 0 \\ x_1 \end{pmatrix}, \quad C = \left\{x \in \mathbb{R}^2 \,:\, x_2 = 0\right\},$$

D empty, and $G(x) = 0$ for all $x \in \mathbb{R}^2$. It is easy to verify that the hybrid system with such data is homogeneous with respect to the dilation

$$M_\lambda = \begin{pmatrix} \lambda & 0 \\ 0 & \lambda^3 \end{pmatrix}$$

and the degree is $d = -2$. We can take $\omega(x) = \left(x_1^6 + x_2^2\right)^{1/6}$. Then

$$\omega^d(x)M(\omega(x)) = \begin{pmatrix} \omega^{-1}(x) & 0 \\ 0 & \omega(x) \end{pmatrix}$$

for $x \neq 0$, and this term blows up as x approaches 0.

It should also be noted that considering a perturbation of G along the lines of what was considered in Definition 6.27, but with $\eta M(\omega(x))$ taking the place of $\rho(x)\mathbb{B}$, i.e., the map

$$x \mapsto \{v \,:\, v \in u + \eta M(\omega(u))\mathcal{B}_\omega, u \in G\left(x + \eta M(\omega(x))\mathcal{B}_\omega\right)\},$$

also leads to a homogeneous mapping. However, this perturbation is not appropriate for the goal at hand. In particular, it would be an obstacle to Lemma 9.10(c).

Theorem 9.8. (Homogeneous perturbations and pAS) *Let M_λ be a dilation and ω be a homogeneous quasinorm. Suppose that the hybrid system $\mathcal{H}$ is well-posed, homogeneous with respect to dilation M_λ with degree $d \in \mathbb{R}$, and that, for that system, $0 \in \mathbb{R}^n$ is pre-asymptotically stable. Then, there exists $\eta > 0$ such that 0 is pre-asymptotically stable for the system $\mathcal{H}_\eta^h$.*

PROOF. As $\mathcal{H}$ is well-posed and 0 is pAS, 0 is uniformly globally pAS thanks to Theorem 7.12 and this property is semiglobally practically robust thanks to Lemma 7.20. Let $\beta \in \mathcal{KL}$ be any function satisfying (7.1). Take an admissible perturbation radius ρ given by $\rho(x) = 1$ for all $x \in \mathbb{R}^n$. By semiglobal practical robustness, there exists $\delta > 0$ such that

$$\omega(\phi(t,j)) \le \beta(\omega(\phi(0,0)), t+j) + 1/4$$

for each solution ϕ to $\mathcal{H}_{\delta\rho}$ with $\omega(\phi(0,0)) \le 1$. Pick $m > 0$ such that $\beta(1,m) \le 1/4$. Now, pick $\eta > 0$ such that

$$\eta M(\omega(x))\mathcal{B} \subset \delta\mathbb{B}$$

for all $x \in \mathbb{R}^n$ with $\omega(x) \le \beta(1,0) + 1/4$ and

$$\eta \omega^d(x) M(\omega(x))\mathcal{B} \subset \delta\mathbb{B}$$

for all $x \in \mathbb{R}^n$ with $1/2 \le \omega(x) \le \beta(1,0)+1/4$. With such η, consider the hybrid system $\mathcal{H}_\eta^h$ and recall that by Proposition 9.6, it is homogeneous with respect to the dilation M_λ with degree d.

Let ϕ be a solution to $\mathcal{H}_\eta^h$ with $1/2 \le \omega(\phi(0,0)) \le 1$. Let $(T,J) \in \operatorname{dom}\phi$ be the "first" element in $\operatorname{dom}\phi$ such that $\omega(\phi(T,J)) \le 1/2$. Then ϕ is also a solution to $\mathcal{H}_{\delta\rho}$, when restricted to $(t,j) \in \operatorname{dom}\phi$ with $t \le T$, $j \le J$. Consequently, for any solution ϕ to $\mathcal{H}_\eta^h$ with $1/2 \le \omega(\phi(0,0)) \le 1$ either there exists $(T,J) \in \operatorname{dom}\phi$ with $T+J \le m$, $\omega(\phi(T,J)) \le 1/2$, and such that $\omega(\phi(t,j)) \le M := \beta(1,0)+1/4$ for all $(t,j) \in \operatorname{dom}\phi$, $t \le T$, $j \le J$; or, $\operatorname{dom}\phi$ is compact, with $t+j \le m$ for all $(t,j) \in \operatorname{dom}\phi$, and for all such (t,j), $\omega(\phi(t,j)) \le M$. Now, Proposition 9.4 finishes the proof. $\qquad\square$

9.3   CONICAL APPROXIMATION AND STABILITY

Given a differential equation $\dot{z} = f(z)$ on $\mathbb{R}^n$, with sufficiently regular f, and a point $\xi \in \mathbb{R}^n$, if ξ is asymptotically stable for the linear differential equation $\dot{z} = f'(\xi)(z - \xi)$, then ξ is asymptotically stable for $\dot{z} = f(z)$. Here and below, $f'(\xi)$ stands for the Jacobian matrix of f at ξ, so that $x \mapsto f(\xi) + f'(\xi)x$ is the affine approximation of f at ξ. A similar conclusion regarding pre-asymptotic stability of a hybrid system can be made from pre-asymptotic stability of a conical approximation of a hybrid system, defined below in Definition 9.9.

Since a point ξ can be pre-asymptotically stable for a hybrid system even if the flow map f is nonzero at ξ, Definition 9.9 deals separately with the case of $f(\xi) = 0$ and $f(\xi) \neq 0$. Note, however, that if ξ is in the jump set then pre-asymptotic stability of ξ requires $g(\xi) = \xi$. Consequently, only approximations having this feature are considered.

Definition 9.9. (Conical approximation of a hybrid system) *Consider a hybrid system $\mathcal{H}$*

$$\begin{cases} x \in C & \dot{x} = f(x) \\ x \in D & x^+ = g(x) \end{cases}$$

and a point $\xi \in \mathbb{R}^n$ such that when $\xi \in \overline{C}$, then either $f(\xi) \neq 0$ and f is continuous at ξ or f is continuously differentiable on a neighborhood of ξ, and when $\xi \in \overline{D}$, then $g(\xi) = \xi$ and g is continuously differentiable on a neighborhood of ξ. A conical approximation of $\mathcal{H}$ at ξ is the hybrid system $\mathcal{CA}(\mathcal{H}, \xi)$

$$\begin{cases} x \in C_{loc} & \dot{x} = f_{loc}(x) \\ x \in D_{loc} & x^+ = g_{loc}(x) \end{cases}$$

where

- *if $\xi \in \overline{C}$ then*

$$f_{loc}(x) = f(\xi) \text{ for all } x \in \mathbb{R}^n \text{ if } f(\xi) \neq 0,$$
$$f_{loc}(x) = f'(\xi)x \text{ for all } x \in \mathbb{R}^n \text{ if } f(\xi) = 0;$$

- *if $\xi \in \overline{D}$ then*

$$g_{loc}(x) = g'(\xi)x \text{ for all } x \in \mathbb{R}^n;$$

- $C_{loc} = T_C(\xi), \qquad D_{loc} = T_D(\xi).$

Recall that a tangent cone to a set S is empty at each point not in $\overline{S}$. Hence, f_{loc} and g_{loc} are irrelevant when $\xi \notin \overline{C}$ and $\xi \notin \overline{D}$, respectively.

A conical approximation of a hybrid system has regular data and is homogeneous. Furthermore, the data of any homogeneous perturbation of a conical approximation $\mathcal{CA}(\mathcal{H}, \xi)$ contains, locally around ξ, the data of the original system $\mathcal{H}$. More precisely, the following result holds.

Lemma 9.10. (Properties of conical approximations) *A conical approximation $\mathcal{CA}(\mathcal{H}, \xi)$ of a hybrid system $\mathcal{H} = (C, F, D, G)$ at a point $\xi \in \mathbb{R}^n$ has the following properties:*

(a) *The data of $\mathcal{CA}(\mathcal{H}, \xi)$ satisfies Assumption 6.5, and hence, $\mathcal{CA}(\mathcal{H}, \xi)$ is well-posed.*

(b) *The system $\mathcal{CA}(\mathcal{H}, \xi)$ is homogeneous with respect to the standard dilation, with degree -1 if $f(\xi) \neq 0$ and with degree 0 if $f(\xi) = 0$.*

(c) *For any $\eta > 0$ there exists a neighborhood U of ξ such that*

$$f(x) \in (f_{loc})_\eta^h(x - \xi) \qquad\qquad \forall\, x \in C \cap U$$
$$C \cap U \subset \xi + (C_{loc})_\eta^h$$
$$g(x) \in g(\xi) + (g_{loc})_\eta^h(x - \xi) \qquad \forall\, x \in D \cap U$$
$$D \cap U \subset \xi + (D_{loc})_\eta^h,$$

where $(f_{loc})_\eta^h$, $(C_{loc})_\eta^h$, $(g_{loc})_\eta^h$, $(D_{loc})_\eta^h$ are homogeneous perturbations of f_{loc}, C_{loc}, g_{loc}, D_{loc}, with degree -1 if $f(\xi) \neq 0$ and 0 if $f(\xi) = 0$, as in Definition 9.5.

PROOF. That $\mathcal{CA}(\mathcal{H}, \xi)$ satisfies Assumption 6.5 is obvious when one recalls that tangent cones are closed. Theorem 6.30 concludes that $\mathcal{CA}(\mathcal{H}, \xi)$ is well-posed. This verifies (a). Verifying part (b) is a straightforward calculation.

Regarding (c), if $f(\xi) \neq 0$, then $f_{loc}(x) = f(\xi)$ for all x and

$$(f_{loc})_\eta^h(x - \xi) = f(\xi) + \eta \omega^{-1}(x - \xi)M(\omega(x - \xi))\mathcal{B} = f(\xi) + \eta\mathcal{B}.$$

Because $\mathcal{B}$ is a neighborhood of 0, $f(x) \in (f_{loc})_\eta^h(x - \xi)$ for all x close enough to ξ. If $f(\xi) = 0$, then $f_{loc}(x) = Ax$ for $A = f'(\xi)$, and

$$(f_{loc})_\eta^h(x - \xi) \supset A(x - \xi) + \eta M(\omega(x - \xi))\mathcal{B} \supset A(x - \xi) + \varepsilon|x - \xi|\mathbb{B}$$

for some $\varepsilon > 0$. Differentiability of f at ξ implies that $f(x) \in (f_{loc})_\eta^h(x - \xi)$ for all x close enough to ξ. The statement for g is shown in the same fashion.

It remains to show the statement regarding C; the one regarding D is analogous. For each $\eta > 0$, $(C_{loc})_\eta^h$ is a cone and contains $C_{loc} \setminus \{0\}$ in its interior. Suppose that there is no neighborhood U of ξ such that $C \cap U \subset \xi + (C_{loc})_\eta^h$. Then there exist points $x_i \in C$, $i = 1, 2, \dots$ with $|x_i - \xi| < 1/i$ and $x_i \notin \xi + (C_{loc})_\eta^h$. Equivalently, $x_i - \xi/|x_i - \xi| \notin (C_{loc})_\eta^h$. Picking a convergent subsequence of $x_i - \xi/|x_i - \xi|$ leads to v with $|v| = 1$ and $v \in T_C(\xi)$. But by construction, this v is not an element of the interior of $(C_{loc})_\eta^h$. This contradiction finishes the proof. $\square$

The following result shows that local pre-asymptotic stability of a point for a hybrid system can be deduced from pre-asymptotic stability of the origin for conical approximation to the hybrid system at the point in question.

Theorem 9.11. (Conical approximation and pAS) *Let $\mathcal{H} = (C, f, D, g)$ be a hybrid system and let $\xi \in \mathbb{R}^n$. If 0 is (globally) pAS for the conical approximation of $\mathcal{H}$ at ξ, then ξ is locally pAS for $\mathcal{H}$.*

PROOF. By Lemma 9.10 (a), the conical approximation $\mathcal{CA}(\mathcal{H}, \xi)$ is well-posed. By (b) of that lemma, $\mathcal{CA}(\mathcal{H}, \xi)$ is homogeneous, so thanks to Proposition 9.4, local and global pAS of ξ for $\mathcal{CA}(\mathcal{H}, \xi)$ are the same. By Theorem 9.8, there exists $\eta > 0$ such that the homogeneous perturbation of $\mathcal{CA}(\mathcal{H}, \xi)$, that is, the system $\mathcal{CA}(\mathcal{H}, \xi)^h_\eta$, has the point 0 pAS. For any such perturbation $\mathcal{CA}(\mathcal{H}, \xi)^h_\eta$, Lemma 9.10 (c) yields a neighborhood U of ξ such that any solution ϕ to $\mathcal{H}$ with $\mathrm{rge}\,\phi \subset U$ is such that $(t, j) \mapsto \phi(t, j) - \xi$ is a solution to $\mathcal{CA}(\mathcal{H}, \xi)^h_\eta$. Hence ξ is locally pAS for $\mathcal{H}$. $\qquad\square$

The following examples illustrate the utility of Theorem 9.11. In Example 9.12, concluding pre-asymptotic stability for the conical approximation is significantly simpler than concluding it directly for the original system. This is even more dramatic in Example 9.13, where the conical approximation does not allow for flow at all and reduces to a simple difference equation, while the original system is truly hybrid.

Example 9.12. ($\mathcal{CA}(\mathcal{H}, 0)$ globally pAS, $\mathcal{H}$ locally pAS) Recall the hybrid system in Example 7.2 and Figure 7.1. Local pAS is established in Example 7.2, via simple but technical argument, and

$$\mathcal{B}^p_A \cap (C \cup D) = \{x \in C \cup D : x_2 < x_1 + 3/16\}.$$

The conical approximation $\mathcal{CA}(\mathcal{H}, 0)$ of this system at $\xi = 0$ is given by

$$C_{loc} = \left\{x \in \mathbb{R}^2 : x_2 \geq 0,\ x_1 \leq 0\right\} \qquad f_{loc}(x) = \begin{pmatrix} 1 \\ 1 \end{pmatrix}$$

$$D_{loc} = \left\{x \in \mathbb{R}^2 : x_2 \geq 0,\ x_1 = 0\right\} \qquad g_{loc}(x) = \begin{pmatrix} -x_2/2 \\ 0 \end{pmatrix}.$$

See Figure 9.1. A straightforward analysis of its solutions indicate that the system $\mathcal{CA}(\mathcal{H}, 0)$ has the point 0 (globally) pAS. This can also be verified using $V(x) = 2x_2 - 3x_1$ as a Lyapunov function in Theorem 3.18. Consequently, by Theorem 9.11, 0 is pAS for $\mathcal{H}$.

Example 9.13. ($\mathcal{CA}(\mathcal{H}, 0)$ globally pAS and "discrete," $\mathcal{H}$ locally pAS) On $\mathbb{R}^2$, with $x = (x_1, x_2)$, consider

$$C = \left\{x \in \mathbb{R}^2 : x_2 \geq 0,\ 0 \leq x_1 \leq x_2^2\right\} \qquad f(x) = \begin{pmatrix} 1 \\ 1 \end{pmatrix}$$

$$D = \left\{x \in \mathbb{R}^2 : x_2 \geq 0, x_1 = x_2^2\right\} \qquad g(x) = \begin{pmatrix} 0 \\ x_2/2 \end{pmatrix}.$$

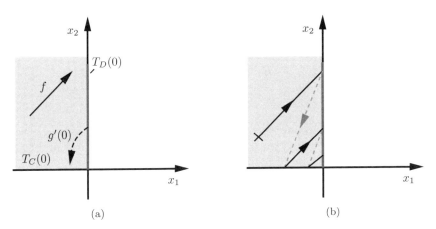

(a)　　　　　　　　　　(b)

Figure 9.1: Approximation of the hybrid system in Figure 7.1. Solutions may flow in the set $\mathbb{R}_{\leq 0} \times \mathbb{R}_{\geq 0}$, which is the tangent cone to C, and they may jump from the set $\{0\} \times \mathbb{R}_{\geq 0}$, which is the tangent cone to D. The solid arrow indicates the direction of flow, which is determined by $f(0) = (1,1)$, and the dotted arrows indicate the jumps, which are determined by $g'(0)x = (-x_2/2, 0)$. A sample solution starting from $\times$ is shown in (b).

The conical approximation $\mathcal{CA}(\mathcal{H}, 0)$ of this system at $\xi = 0$ is given by $f_{loc}(x) = f(x)$, $g_{loc}(x) = g(x)$ for all $x \in \mathbb{R}^2$ and

$$C_{loc} = D_{loc} = \{x \in \mathbb{R}^2 : x_2 \geq 0, x_1 = 0\}.$$

The system $\mathcal{CA}(\mathcal{H}, 0)$ has only discrete solutions and the maximal ones are complete and given by $\phi(0, j) = \phi(0,0)/2^j$ for $j \in \mathbb{N}$. Thus 0 is (globally) pAS for $\mathcal{CA}(\mathcal{H}, 0)$. This implies that 0 is pAS for $\mathcal{H}$. A direct verification shows that, indeed, 0 is locally pAS, with $\mathcal{B}_{\mathcal{A}}^p \cap (C \cup D) = \{x \in C \cup D : x_2 < x_1 + 1/4\}$.

It may happen that conical approximations are not pre-asymptotically stable even when the system is pre-asymptotically stable. Of course, the same issue arises even for differential equations.

Example 9.14. ($\mathcal{H}$ globally pAS, $\mathcal{CA}(\mathcal{H}, 0)$ stable but not pre-attractive) On $\mathbb{R}^2$, consider a system with C and D as in Example 9.13 and

$$f(x) = \begin{pmatrix} x_2 \\ -x_1 \end{pmatrix}, \qquad g(x) = \begin{pmatrix} 0 \\ x_2 \end{pmatrix}.$$

Direct verification shows that 0 is pAS for this system. The conical approximation $\mathcal{CA}(\mathcal{H}, 0)$ is given by $f_{loc}(x) = f(x)$, $g_{loc}(x) = g(x)$ for all $x \in \mathbb{R}^2$ and $C_{loc} = D_{loc}$ as in Example 9.12. The system $\mathcal{CA}(\mathcal{H}, 0)$ has only discrete solutions and the maximal ones are complete and given by $\phi(0, j) = \phi(0,0)$ for $j \in \mathbb{N}$. Thus 0 is stable for $\mathcal{CA}(\mathcal{H}, 0)$ but not attractive or pre-attractive.

Example 9.15. ($\mathcal{H}$ globally pAS, $\mathcal{CA}(\mathcal{H}, 0)$ "unstable") On $\mathbb{R}^2$, consider a system $\mathcal{H}$ given by

$$C = \left\{ x \in \mathbb{R}^2 : x_2 = x_1^2 \right\} \qquad f(x) = x$$
$$D = \emptyset \qquad\qquad\qquad g(x) = x.$$

The only nontrivial solution to $\mathcal{H}$ is given by $\phi(t, 0) = 0$ for $t \in \mathbb{R}_{\geq 0}$. Hence 0 is pAS for $\mathcal{H}$. The conical approximation $\mathcal{CA}(\mathcal{H}, 0)$ has $f_{loc}(x) = x$ and $C_{loc} = \left\{ x \in \mathbb{R}^2 : x_2 = 0 \right\}$; the jump map is irrelevant as the jump set is empty. For initial points in C_{loc}, solutions are given by $\phi(t, 0) = \phi(0, 0)e^t$. Such solutions diverge, and in particular, 0 is not stable for the conical approximation.

A different example of a globally pAS system, in fact globally AS system, for which the conical approximation, appropriately taken, is unstable is the bouncing ball system of Example 1.1. Ignoring the discontinuity in the flow map f at 0, one obtains $f_{loc}(x) = (0, -\gamma)$. Since $C_{loc} = C$, there exist continuous solutions to the conical approximation from every initial condition, including the origin.

Approximations to hybrid systems, as presented above, apply to systems where flow and jump maps are functions and result in approximating systems that are homogeneous with respect to the standard dilation. Approximations of hybrid systems with more general flow and jump maps or approximations relying on general dilations are also possible. Here, only an example on how general dilations can be used in approximations of differential equations is given.

Example 9.16. (Homogeneous approximation with nonstandard dilations) On $\mathbb{R}^2$, consider the differential equation $\dot{z} = f(z)$, where

$$f(x) = \begin{pmatrix} -x_1^3 + x_1^2 x_2 \\ -x_2^{3/2} \end{pmatrix}.$$

Then $f(x) = f_1(x) + f_2(x)$ where f_1 is homogeneous with respect to $M_\lambda = \mathrm{diag}\left\{ \lambda, \lambda^4 \right\}$ with degree $d = 2$ and f_2 is homogeneous with respect to M_λ with degree $d = 5$. Asymptotic stability of the origin for $\dot{z} = f(z)$ can be then deduced from asymptotic stability of the origin for $\dot{z} = f_1(z)$.

9.4 NOTES

The classical result that the origin is asymptotically stable for a differential equation if it is asymptotically stable for its linearization dates back to Lyapunov's thesis [77]. Other early results deducing asymptotic stability for a differential equation from asymptotic stability of its homogeneous approximation appear in Malkin [83], Massera [88], and Hahn [46], where standard homogeneity is considered. Nonstandard dilations are used, for a similar purpose, in Hermes [53] and Rosier [101]. Control designs using homogeneity appear, for example, in Kawski

[59], Grüne [44], and Tuna [120]. Generalized homogeneity for hybrid systems is first studied in Tuna and Teel [121], from where the concept of a homogeneous hybrid system is taken. A common use of tangent cones, which are homogeneous with respect to the standard dilation, to constraint sets is in viability theory for differential equations or inclusions, as briefly discussed in Chapter 5. For an exposition, see Aubin [4]. Homogeneous, with respect to non-standard dilations, approximations of sets were used by Ancona [3] in the study of optimal control problems. Conical approximations of differential inclusions and control systems were used by Frankowska [36] and Smirnov [112] in the study of controllability and stabilizability properties.

Linearization techniques with relation to asymptotic stability have been used in other, related, frameworks; see for example, Liberzon [73] for switching systems or Lakshmikantham et al. [66] for impulsive differential equations. An earlier approximation result applicable to a class of hybrid automata appeared in Lygeros et al. [79]. Earlier results that rely on approximation to conclude Zeno behavior in a hybrid automaton, similar to what is alluded to at the end of Section 9.1, can be found, for example, in Ames et al. [2]. Theorem 9.11 first appeared in Goebel and Teel [41]. A generalization of this result, using generalized homogeneity, appears in Goebel and Teel [42].

Appendix: List of Symbols

$\dot{x}$	The derivative, with respect to time, of the state of a hybrid system
x^+	The state of a hybrid system after a jump
$\mathbb{R}$	The set of real numbers
$\mathbb{R}^n$	The n-dimensional Euclidean space
$\mathbb{R}_{\geq 0}$	The set of nonnegative real numbers, that is, $\mathbb{R}_{\geq 0} = [0, \infty)$
$[0, 1]^n$	All vectors in $\mathbb{R}^n$ such that each component is the interval $[0, 1]$
$\mathbb{Z}$	The set of all integers
$\mathbb{N}$	The set of nonnegative integers, that is, $\mathbb{N} = \{0, 1, \ldots\}$
$\mathbb{N}_{\geq k}$	$\{k, k + 1, \ldots\}$ for a given $k \in \mathbb{N}$
$\varnothing$	The empty set
$\overline{\Sigma}$	The closure of the set Σ
$\mathrm{con}\Sigma$	The convex hull of the set Σ
$\overline{\mathrm{con}}\Sigma$	The closure of the convex hull of a set Σ
$\Sigma_1 \setminus \Sigma_2$	The set of points in Σ_1 that are not in Σ_2
$\Sigma_1 \times \Sigma_2$	The set of ordered pairs (σ_1, σ_2) with $\sigma_1 \in \Sigma_1$, $\sigma_2 \in \Sigma_2$
$x^\top$	The transpose of the vector x
(x, y)	Equivalent notation for the vector $[x^\top \ y^\top]^\top$
$|x|$	The Euclidean norm of a vector $x \in \mathbb{R}^n$
$\mathbb{B}$	The closed unit ball, of appropriate dimension, in the Euclidean norm
$|x|_\Sigma$	$\inf_{y \in \Sigma} |x - y|$ for a set $\Sigma \subset \mathbb{R}^n$ and a point $x \in \mathbb{R}^n$
$\mathbb{S}^n$	The set $\{x \in \mathbb{R}^{n+1} : |x| = 1\}$
$f : \mathbb{R}^m \to \mathbb{R}^n$	A function from $\mathbb{R}^m$ to $\mathbb{R}^n$
$F : \mathbb{R}^m \rightrightarrows \mathbb{R}^n$	A set-valued mapping from $\mathbb{R}^m$ to $\mathbb{R}^n$
$R(\cdot)$	The rotation matrix $R(\phi) = \begin{bmatrix} \cos\phi & -\sin\phi \\ \sin\phi & \cos\phi \end{bmatrix}$
$F(\Sigma)$	$\cup_{x \in \Sigma} F(x)$ for the set-valued mapping $F : \mathbb{R}^m \rightrightarrows \mathbb{R}^n$ and a set $\Sigma \subset \mathbb{R}^m$
$T_\Sigma(\eta)$	The *tangent cone* to the set $\Sigma \subset \mathbb{R}^n$ at $\eta \in \overline{\Sigma}$. $T_\Sigma(\eta)$ is the set of all vectors $w \in \mathbb{R}^n$ for which there exist $\eta_i \in \Sigma$, $\tau_i > 0$, for all $i = 1, 2, \ldots$ such that $\eta_i \to \eta$, $\tau_i \searrow 0$, and $(\eta_i - \eta)/\tau_i \to w$ as $i \to \infty$.
$\mathcal{K}_\infty$	The class of functions from $\mathbb{R}_{\geq 0}$ to $\mathbb{R}_{\geq 0}$ that are continuous, zero at zero, strictly increasing, and unbounded

$V^{-1}(\mu)$ The μ-level set of the function $V : \operatorname{dom} V \to \mathbb{R}$, which is the
                       set of points $\{x \in \operatorname{dom} V : V(x) = \mu\}$

$L_V(\mu)$ The μ-sublevel set of the function $V : \operatorname{dom} V \to \mathbb{R}$, that is,
                       the set of points $\{x \in \operatorname{dom} V : V(x) \leq \mu\}$

$\overline{\operatorname{sign}}$ The set-valued function $\overline{\operatorname{sign}}(r) = -1$ if $r < 0$, 1 if $r > 0$,
                       and $\{-1, 1\}$ if $r = 0$.

Bibliography

[1]   R. Alur, C. Coucoubetis, T. A. Henzinger, P.-H. Ho, X. Nicollin, A. Olivero, J. Sifakis, and S. Yovine. The algorithmic analysis of hybrid systems. *Theoret. Comput. Sci.*, 138(1):3–34, 1995.

[2]   A. D. Ames, A. Abate, and S. Sastry. Sufficient conditions for the existence of Zeno behavior in a class of nonlinear hybrid systems via constant approximations. In *Proc. 46th IEEE Conference on Decision and Control*, pages 4033–4038, 2007.

[3]   F. Ancona. Homogeneous tangent vectors and high-order necessary conditions for optimal controls. *J. Dynam. Control Systems*, 3(2):205–240, 1997.

[4]   J.-P. Aubin. *Viability theory*. Birkhauser, 1991.

[5]   J.-P. Aubin. Impulse differential inclusions and hybrid systems. *Lecture Notes, University of California, Berkeley*, 1999.

[6]   J.-P. Aubin and A. Cellina. *Differential Inclusions*. Springer-Verlag, 1984.

[7]   J.-P. Aubin and H. Frankowska. *Set-Valued Analysis*. Birkhauser, 1990.

[8]   J.-P. Aubin and G. Haddad. Cadenced runs of impulse and hybrid control systems. *Internat. J. Robust Nonlinear Control*, 11(5):401–415, 2001.

[9]   J.-P. Aubin, J. Lygeros, M. Quincampoix, S. Sastry, and N. Seube. Impulse differential inclusions: a viability approach to hybrid systems. *IEEE Trans. Automat. Control*, 47(1):2–20, 2002.

[10]   A. Bacciotti and F. Ceragioli. Stability and stabilization of discontinuous systems and nonsmooth Lyapunov functions. *ESAIM: Control Optim. Calc. Var.*, 4:361–376, 1999.

[11]   A. Bacciotti and F. Ceragioli. Nonpathological Lyapunov functions and discontinuous Carathéodory systems. *Automatica*, 42(3):453–458, 2006.

[12]   A. Bacciotti and L. Mazzi. An invariance principle for nonlinear switched systems. *Systems Control Lett.*, 54(11):1109–1119, 2005.

[13]   D. D. Bainov and P. S. Simeonov. *Systems with Impulse Effect: Stability, Theory, and Applications*. Ellis Horwood Limited, 1989.

[14] E. A. Barbashin and N. N. Krasovskii. On the stability of motion as a whole. *Doklady Akademii Nauk SSSR*, 86:453–456, 1952. In Russian.

[15] O. Beker, C. V. Hollot, Y. Chait, and H. Han. Fundamental properties of reset control systems. *Automatica*, 40(6):905–915, 2004.

[16] G. D. Birkhoff. *Dynamical Systems*. Amer. Math. Soc., 1927.

[17] M. Bohner and A. Peterson. *Dynamic Equations on Time Scales*. Birkhauser, 2001.

[18] M. S. Branicky. *Studies in Hybrid Systems: Modeling, Analysis, and Control.* PhD thesis, Massachusetts Institute of Technology, Cambridge, 1995.

[19] M. S. Branicky. Multiple Lyapunov functions and other analysis tools for switched and hybrid systems. *IEEE Trans. Automat. Control*, 43(4):475 –482, 1998.

[20] B. Brogliato. *Nonsmooth mechanics: models, dynamics, and control.* Communications and control engineering. Springer, 2nd edition, 1999.

[21] B. Brogliato. Some perspectives on the analysis and control of complementarity systems. *IEEE Trans. Automat. Control*, 48(6):918–935, 2003.

[22] C. I. Byrnes and C. F. Martin. An integral-invariance principle for nonlinear systems. *IEEE Trans. Automat. Control*, 40(6):983–994, 1995.

[23] C. Cai, R. Goebel, R. G. Sanfelice, and A. R. Teel. Hybrid systems: limit sets and zero dynamics with a view toward output regulation. In A. Astolfi and L. Marconi, editors, *Analysis and Design of Nonlinear Control Systems*, pages 241–261. Springer-Verlag, 2008.

[24] C. Cai, A. R. Teel, and R. Goebel. Smooth Lyapunov functions for hybrid systems. Part I: existence is equivalent to robustness. *IEEE Trans. Automat. Control*, 52(7):1264–1277, 2007.

[25] C. Cai, A. R. Teel, and R. Goebel. Smooth Lyapunov functions for hybrid systems. Part II: (Pre-)asymptotically stable compact sets. *IEEE Trans. Automat. Control*, 53(3):734–748, 2008.

[26] F. H. Clarke. *Optimization and Nonsmooth Analysis*. Wiley, 1983.

[27] F. H. Clarke, Y. S. Ledyaev, and R. J. Stern. Asymptotic stability and smooth Lyapunov functions. *J. Diff. Eq.*, 149(1):69–114, 1998.

[28] F. H. Clarke, Y. S. Ledyaev, R. J. Stern, and P. R. Wolenski. *Nonsmooth Analysis and Control Theory*. Springer-Verlag, 1998.

[29] P. Collins. A trajectory-space approach to hybrid systems. In *16th International Symposium on Mathematical Theory of Networks and Systems*, 2004.

[30] G. Dal Maso and F. Rampazzo. On systems of ordinary differential equations with measures as controls. *Differential Integral Equations*, 4(4):739–765, 1991.

[31] W. P. Dayawansa and C. F. Martin. A converse Lyapunov theorem for a class of dynamical systems which undergo switching. *IEEE Trans. Automat. Control*, 44(4):751–760, 1999.

[32] R. A. DeCarlo, M. S. Branicky, S. Pettersson, and B. Lennartson. Perspectives and results on the stability and stabilizability of hybrid systems. *Proc. IEEE*, 88(7):1069–1082, 2000.

[33] A. Deshpande. *Control of hybrid systems*. PhD thesis, University of California, Berkeley, 1994.

[34] A. F. Filippov. Differential equations with discontinuous right-hand sides. *Mat. Sb.*, 51:99–128, 1960. In Russian.

[35] A. F. Filippov. *Differential Equations with Discontinuous Right-hand Sides*. Kluwer, 1988.

[36] H. Frankowska. Local controllability and infinitesimal generators of semigroups of set-valued maps. *SIAM J. Control Optim.*, 25(2):412–432, 1987.

[37] R. Goebel, J. Hespanha, A. R. Teel, C. Cai, and R. Sanfelice. Hybrid systems: generalized solutions and robust stability. In *IFAC Symposium on Nonliear Control Systems*, Stuttgart, 2004.

[38] R. Goebel, R. G. Sanfelice, and A. R. Teel. Invariance principles for switching systems via hybrid systems techniques. *Systems Control Lett.*, 57(12):980–986, 2008.

[39] R. Goebel, R. G. Sanfelice, and A. R. Teel. Hybrid dynamical systems. *IEEE Control Systems Magazine*, 29(2):28–93, 2009.

[40] R. Goebel and A. R. Teel. Solutions to hybrid inclusions via set and graphical convergence with stability theory applications. *Automatica*, 42(4):573–587, 2006.

[41] R. Goebel and A. R. Teel. Zeno behavior in homogeneous hybrid systems. In *Proc. 47th Conference on Decision and Control*, pages 2758 – 2763, 2008.

[42] R. Goebel and A. R. Teel. Pre-asymptotic stability and homogeneous approximations of hybrid dynamical systems. *SIAM Rev.*, 52(1):87–109, 2010.

[43] T. B. Goh, Z. Li, B. M. Chen, T. H. Lee, and T. Huang. Design and implementation of a hard disk drive servo system using robust and perfect tracking approach. *IEEE Trans. Control Systems Technology*, 9(2):221–233, 2001.

[44] L. Grüne. Homogeneous state feedback stabilization of homogenous systems. *SIAM J. Control Optim.*, 38(4):1288–1308, 2000.

[45] W. M. Haddad, V. Chellaboina, and S. G. Nersesov. *Impulsive and Hybrid Dynamical Systems: Stability, Dissipativity, and Control.* Princeton University Press, 2006.

[46] W. Hahn. *Stability of motion.* Translated from the German manuscript by Arne P. Baartz. Die Grundlehren der mathematischen Wissenschaften, Band 138. Springer-Verlag, 1967.

[47] O. Hájek. Discontinuous differential equations, I. *J. Diff. Eq.*, 32(2):149–170, 1979.

[48] J. K. Hale, L. T. Magalhães, and W. M. Oliva. *Dynamics in infinite dimensions*, volume 47 of *Applied Mathematical Sciences*. Springer-Verlag, 2nd edition, 2002.

[49] W. P. M. Heemels and B. Brogliato. The complementarity class of hybrid dynamical systems. *European J. Control*, 9(2-3):322–360, 2003.

[50] W. P. M. Heemels, J. M. Schumacher, and S. Weiland. Linear complementarity systems. *SIAM J. Appl. Math.*, 60(4):1234–1269, 2000.

[51] T. A. Henzinger. The theory of hybrid automata. In *Proc. 11th Annual Symp. on Logic in Comp. Science*, pages 278–292. IEEE CS Press, 1996.

[52] H. Hermes. Discontinuous vector fields and feedback control. In *Differential Equations and Dynamical Systems*, pages 155–165. Academic Press, 1967.

[53] H. Hermes. Nilpotent and high-order approximations of vector field systems. *SIAM Rev.*, 33(2):238–264, 1991.

[54] J. P. Hespanha. Uniform stability of switched linear systems: extensions of LaSalle's invariance principle. *IEEE Trans. Automat. Control*, 49(4):470–482, 2004.

[55] J. P. Hespanha, D. Liberzon, D. Angeli, and E. D. Sontag. Nonlinear norm-observability notions and stability of switched systems. *IEEE Trans. Automat. Control*, 50(2):154–168, 2005.

[56] J. P. Hespanha, D. Liberzon, and A. R. Teel. Lyapunov conditions for input-to-state stability of impulsive systems. *Automatica*, 44(11):2735–2744, 2008.

[57] J. P. Hespanha and A. S. Morse. Stability of switched systems with average dwell-time. In *Proc. 38th IEEE Conference on Decision and Control*, volume 3, pages 2655–2660, 1999.

[58] Z-P. Jiang and Y. Wang. Input-to-state stability for discrete-time nonlinear systems. *Automatica*, 37(6):857–869, 2001.

[59] M. Kawski. Homogeneous stabilizing feedback laws. *Control Theory Adv. Tech.*, 6(4):497–516, 1990.

[60] C. M. Kellett and A. R. Teel. Smooth Lyapunov functions and robustness of stability for difference inclusions. *Systems Control Lett.*, 52(5):395–405, 2004.

[61] C. M. Kellett and A. R. Teel. On the robustness of $\mathcal{KL}$-stability for difference inclusions: Smooth discrete-time Lyapunov functions. *SIAM J. Control Optim.*, 44(3):777–800, 2005.

[62] H. K. Khalil. *Nonlinear Systems*. Prentice Hall, 3rd edition, 2002.

[63] N. N. Krasovskii. *Problems of the theory of stability of motion*. Stanford Univ. Press, 1963. Translation of Russian edition, Moscow 1959.

[64] N. N. Krasovskii. *Game-Theoretic Problems of Capture*. Nauka, Moscow, 1970. In Russian.

[65] J. Kurzweil. On the inversion of Ljapunov's second theorem on stability of motion. *Amer. Math. Soc. Transl., Ser. 2*, 24:19–77, 1956.

[66] V. Lakshmikantham, D. D. Bainov, and P. S. Simeonov. *Theory of Impulsive Differential Equations*. World Scientific, 1989.

[67] A. Lamperski and A. D. Ames. Lyapunov-like conditions for the existence of Zeno behavior in hybrid and Lagrangian hybrid systems. In *Proc. 46th IEEE Conference on Decision and Control*, pages 115–120, 2007.

[68] S. Lang. *Real Analysis*. Addison-Wesley, 2nd edition, 1983.

[69] J. P. LaSalle. Some extensions of Liapunov's second method. *IRE Trans. Circuit Theory*, 7(4):520–527, 1960.

[70] J. P. LaSalle. An invariance principle in the theory of stability. In *Differential equations and dynamical systems*. Academic Press, 1967.

[71] J. P. LaSalle. *The Stability of Dynamical Systems*. SIAM's Regional Conference Series in Applied Mathematics, 1976.

[72] D. Liberzon. Hybrid feedback stabilization of systems with quantized signals. *Automatica*, 39(9):1543–1554, 2003.

[73] D. Liberzon. *Switching in Systems and Control*. Systems and Control:
      Foundations and Applications Series. Birkhauser, 2003.

[74] Y. Lin, E. D. Sontag, and Y. Wang. A smooth converse Lyapunov theorem
      for robust stability. *SIAM J. Control Optim.*, 34(1):124–160, 1996.

[75] H. Logemann and E. P. Ryan. Asymptotic behaviour of nonlinear systems.
      Amer. Math. Monthly, 111(10):864–889, 2004.

[76] A. Loria, E. Panteley, D. Popovic, and A. R. Teel. A nested Matrosov
      theorem and persistency of excitation for uniform convergence in stable
      nonautonomous systems. *IEEE Trans. Automat. Control*, 50(2):183–198,
      2005.

[77] A. M. Lyapunov. The general problem of the stability of motion. *Internat.
      J. Control*, 55(3):521–790, 1992. Translated by A. T. Fuller from Édouard
      Davaux's French translation (1907) of the 1892 Russian original.

[78] J. Lygeros. *Hierarchical, hybrid control of large scale systems*. PhD thesis,
      University of California, Berkeley, 1996.

[79] J. Lygeros, K. H. Johansson, S. N. Simić, J. Zhang, and S. S. Sastry.
      Dynamical properties of hybrid automata. *IEEE Trans. Automat. Control*,
      48(1):2–17, 2003.

[80] J. Lygeros, C. Tomlin, and S. S. Sastry. Controllers for reachability spec-
      ifications for hybrid systems. *Automatica*, 35(3):349–370, 1999.

[81] N. Lynch, R. Segala, F. Vaandrager, and H. B. Weinberg. Hybrid i/o au-
      tomata. In *Hybrid Systems III*, volume 1066 of *Lecture Notes in Computer
      Science*. Springer, 1996.

[82] M. Malisoff and F. Mazenc. Constructions of strict Lyapunov functions for
      discrete time and hybrid time-varying systems. *Nonlinear Anal. Hybrid
      Syst.*, 2(2):394–407, 2008.

[83] I. G. Malkin. A theorem on stability in the first approximation. *Doklady
      Akad. Nauk SSSR*, 76:783–784, 1951.

[84] J. L. Mancilla-Aguiar and R. A. Garcia. A converse Lyapunov theorem
      for nonlinear switched systems. *Systems Control Lett.*, 41(1):67–71, 2002.

[85] J. L. Mancilla-Aguilar and R. A. Garcia. An extension of LaSalle's invari-
      ance principle for switched systems. *Systems Control Lett.*, 55(5):376–384,
      2006.

[86] M. D. P. Monteiro Marques. *Differential inclusions in nonsmooth mechan-
      ical problems: Shocks and dry friction*, volume 9 of *Progress in Nonlinear
      Differential Equations and their Applications*. Birkhauser Verlag, 1993.

[87] J. L. Massera. On Liapounoff's conditions of stability. *Ann. of Math.*, 50:705–721, 1949.

[88] J. L. Massera. Contributions to stability theory. *Ann. of Math. (2)*, 64:182–206, 1956.

[89] V. M. Matrosov. On the stability of motion. *J. Appl. Math. Mech.*, 26:1337–1353, 1963.

[90] C. G. Mayhew, R. G. Sanfelice, and A. R. Teel. Quaternion-based hybrid controller for robust global attitude tracking. *IEEE Transactions on Automatic Control*, 56(11):2555–2566, November 2011.

[91] V. S. Melnik and J. Valero. On attractors of multivalued semi-flows and differential inclusions. *Set-Valued Anal.*, 6(1):83–111, 2004.

[92] R. E. Mirollo and S. H. Strogatz. Synchronization of pulse-coupled biological oscillators. *SIAM J. Appl. Math.*, 50(6):1645–1662, 1990.

[93] J.-J. Moreau. Unilateral contact and dry friction in finite freedom dynamics. In *Non-smooth Mechanics and Applications*, volume 302 of *International Centre for Mechanical Sciences, Courses and Lectures*, pages 1–82. Springer-Verlag, 1988.

[94] D. Nešić and A. R. Teel. Matrosov theorem for parameterized families of discrete-time systems. *Automatica*, 40(6):1025–1034, 2004.

[95] D. Nešić, L. Zaccarian, and A. R. Teel. Stability properties of reset systems. *Automatica*, 44(8):2019–2026, 2008.

[96] B. Paden and R. Panja. Globally asymptotically stable 'PD+' controller for robot manipulators. *Internat. J. Control*, 47(6):1697–1712, 1988.

[97] C. Prieur. Uniting local and global controllers with robustness to vanishing noise. *Math. Control Signals Systems*, 14(2):143–172, 2001.

[98] A. Puri and P. Varaiya. Decidability of hybrid systems with rectangular differential inclusions. In *Computer Aided Verification*, Lecture Notes in Computer Science Series, pages 95–104. Springer, 1994.

[99] T. Raff and F. Allgöwer. An impulsive observer that estimates the exact state of a linear continuous-time system in predetermined finite time. In *Proc. Mediterranean Conference on Control and Automation, Athens, Greece*, pages 1–3, 2007.

[100] R. T. Rockafellar and R. J-B Wets. *Variational Analysis*. Springer, 1998.

[101] L. Rosier. Homogeneous Lyapunov function for homogeneous continuous vector fields. *Systems Control Lett.*, 19(6):467–473, 1992.

[102] E. P. Ryan. An integral invariance principle for differential inclusions with applications in adaptive control. *SIAM J. Control Optim.*, 36(3):960–980, 1998.

[103] R. G. Sanfelice. *Robust Hybrid Control Systems*. PhD thesis, University of California, Santa Barbara, 2007.

[104] R. G. Sanfelice, R. Goebel, and A. R. Teel. Invariance principles for hybrid systems with connections to detectability and asymptotic stability. *IEEE Trans. Automat. Control*, 52(12):2282–2297, 2007.

[105] R. G. Sanfelice, R. Goebel, and A. R. Teel. Generalized solutions to hybrid dynamical systems. *ESAIM Control Optim. Calc. Var.*, 14(4):699–724, 2008.

[106] R. G. Sanfelice and A. R. Teel. A "throw-and-catch" hybrid control strategy for robust global stabilization of nonlinear systems. In *Proc. 26th American Control Conference*, pages 3470–3475, 2007.

[107] R. G. Sanfelice and A. R. Teel. Asymptotic stability in hybrid systems via nested Matrosov functions. *IEEE Trans. Automat. Control*, 54(7):1569–1574, 2009.

[108] R. G. Sanfelice and A. R. Teel. Dynamical properties of hybrid systems simulators. *Automatica*, 46(2):239–248, 2010.

[109] R. G. Sanfelice and A.R. Teel. On singular perturbations due to fast actuators in hybrid control systems. *Automatica*, 47(4):692–701, 2011.

[110] D. Shevitz and B. Paden. Lyapunov stability theory of nonsmooth systems. *IEEE Trans. Automat. Control*, 39(9):1910–1914, 1994.

[111] G. N. Silva and R. B. Vinter. Measure driven differential inclusions. *J. Math. Anal. Appl.*, 202(3):727–746, 1996.

[112] G. V. Smirnov. Stabilization by constrained controls. *SIAM J. Control Optim.*, 34(5):1616–1649, 1996.

[113] E. D. Sontag. Comments on integral variants of ISS. *Systems Control Lett.*, 34(1-2):93–100, 1998.

[114] D. Stewart. Rigid-body dynamics with friction and impact. *SIAM Rev.*, 42(1):3–39, 2000.

[115] H. D. Taghirad and E. Jamei. Robust performance verification of adaptive robust controller for hard disk drives. *IEEE Trans. Industrial Electronics*, 55(1):448–456, 2008.

[116] L. Tavernini. Differential automata and their discrete simulators. *Nonlinear Anal.*, 11(6):665–683, 1987.

[117] A. R. Teel and D. Nešić. Averaging for a class of hybrid systems. *Dynamics of Continuous, Discrete and Impulsive Systems Series A: Mathematical Analysis*, 17:829–851, 2010.

[118] A. R. Teel and L. Praly. A smooth Lyapunov function from a class-$\mathcal{KL}$ estimate involving two positive semidefinite functions. *ESAIM Control Optim. Calc. Var.*, 5:313–367 (electronic), 2000.

[119] A. R. Teel and L. Zaccarian. On "uniformity" in definitions of global asymptotic stability for time-varying nonlinear systems. *Automatica*, 42(12):2219–2222, 2006.

[120] S. E. Tuna. Optimal regulation of homogeneous systems. *Automatica*, 41(11):1879–1890, 2005.

[121] S. E. Tuna and A. R. Teel. Homogeneous hybrid systems and a converse Lyapunov theorem. In *Proc. 45th IEEE Conference on Decision and Control*, pages 6235–6240, 2006.

[122] A. van der Schaft and H. Schumacher. Complementarity modeling of hybrid systems. *IEEE Trans. Automat. Control*, 43(4):483–490, 1998.

[123] A. van der Schaft and H. Schumacher. *An Introduction to Hybrid Dynamical Systems*, volume 251 of *Lect. Notes in Contr. and Inform. Sci.* Springer, 2000.

[124] V. Venkataramanana, B. M. Chena, T. H. Leea, and G. Guo. A new approach to the design of mode switching control in hard disk drive servo systems. *Control Engineering Practice*, 10(9):925–939, 2002.

[125] M. Vidyasagar. *Nonlinear Systems Analysis*. Prentice Hall, 2nd edition, 1993.

[126] E. Vinograd. Inapplicability of the method of characteristic exponents for the study of non-linear differential equations. *Math. Sbornik*, 41:431–438, 1957.

[127] F. W. Wilson. Smoothing derivatives of functions and applications. *Trans. Amer. Math. Soc.*, 139:413–428, 1969.

[128] H. S. Witsenhausen. A class of hybrid-state continuous-time dynamic systems. *IEEE Trans. Automat. Control*, 11(2):161–167, 1966.

[129] T. Yang. *Impulsive control theory*, volume 272 of *Lecture Notes in Control and Information Sciences*. Springer-Verlag, 2001.

[130] H. Ye, A. N. Michel, and L. Hou. Stability theory for hybrid dynamical systems. *IEEE Trans. Automat. Control*, 43(4):461–474, 1998.

Index

basic conditions, 120, 132
basin of pre-attraction, 141, 152
    nominally well-posed system, 141
bouncing ball, 2, 35, 54, 61, 82, 125,
    173, 178, 186

class-$\mathcal{K}_\infty$ function, 45
class-$\mathcal{KL}$ function, 68
    exponential decay, 69
conical approximation, 192

data of a hybrid system, 25
dilation, 185
discrete states, 15
distance to a closed set, 45

flashing fireflies, 4, 82
flow map, 2, 25
flow set, 2, 25

generalized solutions to differential
    equations, 74

hybrid arc, 28
    compact, 28
    complete, 28
    continuous, 28
    discrete, 28
    eventually continuous, 28
    eventually discrete, 28
    graph, 108
    graphical convergence, 108
    nontrivial, 28
    types of, 28
    Zeno, 28
hybrid automata, 17, 121, 159
hybrid system, 1

conical approximation, 192
    homogeneous, 185
    homogeneous perturbation, 189
    Krasovskii regularization, 81
    nominally well-posed, 117
    perturbation, 132
    well-posed, 133
hybrid time domain, 27
    convergence, 99

impulsive differential equations, 20

jump map, 2, 25
jump set, 2, 25

$\mathcal{KL}$ pre-asymptotic stability, 145

local eventual boundedness, 110
logical modes, 15
Lyapunov function, 52
    candidate, 52
    hybrid automata, 159
    smooth, 157

meagre function, 178

ω-limit of a set, 130, 144
ω-limit of an arc, 127

pre-asymptotic stability, 45, 176
    exponential decay, 70
    from conical approximation, 194
    homogeneity, 188
    $\mathcal{KL}$ characterization, 68, 146
    local, 139
    relaxed Lyapunov conditions, 60

semiglobal practical robustness, 149
sufficient Lyapunov conditions, 52
uniform global, 45
pre-attractivity
    local, 139
    uniform global, 45
proper indicator, 145

quantized control system, 9

reachable set, 127
reset control system, 10
robust zero-crossing detection, 83
robustness of pre-asymptotic stability, 148
    well-posed system, 152
    $\mathcal{KL}$ characterization, 149

sample-and-hold control, 7, 44
sampled-data systems, 56, 61
set convergence, 97
set-valued mapping, 25
    domain, 25
    graph, 101
    graphical convergence, 106
    local boundedness, 105
    outer semicontinuity, 102
    range, 101
solution to a hybrid system, 29
    compact, 30
    complete, 30
    continuous, 30
    dependence on initial conditions, 126, 135
    dilation, 187
    discrete, 30
    eventually continuous, 30
    eventually discrete, 30
    existence, 33, 124
    existence under state perturbations, 79
    Hermes, 81, 134
    Krasovskii, 81, 134, 148

maximal, 30
nontrivial, 30
types of, 30
uniqueness, 34
Zeno, 30
stability
    local, 139
    uniform global, 45
state perturbation, 76
strong forward pre-invariance, 131, 142
switching systems, 21, 44, 57
    arbitrary switching and differential inclusions, 93
    classes of switching signals, 37
    invariance principle, 182

tangent cone, 103

uniform pre-attractivity, 131, 144
uniting local and global controllers, 58, 64

weak invariance, 128, 169